TRANSGENIC PLANTS

Volume 2

Present Status and Social and Economic Impacts

TRANSGENIC PLANTS

Volume 2

Present Status and Social and Economic Impacts

EDITED BY

Shain-dow Kung

The Hong Kong University of Science and Technology
Kowloon, Hong Kong
and
Center for Agricultural Biology
Maryland Biotechnology Institute
Department of Botany
The University of Maryland
College Park, Maryland

Ray Wu

Section of Biochemistry
Molecular and Cell Biology
Cornell University
Ithaca, New York

Academic Press, Inc.
Harcourt Brace Jovanovich, Publishers
San Diego New York Boston
London Sydney Tokyo Toronto

This book is printed on acid-free paper. ♾

Academic Press, Inc.
1250 Sixth Avenue, San Diego, California 92101-4311

United Kingdom Edition published by
Academic Press Limited
24–28 Oval Road, London NW1 7DX

Library of Congress Cataloging-in-Publication Data

Transgenic plants / edited by Shain-dow Kung, Ray Wu.
p. cm.
Includes indexes.
Contents: v. 1. Engineering and utilization – v. 2. Present status and social and economic impacts.
ISBN 0-12-428781-6 (v. 1). – ISBN 0-12-428782-4 (v. 2)
1. Transgenic plants. 2. Crops–Genetic engineering. 3. Plant genetic engineering. I. Kung, Shain-dow. II. Wu, Ray.
SB123.57.T73 1992
631.5'23–dc20 92-11016
CIP

PRINTED IN THE UNITED STATES OF AMERICA
92 93 94 95 96 97 BB 9 8 7 6 5 4 3 2 1

Contents

Part III

Social and Economic Impacts of Transgenic Plants

Part IV

Concluding Remarks

Contributors

Thomas R. Adams, Department of Discovery Research, DEKALB Plant Genetics, Groton, Connecticut 06340

Thomas Altmann, Institut für Genbiologische Forschung GmbH, 1000 Berlin 33, Germany

Benny Bytebier, Department of Botany, University of Nairobi, Nairobi, Kenya

Pierre J. Charest, Petawawa National Forestry Institute, Forestry Canada, Chalk River, Ontario, Canada K0J 1J0

E. C. Cocking, Plant Genetic Manipulation Group, Department of Botany, University of Nottingham, University Park, Nottingham NG7 2RD, United Kingdom

Brigitte Damm, MOGEN International nv, 2333 CB Leiden, The Netherlands

Abhaya M. Dandekar, Department of Pomology, University of California, Davis, California 95616

M. R. Davey, Plant Genetic Manipulation Group, Department of Botany, University of Nottingham, Nottingham NG7 2RD, United Kingdom

David L. DeBoer, Monsanto Corporation, St. Louis, Missouri 63198

D. Facciotti, Calgene Inc., Davis, California 95616

R. P. Finch, Plant Genetic Manipulation Group, Department of Botany, University of Nottingham, Nottingham NG7 2RD, United Kingdom

Ebrahim Firoozabady, DNA Plant Technology Corporation, Oakland, California 94608

John F. Gerber, Institute of Food and Agricultural Sciences, University of Florida, Gainesville, Florida 32601

R. M. Goodman, Russell Laboratories, University of Wisconsin, Madison, Wisconsin 53705

William J. Gordon-Kamm, Department of Discovery Research, DEKALB Plant Genetics, Groton, Connecticut 06340

Ursula Halfter, The Rockefeller University, New York, New York 10021

Timothy C. Hall, Department of Biology, Texas A&M University, College Station, Texas 77843

Ralph W. F. Hardy, Boyce Thompson Institute for Plant Research, Ithaca, New York 14853

Jean-Pierre Hernalsteens, Laboratorium Genetische Virologie, Vrije Universiteit Brussel, B-1640 Sint-Genesius-Rode, Belgium

C. M. Houck, Calgene, Inc., Davis, California 95616

Bin Huang, Allelix Crop Technologies, Mississauga, Ontario, Canada L4V 1P1

Darrell Hueth, Department of Agriculture and Resource Economics, University of Maryland, College Park, Maryland 20742

David J. James, Horticulture Research International, East Malling, Maidstone, Kent, ME19 6BJ, United Kingdom

Albert P. Kausch, Department of Discovery Research, DEKALB Plant Genetics, Pfizer Central Research, Groton, Connecticut 06340

Edward L. Korwek, Hogan & Hartson Law Offices, Washington, D. C. 20004

S. L. Kothari, Plant Genetic Manipulation Group, Department of Botany, University of Nottingham, Nottingham NG7 2RD, United Kingdom

Roger W. Krueger,[1] Department of Discovery Research, DEKALB Plant Genetics, Pfizer Central Research, Groton, Connecticut 06340

Shain-dow Kung, The Hong Kong University of Science and Technology, Kowloon, Hong Kong, and Center for Agricultural Biotechnology, Maryland Biotechnology Institute, Department of Botany, The University of Maryland, College Park, Maryland 20742

Peggy G. Lemaux,[2] Department of Discovery Research, DEKALB Plant Genetics, Pfizer Central Research, Groton, Connecticut 06340

P. T. Lynch, Plant Genetic Manipulation Group, Department of Botany, University of Nottingham, Nottingham NG7 2RD, United Kingdom

Catherine J. Mackey, Department of Discovery Research, DEKALB Plant Genetics, Pfizer Central Research, Groton, Connecticut 06340

Gale H. McGranahan, Department of Pomology, University of California, Davis, California 95616

Brian L. A. Miki, Plant Research Centre, Agriculture Canada, Ottawa, Ontario, Canada K1A 0C6

Elizabeth E. Murray,[3] McArdle Laboratory for Cancer Research, University of Wisconsin, Madison, Wisconsin 53705

Douglas D. Parker, Center for Public Issues in Biotechnology, Maryland Biotechnology Institute, University of Maryland, College Park, Maryland 20742

T. Michael Spencer, Department of Discovery Research, DEKALB Plant Genetics, Pfizer Central Research, Groton, Connecticut 06340

Marc Van Montagu, Laboratorium voor Genetica, Universiteit Gent, B-9000 Gent, Belgium

Lothar Willmitzer, Institut für Genbiologische Forschung GmbH, 1000 Berlin 33, Germany

Ning-Sun Yang, Department of Mammalian Genetics, Agracetus, Inc., Middleton, Wisconsin 53562

Alvin L. Young, Office of Agricultural Biotechnology, United States Department of Agriculture, Washington, D. C. 20250

[1]Current affiliation: Agricultural Research Division—Plant Biology, American Cyanamid Company, Princeton, New Jersey 08543

[2]Current affiliation: Department of Plant Biology, University of California, Berkeley, California 94710

[3]Current affiliation: Promega Corporation, Madison, Wisconsin 53711

Preface

Biotechnology offers new ideas and techniques applicable to agriculture. It uses the conceptual framework and the technical approaches of molecular biology and plant cell culture systems to develop commercial processes and products. With the rapid development of biotechnology, agriculture has moved from a resource-based to a science-based industry. Indeed, plant breeding has been dramatically broadened by the introduction of genetic engineering techniques based on knowledge of gene structure and function. This has ushered agriculture into a new era where it joins ranks with the most sophisticated of biological sciences in using molecular and cell culture approaches that promise to yield a new generation of plants of almost any desirable trait. Biotechnology has taken us from the era of hybrid plants to the era of transgenic plants. We have advanced far beyond the "Green Revolution" to the "Gene Revolution."

Since 1985, more than fifty species of transgenic plants have been engineered. However, reports on this work have been published in different journals and in various formats. We have attempted to harvest the first group of transgenic crops by assembling the important information in a two-volume work on "Transgenic Plants." These volumes represent a significant milestone in plant/agricultural biology, promote the practical application of recombinant DNA technology, and assist in transforming the agricultural industry.

Reports on transgenic plants that have been engineered to date include the development of new techniques for transformation, basic study on a specific gene, and crop improvement for a specific cause. A number of publications on transgenic plants focus mainly on the development of efficient and reproducible methods for the transformation and regeneration of major crops. To date, the plants transformed have been predominantly dicot, particularly from the family Solanaceae, with only a few monocots and woody plants. Thus, the utilization of transgenic plants to improve agriculture is still in its infancy.

Volume 2 has been organized into three major parts. Part I covers transgenic monocotyledonous plants: rice, maize, and asparagus. Part II is directed to the transgenic dicotyledonous plants, including the most popular family of Solanaceae, legumes, oilseeds and vegetables, woody plants, cotton, and *Arabidopsis*. Part III focuses on the social and economic impacts of transgenic plants. In Volume 1, Engineering and Utilization, organized into three parts, Part I provides an overview of basic techniques for engineering transgenic plants. Part II presents selected examples of the value of transgenic plants to basic research. Part III includes some of the most significant contributions transgenic plants have made thus far to crop improvement, ranging from herbicide tolerance to the improvement of seed storage proteins. These volumes were designed so that they will be suitable for use in classrooms as well as in research laboratories.

We thank the authors for their valuable contributions. Their generosity, efforts, and patience are deeply appreciated. We also thank the staff of Academic Press for their help in making these volumes possible.

Shain-dow Kung
Ray Wu

Contents of Transgenic Plants, Volume 1

ENGINEERING AND UTILIZATION

Part II

Transgenic Plants and Basic Research

Part III

Transgenic Plants and Crop Improvement

Introduction: From Green Revolution to Gene Revolution

Shain-dow Kung

The Hong Kong University of Science and Technology
Kowloon, Hong Kong
and
Center for Agricultural Biology
Maryland Biotechnology Institute
Department of Botany
The University of Maryland
College Park, Maryland 20742

When farmers began harvesting the first domesticated plants in about 8000 B.C. the earth's population was estimated to be four million people. Today, world population exceeds five billion. We will have to grow as much food in the first two decades of the new century as was produced over the past 10,000 years. Techniques for selecting and upgrading plant products have been refined over centuries of human history to current standards of plant breeding. After the rediscovery of Mendel's laws of genetics in the first decade of the twentieth century (Borlaug, 1983), plant breeding became a science-based endeavor.

Conventional plant breeding techniques have resulted in dramatic crop improvements (Borlaug, 1983; Goodman *et al.*, 1987) and will continue in the future. The total number of plant species is estimated to be 250,000. Of these, 3000 species are used for food, and only 29 species are major crops. These species include cereals, legumes, oilseeds, root crops, sugar crops, and tree crops (Harlan, 1976). Supplementing these 29 basic food crops are about 15 major species of vegetables and 15 major species of fruits (Borlaug, 1983), which collectively constitute the human diet. Four major crops (wheat, rice, maize, and potato) contribute more tonnage to the world total than the 25 other crops combined.

Since our food supply now depends on the success of a small number of plant species, the failure of one may mean starvation for millions of people. The 1970 corn leaf blight epidemic, which caused the loss of an estimated 15% of the U.S. corn crop, provides a compelling example of this dependency. By definition, domesticated plants are dependent on man for survival since domestication is the result of selections that make plants better adapted to man-made environments and less well adapted to natural environments. For example, once the natural mechanism for seed dispersal is lost, plants are dependent on man for reproduction.

Strong pressures for further improvements in crop quality and quantity are exerted by population growth, social demands, health requirements, environmental stresses, and ecological considerations. World population is projected to reach eight billion by the year 2010, an alarming doubling in only 35 years. Ensuring the feeding of three billion additional people in the next 20 years will require dramatic increases in crop production. This in turn will require continuous improvement of technology applicable to agriculture to achieve this goal.

I would like to descibe recent steps taken from the era of the Green Revolution to that of the Gene Revolution from the perspective of a plant molecular biologist. Many scientific and technological advances underlie this history. In my view, the present era of biotechnology-based agricultural science, the Gene Revolution, has emerged in response to the limitations of conventional plant breeding techniques which have been unable to overcome the barrier of sexual incompatibility, limiting the genetic pool available for crop improvements. The later development of somatic hybrid plants, while transcending this limit (Carlson, 1973; Shepard *et al.*, 1983; Glimelius *et al.*, 1991), has not yielded results with immediate economic value. However, the Gene Revolution holds the promise of introducing crop improvements at the molecular level, with a power and control previously inconceivable, thereby producing plants with far greater ranges of genetic variability.

I.1 Plant Hybridization and the Green Revolution

Conventional plant breeding's principal achievement has been in the discovery and utilization of heterosis. The application of this discovery, refined over a 100-year period, culminated in the Green Revolution. The contributions of plant hybridization and the impact of the Green Revolution during that period will be described.

I.1.1 PLANT HYBRIDIZATION

The most frequently employed plant breeding technique is hybridization. When two lines are cross-pollinated and the resulting seeds are grown, the F_1 hybrid generation is obtained. All plants in this F_1 generation are identical. If they are inbred by self-pollination, a great range of diversity appears in the F_2 generation. The F_2 generation contains an enormous number of new genotypes. All possible combinations of parental traits are represented. The desirable varieties are selected. By repeated self-pollination, the less desirable varieties are eliminated. The end result is the selection of a single genotype respresenting the optimal combination of parental characteristics. The limitation of this method is that sexual incompatibility prevents cross-fertilization between distantly related species.

Our forefathers recognized very early that both animals and plants could be selected and crossed to reproduce and upgrade species. The mule, a cross between horse and donkey, produced over 2500 years ago, was probably the first hybrid domestic animal. Although I have not found a historical record of the date the first hybrid plant was made, a sexual hybrid between carnation and Sweet William was produced in 1717 by Thomas Fairchild (Cocking *et al.*, 1984). From 1856 to 1864 Gregor Mendel conducted the first systematic and scientific plant hybridization experiments. His analysis of over 10,000 hybrid peas resulted in the formulation of his laws of genetics, the scientific basis for modern plant breeding.

Plant breeding has been described as the selection of plants with desired traits after sexual exchange of genes by cross-fertilization between two parents (Goodman *et al.*, 1987). This definition clearly indicates that hybridization is an essential technique in plant breeding. There are two broad types of hybrid plants: interspecific and intergeneric hybrids. Due to sexual incompatibility, interfamilial hybrid plants have not been produced. Successful examples of interspecific hybridizations are the breeding of wheat (McFadden, 1930), tomato (Bohn and Tucker, 1939), and soybean (Newell and Hymowitz, 1982). It is believed that the origin of many tobacco species was the result of interspecific hybridizations followed by chromosome doubling (Gray *et al.*, 1974). Some modern crop species, such as rapeseed and certain wheat strains, originated in nature by hybridization between different species or genera (Simmonds, 1979). A dramatic example of an intergeneric hybrid was the Nicotunia, produced by L. Burbank by crossing thousands and thousands of large flowering nicotianas with petunias (Cocking *et al.*, 1984). Nicotunias were generally slender with purple stalk, large leaves, and handsome flowers. No seed was ever produced, but the plants were readily multipled by cuttings.

The history of improvement in hybrid maize is an outstanding illustration of the successful application and refinement of plant hybridization techniques. The first hybrid maize plant was developed after recognition that inbreeding in maize leads to reduced vigor in the following generation and that vigor can be restored by crossing. In 1908, Shull reported that the hybrid plants produced by a cross between two different pure lines quadrupled the yield per acre (Chrispeels and Sadava, 1977). Yields increased roughly sixfold in less than 20 years (Jennings,

1976). Currently, hybrid maize plants produced in the United States can have yields as high as 130 bushels per acre.

I.1.2 GREEN REVOLUTION

The success in applying conventional plant breeding principles and agricultural practices to crop improvement reached its peak when high-yielding wheat and rice varieties were cultivated in the 1960s, with a profound impact on agricultural production (Borlaug, 1983). Before the introduction of high-yielding semi-dwarf varieties of Mexican wheat into India between 1966 and 1968, Indian annual wheat production was 11.39 million metric tons. By 1981, after widespread adoption of the high-yielding varieties, annual wheat production increased to 36.5 million metric tons. India's increase in production in 15 years provided sufficient grain to feed 184 million additional people at 375 g of wheat per person per day (Borlaug, 1983). The wheat production gains in Argentina, Bangladesh, China, Pakistan, and Turkey were equally impressive. In China, agricultural production has risen 8% yearly for the past 10 years. This gives China the distinction of being the largest food producer in the world. For this remarkable accomplishment, He Kang, Minister of Agriculture of the People's Republic of China from 1983 to 1990, was awarded an honorary doctoral degree by the University of Maryland at College Park in 1986. In the developing countries as a whole, wheat and rice production increased by about 75% between 1965 and 1980. Individually, some countries did better than others. For example, from 1950 to 1984 wheat production in Mexico increased 400% while Indonesia doubled its rice production. Furthermore, improved seed and agricultural practices have brought a 300–400% increase in the yield of maize, sorghum, and millet in the Sudan, Ghana, Tanzania, and Zambia.

The doubling or tripling of rice and wheat production in Asia was termed the Green Revolution to describe the social, economic, and nutritional impact (Chrispeels and Sadava, 1977). Norman Borlaug was awarded the Nobel Peace Prize in 1970 for his remarkable contributions.

Since the 1960s, the Green Revolution has been both praised and damned. As use of the new varieties spread rapidly in Asia, some experts claimed that many underdeveloped countries would soon be self-sufficient in cereal grains. However, the new varieties were highly responsive and required far more intensive fertilization and irrigation than traditional crops. Consequently, poor weather, high energy costs, and global economic constraints slowed the progress of the revolution considerably. Critics argued that the Green Revolution's emphasis on intensive agriculture, large farms, and prime crop land damaged the environment and offered little to the poor farmers, the intended beneficiaries. Thus, the history of the Green Revolution illustrates both the potential and limitations of conventional plant breeding technology.

I.2 Link between Green Revolution and Gene Revolution—Protoplast Fusion and Somatic Hybrid Plants

The development of new plant breeding technologies resulted from new advances in the understanding of plants at the cellular level. Two areas, protoplast fusion and somatic hybrid plants, will be discussed.

I.2.1 PROTOPLAST FUSION

In contrast to animal cells, plant cells are totipotent. Each living cell is capable of regenerating into an entire plant identical to the one from which the cell was obtained. The term "cell culture" is used to describe *in vitro* culture of plant cells, cell suspension, and protoplast. In the 1950s, F. C. Stewart and colleagues pioneered plant cell culture and regeneration techniques based on their discovery that plants could be developed from single cells (Stewart *et al.,* 1983). For example, they produced seed-forming plants using specially treated cells that broke away from pieces of cells derived from carrot root phloem.

Perhaps the greatest contributions of cell culture has been its subsequent role in enabling the genetic engineering of plants. Cell and tissue cultures allow the rapid propagation and regeneration of genetically engineered cells, an essential step in producing transgenic plants.

Successful cell culture techniques were soon extended to the production of protoplasts from plant cells by stripping off their walls (Shepard *et al.,* 1983). Protoplasts are studied as simple cellular entities, like microorganisms. Isolated protoplasts can be used for fusion, DNA uptake, cell wall studies, and other cellular investigations. In the 1960s, Takebe and co-workers prepared tobacco protoplasts for efficient viral infection experiments. Protoplasts were quickly adapted for regeneration (Nagata and Takebe, 1971). Success in regenerating complete plants from protoplasts eventually led to attempts to combine cells with different genetic backgrounds. In 1972, Carlson *et al.* were the first group to succeed in fusing tobacco protoplasts from two genetically compatible *Nicotiana* species *N. glauca* and *N. langsdorffii*. Initially the fused products were termed parasexual hybrids (Carlson *et al.,* 1972). Later the term somatic hybrid was adopted (Shepard *et al.,* 1983).

I.2.2 SOMATIC HYBRID PLANTS

Somatic hybrid plants are hybrid plants derived from the fusion of somatic cells. Many somatic hybrids were produced after the successful attempts of protoplast fusions. Since all fusions are possible in principle, regardless of the phylogenetic relatedness, somatic hybrid plants theoretically offered unlimited possibilities for genetic exchange, a prospect which generated great excitement and great expectations during the 1970s. Many fusions between protoplasts of phylogenetically unrelated species were attempted. Notable examples include fusion between soybean and tobacco, soybean and barley, soybean and maize, soybean and clover, soybean and alfalfa, soybean and rapeseed, carrot and barley, potato and tomato, carrot and parsley, sorghum and maize, and even between plant and human cells (Quellette *et al.*, 1985; Lima-De-Faria *et al.*, 1977). Not surprisingly, most of these attempts failed. There are notable exceptions:

1. The fusion produced between two sexually incompatible *Petunia* species, *P. parodi* and *P. parviflora* (Power *et al.*, 1980).
2. The fertile integeneric somatic hybrid plants generated in the Brassicaceae family (Glimelius *et al.*, 1991).
3. Pomato, an intergeneric hybrid produced between potato and tomato.

It should be noted that potato and tomato are not sexually compatible, and repeated attempts to hybridize these two plants belonging to different genera failed. However, Burbank was able to graft potato onto tomato or vice versa and obtained a new fruit from this chimera which he named the "pomato" (Howard, 1945). Almost a century later, in 1978, Melchers *et al.* were the first to produce an intergeneric somatic hybrid between potato and tomato. The hybrid plants were also named "pomato" (Melchers *et al.*, 1978). Some of these somatic hybrid plants formed "tuberlike stolons"; however, none set fertile flowers, or fruits, or produced true tubers (Shepard *et al.*, 1983).

The advance from the era of hybrid plants to the era of somatic hybrid plants expanded applications of plant breeding technology from the organismic to the cellular level. Further advances in plant breeding technology—to the molecular level—are the hallmark of the modern era of biotechnology or genetic engineering.

I.3 Genetic Engineering and the Gene Revolution

The engineering of transgenic plants to develop higher yielding, more nutritious, disease- and pest-resistant crop varieties signals a revolution in plant breeding—the Gene Revolution.

I.3.1 GENETIC ENGINEERING

It was only in 1983 that scientists inserted the first foreign genes into petunia and tobacco, the "*Escherichia coli*" of the plant world (Kung, 1984). Independently, several transgenic tobacco plants were produced to express foreign genes engineered by the *Agrobacterium tumefaciens* vectors (Murai *et al.*, 1983; Horsch *et al.*, 1984; and De Block *et al.*, 1984). Early transformation experiments often utilized plant protoplasts as the recipient cells; the subsequent development of transformation methods based on a regenerable explant (Horsch *et al.*, 1985) such as leaves, stems, and roots significantly simplified transformation techniques and are widely used today. Recent breakthroughs in the regeneration of the transformed monocot plants, rice and maize, have removed a major obstacle to improving cereal crops. In this volume many successful examples of producing genetically engineered transgenic plants are described (Chapters 1–9). Thus, only a few cases are illustrated here.

Transgenic plants resistant to pathogens have been successfully produced. By recognizing the cross-protection phenomenon in plants, cDNA encoding the coat protein (CP) gene of tobacco mosaic virus (TMV) was introduced into tobacco and tomato cells, and plants were regenerated. The expression of CP in transgenic plants provided maximum protection when challenged with viruses (Powell-Abel *et al.*, 1986). Field tests demonstrated the effectiveness of this protection: less than 5% of the inoculated transgenic tomatoes containing the chimeric CP gene displayed symptoms. In contrast, 100% of the control plants exhibited virus symptoms.

Herbicide resistance research has focused on developing crops that are resistant to herbicides (Fraley, 1989) such as glyphosate. Two general approaches have been taken in using agricultural biotechnologies to obtain herbicide resistance: (1) altering the level of sensitivity of the target enzyme for the herbicide resistance and (2) incorporating a gene that will detoxify the herbicide. For example, the gene responsible for resistance to glyphosate has been inserted into tomatoes, tobacco, soybeans, cotton, and poplar trees.

Frost protection has been approached by genetically altering microorganisms that interact with plants. Field trails with this "ice minus" organism have been conducted successfully on strawberries and potatoes (Lindow, 1989). Furthermore, attempts have even been made to produce antibodies in transgenic plants (Hiatt *et al.*, 1990).

One of the major difficulties in genetically engineering plants is that many of the more important economic traits in plants are controlled by multiple genes. Current genetic transformation techniques are most effective when dealing with one gene at a time.

I.3.2 GENE REVOLUTION

The Gene Revolution refers to the manipulation of the structure and function of genes at the sequence level aimed to benefit the human race. This is a remarkable advancement since the term "gene" was first used to describe the hereditary factors in 1909. The success of the Gene Revolution rests on a solid foundation of basic research (Watson and Crick, 1953; Meynell, 1973; and Smith *et al.*, 1971), which led to the development of recombinant DNA technology (Jackson *et al.*, 1972). The manipulation of plant genes includes isolation, identification, and analysis of structure and regulation of beneficial genes, followed by the transferring of these genes into plants. Selected examples of achievements in plant gene identification and gene regulation are discussed below for quick illustration, not for comprehensive review.

I.3.2.1 Gene Identification

A search of plant gene sequences in the GenBank database (updated through 1/91) revealed that over 2780 entries of gene or sequences of plant origin have been identified and deposited. Many of these entries are undoubtedly redundant. For example, there are over 60 entries for the small subunit of Rubisco (rbcS) and more than 50 entries for chlorophyll a/b binding protein (Cab).

About 200 distinct nuclear genes of higher plants were identified from 1500 entries and are listed in Table I.1. Genes of organellar origins, mobile genetic elements, satellite and repeated sequences, many rRNA, tRNA, and ribosomal proteins are not included. The most frequently studied nuclear genes from higher plants are rbcS, Cab, and leghemoglobin followed by zein, storage proteins, nodulin, and amylase. The most frequently used plants are maize (198 entries), soybean (111 entries), and wheat, followed closely by rice, pea, tobacco, and tomato. Plant nuclear genes exhibit the same salient structural features as those of other eukaryotes.

In addition to plant nuclear genes, a great number of plant organellar genes from chloroplast and mitochondria have been studied. It was estimated that over 250 species of chloroplast genomes belonging to 90 genera and 33 families of higher plants have been analyzed and characterized (Kung, 1989). The number of organellar genes sequenced is growing at a remarkable rate. To date three complete nucleotide sequences—tobacco (Shinozaki *et al.*, 1986), liverwort (Ohyama *et al.*, 1986), and rice (Hiratsuka *et al.*, 1989)—and more than 40 physical maps of chloroplast genomes have been published.

These three sequenced chloroplast genomes represent three distinct phylogenetic groups of plants: dicot (tobacco), monocot (rice), and a fern (liverwort). Their chloroplast genome sizes vary greatly: 155,844 bp for tobacco, 134,525 bp for rice, and 121,024 bp for liverwort. Chloroplast genomes studied so far probably code for all of their own tRNA and rRNA, and some, but certainly not all, of the chloroplast proteins. Tobacco and liverwort chloroplast genomes contain an estimated number of about 146 and 128 genes, respectively. These estimates may be conservative (Hiratsuka *et al.*, 1989).

TABLE I.1
Plant Nuclear Genes

Acetolactate synthase
Actin
Actinidin
Acyl carrier protein adenine nucleotide translocator
ADP-glucose pyrophosphorylase
Agglutinin
Albumin
Alcohol dehydrogenase
Aldolase
Amine oxidase
1-Aminocyclopropane-1-carboxylate synthase
α-Amylase
Amylase/protease inhibitor
α-Amylase/subtilis inhibitor
α-Amylase/trypsin inhibitor
Anionic peroxidase
Anther specific
Arcelin
Ascorbate oxidase
Aspartic protease
Aspartic proteinase
ATP/ADP translocator
ATP-dependent protease
Auxin-binding protein
Auxin-regulated protein
Avenin
Betaine-aldehyde dehydrogenase
Bowman-Birk trypsin
Calmodulin
Canavalin
Capestris acyl carrier protein
Carboxypeptidase
Catalase
Cellobiohydrolase II
Cellulase
Chalcone flavanone isomerase
Chalcone synthase
Chitinase
Chloroperoxidase
Chlorophyll a/b-binding protein
Chymotrypsin inhibitor cinnamyl-alcohol dehydrogenase
Germ agglutinin isolectin
Germin
Gibberellin responsive protein
Gliadin
α-/β-Gliadin
α-Globulin
β-Globulin
β-Glucan endohydrolase
β-Glucanase
β-1-3-Glucanase
Glucoamylase
β-Glucosidase
Glutamate-1-semialdehyde aminotransferase
Glutamine synthetase
Glutathione-*S*-transferase
Glutelin
Glyceraldehyde-3-phosphate dehydrogenase
Glycerol-3-phosphate acyltransferase
Glycine decarboxylase
Glycine-rich protein
Glycinin
Glycoprotein
Glyoxysomal malate dehydrogenase
Glyoxysomal malate synthase
H^+-ATPase
Heat shock protein
Hevein
Histone
Hordein
Hydroxyproline-rich glycoprotein
Isocitrate lyase
Isopenicillin N synthetase
α-Isopropylmalate isomerase
Kafirin
Kunitz trypsin
Latax protein
Lectin
Leghemoglobin
Legumin
Light-inducible tissue-specific ST-LS1
Ligninase
Lignin peroxidase
Opaque-2
Orotidylic acid pyrophosphorylase
Oryzacystatin
Osmatin
Papain
Patatin
Pectin esterase
Peroxidase
α-Type phaseolin
β-Type phaseolin
Phenylalanine ammonia-lyase
Phosphate translocator protein
Phosphoenolpyruvate carboxylase
Phosphoglucose isomerase
Phosphoglycerate kinase
Phosphoribulokinase photosystem I protein
Phytochrome
Phytohemagglutinin
Plastocyanin
Pollen allergen
Polygalacturonase
Preproglutelin
Prolamine
Proline-rich cell wall protein
Protease inhibitor
Proteinase
Pyruvate orthophosphate dikinase
Rennin
Ribulose-1,5-bisphosphate carboxylase/oxygenase L
Ribulose-1,5-bisphosphate carboxylase S
Ricin
Ripening-related protein
Seed acyl-carrier protein
Seed lipoxygenase
Sexual inducer glycoprotein
Sporamin
S-rich γ-secalin-related storage protein
Starch synthase
Storage protein
Sucrose synthetase
Sulfhydryl endopeptidase

(*continued*)

TABLE I.1 *(Continued)*

Concanavalin	Lipoxygenase	Sulfur-rich water-soluble seed protein
Copper/zinc-superoxide dismutase	Low-temperature-induced thiol protease	Superoxide dismutase
4-Coumarate: coenzyme A ligase	L-Phenylalanine ammonia-lyase	Thaumatin
Cruciferin	Megasperma actin	α-Trichosanthin
Cyclin	Malate synthase	Thiol protease aleurain
Cysteine proteinase	Manganese peroxidase	Thionin
Cytochrome	Manganese superoxide dismutase	Thioredoxin f
Dehydrin	NADP-dependent malic enzyme	Toxin α-hordothionin
Disease-resistance response protein	NADP malate dehydrogenase	Trichodiene synthase
Elongation factor 1-alpha	$NADP^+$ oxidoreductase	Triosephosphate isomerase
Endochitinase	NADPH-protochlorophyllide oxidoreductase	Tryptophan decarboxylase
Endoglucanase III	NAPH-dependent hydroxy-pyruvate reductase	Tryptophan synthetase
Endosperm glutelin-2 protein	Napin	β-Tubulin
5-Enolpyruvylshikimate-3-phosphate synthase	Negatively light-regulated protein	Tubulinlike protein
Ethylene-responsive fruit ripening portein	Nif	Ubiquitin
Extensin	Nitrate reductase	Ubiquitin carrier protein
Ferredoxin	Nitrite reductase	UDPG-flavonal glucosyltransferase
Ferredoxin-$NADP^+$ reductase	Niveus RNase Rh	UDP-glucose flavonoid glycosyltransferase
Fructose-1,6-bisphosphatase	Nodulin oleosin KD18	Vicilin
γ-Gliadin		Water-stress inducible protein
		Wound-inducible proteinase inhibitor
		Zein

I.3.2.2 Gene Regulation

After establishing that plant genes share a similar general structure with that of other eukaryotic genes, the next step is to identify regulatory sequences. The well-studied examples are also rbcS and Cab (Schell, 1987; Benfey and Chua, 1989). A "Light Responsive Element" (LRE) upstream from the TATA box responds to light (Moses and Chua, 1987). It was clearly demonstrated that this LRE can regulate many foreign genes placed under its control. It also exhibits organ specificity: a high expression in leaves, low expression in stems, and no expression in roots (Benfey and Chua, 1989).

A similar analysis of the upstream regulatory sequences of the pea Cab gene also demonstrated the existence of LRE and its light responsiveness and organ specificity (Schell, 1987). Another example of the existence of regulatory sequence is the chalcone synthase (chs) gene which is induced by strong and prolonged UV light (Schell, 1987). A UV light-responsive sequence upstream of the TATA box has also been identified (Schell, 1987).

The genes identified, isolated, cloned, and sequenced to date are mostly involved in the metabolic pathways such as from the carbon and nitrogen cycle, and very few, such as plant hormone genes, are involved in regulation. Major efforts are now underway to identify plant regulatory genes including those conferring resistance to diseases, pests, and other stresses. It is quite conspicuous that genes listed in Table I.1 lack these properties.

I.4 Milestones from the Green Revolution to the Gene Revolution

Table I.2 presents a chronology of the milestones discussed in this introduction. Figure I.1 shows a striking difference between the Green Revolution and the Gene Revolution. The Green Revolution was based on the application and subsequent refinement of a single scientific discovery made in 1864, Mendel's laws of genetics. Advances in the application of Mendel's laws of genetics resulted in the Green Revolution. This was achieved as the result of countless observations and selections of genetic recombination of many crops. Prior to the availability of modern tools of agricultural biotechnology, there has been gradual progress in crop improvement based on this single scientific discovery. It took approximately one entire century to advance from the discovery of genetic laws to the peak of its application—the Green Revolution (Fig. I.1). In contrast, the Gene Revolution is the result of a profound transformation in our understanding of plant biology and

TABLE I.2

Milestones of Scientific and Technological Developments Which Were Essential to the Advancement from the Green Revolution to the Gene Revolution

Year	Development	Developers
1864	Genetic laws	G. Mendel
1908–1920	Hybrid corn	G. Shull
1950	Cell culture	F. Stewart
1953	DNA	J. Watson, F. Crick
1959	Plasmid	T. Akiba, K. Ochiai
1960s	Green Revolution	N. Borlaug
1970	Protoplasts and regeneration	I. Tekebe
1970	Restriction enzyme	H. Smith, D. Nathans
1972–1973	Recombinant DNA	S. Cohen, P. Boyer, P. Berg
1972	Somatic hybrid	P. Carlson
1974	Ti plasmid	J. Schell
1978	Pomato	G. Melchers
1982	Direct gene transfer	
1983	Transgenic plants	

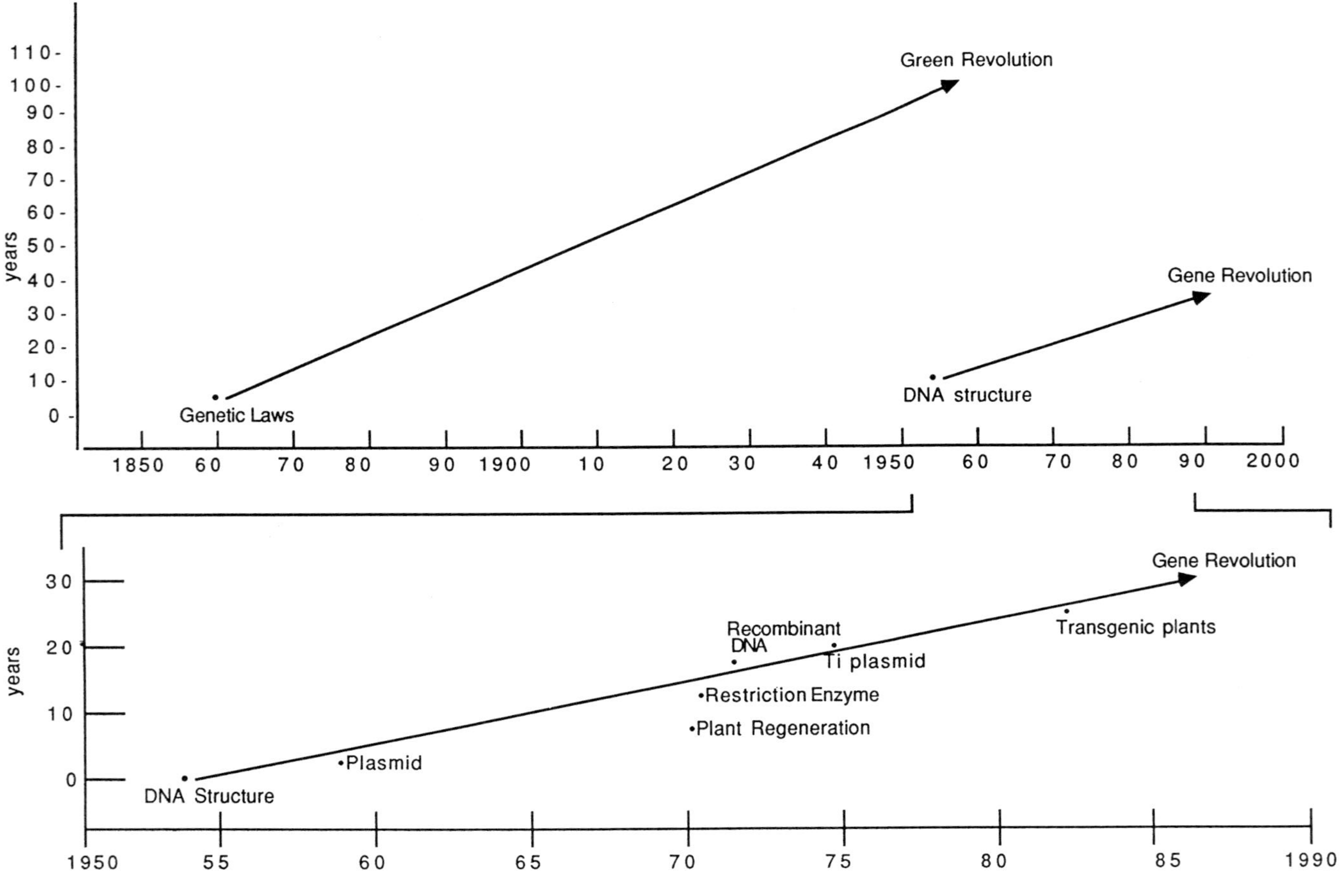

Figure I.1. A striking difference between the Green Revolution and the Gene Revolution. Green Revolution was based on a single discovery and progressed gradually. Gene Revolution involves several discoveries and progresses rapidly.

has progressed over a much shorter time period, involving many important discoveries (Fig. I.1).

The first transgenic plant (1983) was engineered only 29 years after the fundamental discovery of DNA structure (1953). The subsequent discoveries of the plasmid (1959), restriction enzymes (1970), plant regeneration (1970), recombinant DNA technology (1971), and Ti plasmid (1974) collectively made the genetic transformation of plants possible. This illustrates an important aspect of crop improvement technology: the lag from scientific discovery to technology transfer is becoming progressively narrower (Fig. I.1). Furthermore, the technologies used to improve crops are no longer dependent on only a single but on multiple scientific discoveries (Fig. I.1).

I.5 Concluding Remarks

The history and future of crop improvement practices represent a continuous effort to develop novel approaches which overcome the limitations of existing methods. The application of genetic principles to breeding programs has been and will continue to be very important as evidenced by the "Green Revolution." The success of this application was enhanced by the development of tissue culture and protoplast fusion techniques. The combination of conventional breeding approaches and tissue culture techniques opened many new avenues for crop improvements. The current application of agricultural biotechnology to crop improvement has already transformed plant breeding and offers the opportunity to engineer ideal crops with unprecedented power and precision.

Acknowledgment

Greg Silsbee's and David Swartz's help in preparing this manuscript and J. R. Wu and Shyam Dube's review are greatly appreciated.

References

Benfey, P. N. and Chua, N.-H. (1989). *Science* 244, 174–181.

Bohn, G. W. and Tucker, C. M. (1939). *Science* 89, 603–605.

Borlaug, N. E. (1983). *Science* 219, 689–693.

Carlson, P. S., Smith, H. H., and Dearing, R. D. (1972). *Proc. Natl. Acad. Sci. U.S.A.* 69, 2292–2294.

Carlson, P. S. (1973). *Proc. Natl. Acad. Sci. U.S.A.* 70, 598–602.

Chilton, M.-D., Drummond, H. M., Merlo, D. J., Sciaky, D., Montoya, A. L., Gordon, M. P., and Nester, E. W. (1977). *Cell,* 11, 263–271.

Chrispeels, M. J. and Sadava, D. (1977). In *Plants, Food and People,* W. H. Freeman, San Francisco, p. 192.
Cocking, E. C. (1984). In *Handbook of Plant Cell Culture* (P. V. Ammirato, D. A. Evans, W. R. Sharp and Y. Yamada, eds.) Vol. 3, 1–8.
Cohen, S. N., Chang, A. C. Y., Boyer, H. W., and Hellings, R. B. (1973). *Proc. Natl. Acad. Sci. U.S.A.* 70:3240–3244.
Danna, K. J. and Nathans, D. (1971). *Proc. Natl. Acad. Sci. U.S.A.* 68:2913–2917.
De Block, M., Herrera-Estrella, L., Van Montagu, M., Schell, J., and Zambroyski, P. (1984). *EMBO J.* 3, 681–686.
Fraley, R. (1989). In *Plant Biotechnology* (S. D. Kung and C. J. Arntzen, eds.). Butterworths, Boston, pp. 395–407.
Glimelius, K., Fahlesson, J., Landgren, M., Sjödin, C., and Sundberg, E. (1991). *Trends in Biotech.* 9, 24–30.
Goodman, R. M., Hauptli, H., Crossway, A., and Knauf, V. C. (1987). *Science* 236, 48–54.
Gray, J. C., Kung, S. D., and Wildman, S. G. (1974). *Nature* 252, 226–227.
Harlan, J. R. (1976). *Sci. Amer.* 57–65.
Hiatt, A. H., Cafferkey, R., and Bowdish, K. (1990). *Nature,* 344, 469–470.
Hiratsuka, J. *et al.* (1989). *Mol. Gen. Genet.* 217, 185–191.
Horsch, R. B., Fraley, R. T., Rogers, S. G., Sanders, P. R., Lloyd, A., and Hoffman, N. (1984). *Science* 223, 496–498.
Horsch, R. B., Frey, J. E., Hoffmann, N. L., Eichholtz, D., Rogers, S. G., and Fraley, R. T. (1985). *Science* 227, 1229–1231.
Howard, W. L. (1945). *Chron. Bot.* 9, 300–306.
Jackson, D. A., Symons, R. H., and Berg, P. (1972). *Proc. Natl. Acad. Sci. U.S.A.* 69:2904–2909.
Jennings, P. R. (1976). *Sci. Amer.* 125–133.
Kung, S. D. (1984). *CRC Critical Rev. in Plant Sci.,* 1, 227–267.
Kung, S. D. (1989). In *Biotechnology in Agriculture and Forestry* (Y. P. S. Bajaj, ed.). Pp. 283–303.
Lima-De-Faria, A., Eriksson, T., and Kjellen, L. (1977). *Hereditas* 87, 57–61.
Lindow, S. (1989). *Phyto. Path.* 79, 619–627.
McFadden, E. S. (1930). *J. Am. Soc. Agron.* 22, 1050–1051.
Melchers, G., Sacristan, M. D., and Holder, A. A. (1978). *Carlsberg Res. Commen.* 43, 203–208.
Meynell, G. G. (1973). *Bacterial Plasmids.* MIT Press, Cambridge.
Moses, P. B. and Chua, N.-H. (1988). *Sci. Am.* 88–93.
Murai, N., Sutton, D. W., Murray, M. G., Slighton, J. L., Merlo, D. J., Reichert, N. A., Sengupta-Gopalan, C., Stock, C. A., Barker, R. F., Kemp, J. D., and Hall, T. C. (1983). *Science* 22, 476–482.
Nagata, T. and Takebe, I. (1971). *Planta* 99, 12–16.
Newell, C. A. and Hymowitz, R. (1982). *Crop Sci.* 22, 1062–1066.
Ohyama, K. *et al.* (1986). *Nature,* 322, 572–574.
Powell-Abel, P., Nelson, R. S., De, B., Hoffmann, N., Rogers, S. G., Fraley, R. T., and Beachy, R. N. (1986). 232, 738–743.
Power, J. B., Berry, S. F., Champman, J. V., and Cocking, E. C. (1980). *Theor. Appl. Genet.* 57, 1–6.
Quellete, R. P. and Cheremisinoff, P. N. (1985). In *Applications of Biotechnology,* Technormic Pub. Co., Lancaster, Pennsylvania.
Schell, J. (1987). *Science,* 237, 1176–1183.
Shepard, J. F., Bidney, D., Barsby, T., and Kemble, R. (1983). *Science* 219, 683–688.
Shinozaki, A. Ohme, M. Tomaka, M. (1986). *EMBO J.* 5, 2043–2049.
Simmonds, N. W. (1979). In *Principal of Crop Improvements* (N. W. Simmonds, ed.). Longman, New York.

Smith, H. D. and Wilcox, K. W. (1970). *J. Mol. Biol.* 51:371–391.
Stewart, F. C. (1983). In *Handbook of Plant Cell Culture* (P. V. Ammirato, D. A. Evans, W. R. Sharp, and Y. Yamada, eds.). Vol. 1, 1–12.
Watson, J. D. and Crick, F. H. C. (1953). *Nature,* 171, 964–967.
Zaenen, I., Van Larebeke, N., Teuchy, H., Van Montagu, M., and Schell, J. (1974). *J. Mol. Biol.* 86, 109–127.

Part I

Transgenic Monocotyledonous Plants

1

Transgenic Rice

S. L. Kothari, M. R. Davey,
P. T. Lynch, R. P. Finch,
and E. C. Cocking

Plant Genetic Manipulation Group
Department of Botany
University of Nottingham
Nottingham, NG7 2RD, United Kingdom

1.1 Introduction

During the past decade, we have witnessed the genetic manipulation of a wide range of plants, using both recombinant DNA and protoplast fusion technology. For cereal plants, direct interaction of isolated protoplasts with plasmids, coupled with improved procedures for the regeneration of plants from protoplasts, has provided a procedure for producing transgenic plants (Cocking and Davey, 1987). Rice (*Oryza sativa* L.) is the staple food of more than half the population of the world and, as such, is a prime target for genetic manipulations leading to the production of transgenic plants. Many of the advances in plant genetic manipulation have followed advances made in the animal field. In the latter case, the use of a chimeric gene for the production of transgenic individuals first was demonstrated in mice by microinjecting eggs. The imagination of animal scientists was captivated by this success, since it was straightforward conceptually and appealing from the standpoint of both the speed and the precision with which phenotypic changes could be induced (Wall *et al.*, 1990). However, it soon became clear that, before transgenic farm animals would become common in the livestock production industry, a number of formidable hurdles needed to be overcome, including the development of mass gene transfer technology and mass screening procedures. A close parallel exists in the development of procedures for the production of transgenic rice plants, which have, since the late 1980s, similarly captivated the imagination of plant investigators. Among the cereals, rice has been the predominant subject, because of the availability of a somatic cell protoplast system that is capable of efficient regeneration into plants through somatic embryogenesis (Abdullah *et al.*, 1986; Finch *et al.*, 1990). This system has facilitated direct interaction with chimeric genes, coupled with stringent selection procedures, for the production of transgenic rice plants. In this chapter, we will highlight the successes achieved in rice, which clearly indicate that major advances in other cereals soon will be forthcoming.

1.2 Transient Expression of Foreign Genes in Rice

1.2.1 DNA DELIVERY INTO RICE PROTOPLASTS

Gene transfer into plant cells can be assessed within several hours of the transformation process by assaying the activity of the gene product. These transient expression assays use specific reporter genes that transcribe enzymes such as β-glucuronidase (GUS), chloramphenicol acetyltransferase (CAT), and luciferase, the activity of which can be monitored readily. In addition, neomycin phosphotransferase II activity (NPTII), conferred by the *npt*II gene, also has been employed in transient expression studies as well as, more extensively, in stable transformation studies. Transient gene expression has been used to optimize the parameters that regulate DNA delivery into protoplasts and intact cells, and to assess the nature and functional activity of different promoter sequences (Ou-Lee *et al.*, 1986; Werr and Lörz, 1986; Junker *et al.*, 1987; Dekeyser *et al.*, 1989). The information obtained from transient expression experiments can provide a basis for studies on stable gene introgression.

Since the first reported use of polyethylene glycol (PEG) to introduce plasmids into tobacco protoplasts (Krens *et al.*, 1982), this agent has been used extensively to mediate plasmid uptake into a number of protoplast systems, including that of rice (Davey *et al.*, 1989). Junker *et al.* (1987) studied the transient expression of the *npt*II gene, fused to a range of promoter sequences, by applying the PEG transformation procedure to protoplasts isolated from several cereals, including the Japonica rice variety 'Taipei 309.' The level of NPTII activity in extracts from transformed protoplasts was dependent on the amount of DNA used for transformation, and also varied according to the promoter sequence. The chimeric constructs were rated, in descending order, as 35S-NPT, 1′,2′-NPT, Nos-NPT, and SuSy-NPT. The authors concluded that any type of cereal protoplast could be used for transient expression studies, as long as protoplast yields were adequate and their viability in culture could be maintained for several days. The investigators stated that, in the case of the *npt*II gene, the minimum number of protoplasts required for measurable gene expression was 5.0×10^6.

Activity of the *gus* gene has been reported in two transient expression studies after PEG-induced plasmid uptake (Marcotte *et al.*, 1988; McElroy *et al.*, 1990). The uptake procedure used by Marcotte and colleagues was essentially the same as that reported by Krens *et al.* (1982). In their studies, Marcotte *et al.* (1988) demonstrated expression and normal regulation of an ABA-inducible promoter from wheat in rice cell suspension protoplasts; the 650-bp *Em* promoter was sufficient for ABA induction of the *gus* gene. Such results also provided evidence for rapid phytohormone activation of gene transcription. In studies using protoplasts from cell suspensions of the Japonica rice variety 'Lemont,' McElroy *et al.* (1990) characterized the 5′ region of the rice actin 1 (*Act*1) gene, and showed that it was an efficient promoter for regulating the constitutive expression of the *gus* gene in protoplasts transformed using the PEG method. These studies with isolated protoplasts complemented investigations using the same plasmid constructs delivered to suspension-cultured rice cells by the biolistic method (Wang *et al.*, 1988; Cao *et al.*, 1990).

In experiments to optimize the parameters for stable cotransformation, Peng *et al.* (1990) studied PEG-induced uptake of pPUR into protoplasts of the Indica rice variety 'IR54.' The composition of the incubation solution influenced plasmid uptake, and heat-shock treatment of the recipient protoplasts also stimulated gene expression. Although PEG alone has been employed by many workers to induce plasmid uptake into isolated protoplasts, Zhu *et al.* (1990) report its use in combination with cationic liposomes. They prepared a complex of pAP2034, encoding the *npt*II gene, with a commercial preparation of cationic liposomes (Lipofectin). This complex was mixed with protoplasts isolated from suspension-cultured cells of the Chinese rice variety 'DGZ' prior to treatment with PEG of molecular weight 6000. Of 133 calli arising from these experiments, 19 showed NPTII activity, giving a transformation frequency higher than that reported by other workers using reversed-phase evaporation or pH-sensitive liposomes in dicotyledonous protoplast systems.

Concurrent with the use of chemical procedures has been the development of electroporation to mediate DNA delivery into protoplasts. Ou-Lee *et al.* (1986) reported expression of the *cat* gene by rice protoplasts after electroporation. In this study, the efficiency of the cauliflower mosaic virus (CaMV) 35S promoter was compared with that of the long terminal repeat promoter of *Drosophila*, but no significant difference in the level of gene expression was observed with these different promoters. Dekeyser *et al.* (1989) compared transient expression of the *npt*II gene carried on pLD1 (8.4 kb), driven by constitutive and regulated promoters, after electroporation of rice protoplasts. In these studies, the 2′ promoter of the octopine T-DNA was 3–4 times more efficient than the CaMV 35S promoter and 10 times more efficient than the *nos* or the 1′ promoter.

Direct comparison of the two methods of DNA uptake into protoplasts is difficult, since other parameters are involved, including the type of gene construct employed, the amount of DNA used, and the nature of the tissue used as the source of protoplasts. Zhang *et al.* (1992) compared electroporation with PEG-induced uptake of the plasmid pDW2, carrying the *cat* gene, into rice protoplasts. In this study, CAT activity of PEG-treated protoplasts was significantly higher than that of the electroporated protoplasts. However, this result was not in agreement with previous results on stable transformation in rice by the same authors (Zhang *et al.*, 1988). These contradictory results may have resulted from the different ages of the cell suspension cultures used as the source of protoplasts, and the deleterious effects of PEG on subsequent protoplast culture.

Blackhall *et al.* (1990) have established a flow cytometric technique to optimize conditions for the electroporation of rice protoplasts, without the use of a transient expression assay. Protoplasts were electroporated in the presence of the fluorescent macromolecule fluorescein isothiocyanate-labeled dextran (FITC-D), followed by incubation with propidium iodide and analysis by flow cytometry. Optimal electroporation parameters gave a population of protoplasts that fluoresced green as a result of the uptake of FITC-D, and did not stain with propidium iodide. Such protoplasts were assumed to be viable, since their intact plasma membranes excluded propidium iodide.

1.2.2 DNA DELIVERY INTO RICE TISSUES

Transient expression assays have been used to assess the efficiency of a number of innovative methods for delivering foreign DNA into intact rice cells; some of these methods are discussed later in this chapter. Such techniques include particle bombardment of rice callus by explosive discharge (Wang *et al.*, 1988) or compressed air (Oard *et al.*, 1990), imbibition of mature dry rice seeds in DNA solutions (Aryan *et al.*, 1991), and electroporation of leaf tissues (Dekeyser *et al.*, 1990). The particle bombardment technique already has permitted rapid assessment of novel gene constructs in transient expression assays. However, in such systems, the *gus* gene is often the transient expression marker employed, which can lead to misinterpretation of results, since a number of rice tissues, including embryos, exhibit intrinsic GUS-like activity (Hu *et al.*, 1990). Further, older rice tissues can accumulate contaminating fluorogenic compounds, resulting in endogenous fluorescence that is not the result of GUS activity (Raineri *et al.*, 1990). Consequently, experimental plant tissues must be chosen with caution, and improved marker systems remain a necessity.

1.3 Selectable Markers for Stable Transformation of Rice

Several selective agents and antibiotic resistance genes have been investigated to obtain stably transformed tissues in the Gramineae, including rice. The *npt*II gene, encoding resistance to kanamycin, has been used most commonly to select transformed protoplast-derived rice tissues. Uchimiya *et al.* (1986) recovered transformed callus from protoplasts of the rice variety C5924, treated with pCT2T3 (18.3 kb) in the presence of PEG, that showed resistance to 100 μg/ml kanamycin 7 days after DNA uptake. In subsequent experiments in the same laboratory, employing the smaller plasmid pCW (5.4 kb), Toriyama *et al.* (1988) used geneticin 418 (G418) instead of kanamycin for selection. The transformed colonies were selected on 20 μg/ml G418 after 3 wk of culture. One month after electroporation, the transformed colonies were rechallenged with the same concentration of G418. These authors considered G418 to be more effective than kanamycin for selecting transformed protoplast-derived rice tissues. In experiments to assess PEG, electroporation, and a combination of these procedures for delivering pCaMVNEO (5.4 kb) to cell-suspension protoplasts of the variety 'Taipei 309,' Yang *et al.* (1988) exposed plasmid-treated protoplasts to 100 μg/ml kanamycin after 14 days of culture, and replaced the antibiotic every 7 days over a period of 4 wk. Peng *et al.* (1990) selected transformed colonies of the Indica rice variety 'IR54' on the same level of kanamycin 3 wk after PEG-induced cotransformation of protoplasts with pKAN (5.78 kb) and pPUR (5.68 kb). These plasmids carried the CaMV 35S promoter and the *neo* and *gus* genes, respectively. After 4 wk, colonies were transferred to medium without the antibiotic. Of the protoplast-derived kanamycin resistant tissues, 20–30% also showed *gus* gene expression, although this result varied among calli. Southern blot analysis, using a probe for *gus*A, indicated integration of several copies of the gene, often as head-to-tail tandem repeats, but there was no clear relationship between foreign gene copy number and GUS activity. In a detailed evaluation of selectable markers for rice transformation, Dekeyser *et al.* (1989) tested the sensitivity of protoplast-derived rice tissues to the antibiotics kanamycin, G418, hygromycin, and bleomycin, to the drug methotrexate, and to the herbicide phosphinothricin. They reported that the growth of rice callus was inhibited by 2.0 mg/liter methotrexate, 10 mg/liter phosphinothricin, 20 mg/liter bleomycin, 50 mg/liter hygromycin, and 100 mg/liter G418. Kanamycin at 500 mg/liter only partially inhibited callus growth. These workers evaluated kanamycin and G418 resistance conferred on protoplast-derived rice tissues by the *npt*II gene and phosphinothricin resistance conferred by the bialaphos resistance (*bar*) gene. Colonies derived from protoplasts transformed by pLDI (8.4 kb) carrying the *npt*II gene or by pGSFR280 (10.1 kb) carrying the *bar* gene could be selected on 200 mg/liter kanamycin (or 100 mg/liter G418) and 10 mg/liter phosphinothricin, respectively, 7 days after gene uptake into protoplasts. Thus, gene products that inactivated G418 or phosphinothricin permitted transformed calluses to grow almost unimpeded on toxic concentrations of these selective agents.

The apparent "leakiness" of kanamycin selection reported by some workers has encouraged the use of hygromycin as an agent to select transformed rice cells. Shimamoto *et al.* (1989) reported that hygromycin was more effective than kanamycin for selecting protoplast-derived tissues of the Japonica rice variety 'Nipponbare' that were transformed by pGL2 carrying the bacterial hygromycin phosphotransferase (*hph*) gene flanked by the CaMV 35S promoter and a polyadenylation site from CaMV. Protoplasts were selected by addition of 20 μg/ml hygromycin to the culture medium 2 wk after electroporation-induced uptake of plasmid in the presence of calf thymus carrier DNA. Few background colonies formed on the selection plates. Most of the hygromycin-resistant colonies continued to grow after transfer to medium containing the same concentration of hygromycin; 3–4 wk later, the colonies were transferred to medium lacking the antibiotic. Datta *et al.* (1990) used the same plasmid pGL2 to transform protoplasts isolated from a microspore-derived embryogenic cell suspension of the "Indica-type" variety 'Chinsurah Boro II.' Transgenic protoplast-derived colonies were selected on 25 μg/ml hygromycin 2 wk after PEG-induced gene uptake. Colonies that appeared after 4–5 wk of selection were transferred to medium with the same concentration of hygromycin for an additional 2–4 wk.

1.4 Integration and Expression of Foreign Genes in Transformed Rice Tissues and Plants

Foreign genes introduced into protoplasts may become integrated into the chromosomes of recipient cells. Several genes, such as the *cat, gusA, npt*II, and *hph* genes, have been analyzed in transformed tissues and transgenic rice plants. In the earliest reports, Uchimiya *et al.* (1986) and Morota and Uchimiya (1987) provided evidence for the presence of the aminophosphotransferase(3′)II gene [*aph(3′)*II; *npt*II] in transformed rice callus using Southern hybridization. One transformed clone showed more than a 10-fold higher gene activity than other colonies because of higher copy numbers of the gene located at multiple sites in the chromosome(s). The DNA (pCT2T3) used in these experiments also contained the nopaline synthase gene and *ars* from tobacco chromosomal DNA, but the authors were unable to conclude whether the *ars* sequence had any role in rice protoplast transformation. Such transformed colonies were selected at a frequency of 2–3% relative to the number of colonies that developed in the absence of kanamycin selection. In experiments described by Toriyama *et al.* (1988), 5 transgenic plants were regenerated from 19 G418-resistant colonies. In this report, the authors also emphasized the advantage of G418 for selection, since colonies selected on kanamycin-containing medium failed to produce plants. APH(3′)II activity was detected in leaf extracts of regenerated plants; the chimeric *aph(3′)*II gene, used as a probe, hybridized to a 2.7-kb fragment of *Eco*RI/*Hin*dIII digested plant DNA corresponding to the intact *aph(3′)*II gene of pCN. Cytological analysis of root tip preparations showed one plant to be diploid and three to be tetraploid, whereas one had an unidentified chromosome complement.

Studies to compare the delivery of pCaMVNEO to cell suspension protoplasts of 'Taipei 309' by PEG, electroporation, or a combination of these procedures (Yang *et al.*, 1988) showed that electroporation gave the highest absolute transformation frequency of 19.9×10^{-6} (defined as the number of kanamycin-resistant protoplast-derived colonies compared with the number of protoplasts electroporated with DNA) and a relative frequency of 26.1% (the number of colonies that developed in the presence of kanamycin from plasmid-treated protoplasts compared with the number of colonies derived from plasmid-treated protoplasts that developed in the absence of the antibiotic). NPTII activity was confirmed in 20 randomly selected kanamycin-resistant colonies obtained using the three plasmid uptake procedures. The presence of plasmid DNA sequences containing the *npt*II gene also was confirmed in 20 protoplast-derived colonies by Southern hybridization, but only 1 of the 20 clones analyzed had an unmodified form of the *npt*II gene. This particular clone arose after electroporation. Other clones showed a series of bands additional to that of the *npt*II gene, some of which were common among different clones. Thus, some clones contained common bands of 10.0, 4.1, 3.9, and 2.8 kb that were probably the result of rearrangement of plasmid in the transformed cells. Reconstruction experiments indicated that the clones analyzed contained 1–4 copies of the *npt*II gene per plant genome. Such DNA insertion patterns are similar to those reported in transformed maize tissues (Fromm *et al.*, 1986) and transgenic maize plants (Rhodes *et al.*, 1988).

Transformed rice tissues have been produced after the uptake of plasmid molecules of different configurations into isolated protoplasts. For example, Zhang *et al.* (1988) and Shimamoto *et al.* (1989) used supercoiled DNA in their transformation experiments. Uchimiya *et al.* (1986) and Toriyama *et al.* (1988) employed circular plasmid molecules, whereas Matsuki *et al.* (1989) and Datta *et al.* (1990) treated plasmids with restriction enzymes that cut the DNA on either side of the foreign gene prior to DNA uptake into protoplasts. Some workers also have mixed plasmids with carrier DNA, such as calf thymus DNA sheared to about 4 kb (Yang *et al.*, 1988; Shimamoto *et al.*, 1989; Datta *et al.*, 1990), but others have reported this procedure to be unnecessary (Zhang *et al.*, 1988).

The experiments performed by Yang *et al.* (1988) formed the basis for subsequent work that resulted in the production of transgenic plants of the variety 'Taipei 309' (Zhang *et al.*, 1988). After electroporation of protoplasts with pCaMVNEO, 14 plants, including 2 albinos, were regenerated through somatic embryogenesis from 400 kanamycin-resistant colonies. DNA from 6 of the green regenerated plants and 9 randomly chosen kanamycin-resistant protoplast-derived colonies hybridized to the 1.0-kb *Bam*HI fragment of pCaMVNEO that contained the *npt*II gene. The transferred gene apparently was unmodified in transformed tissues and regenerated plants. Interestingly, the 9 kanamycin-resistant colonies expressed NPTII activity, but only 2 of the regenerated transgenic plants (defined as the R_0 generation) gave a positive enzyme assay. Transformed tissues showed gene activity after several passages on kanamycin-free medium, thereby confirming the constitutive expression of the engineered gene. Zhang *et al.* (1988) also observed that heat-shock treatment of freshly isolated protoplasts prior to DNA uptake increased the relative and absolute transformation frequencies, as assessed by the number of kanamycin-resistant colonies, from 16 to 25% and 2.5×10^{-6} to 28×10^{-6}, respectively. Plant regeneration was inhibited by the presence of kanamycin in the medium used for the induction of this process. Using protoplasts of the same rice variety ('Taipei 309') used by Zhang *et al.* (1988), Battraw and Hall (1990) produced transgenic plants cotransformed by the *npt*II and *gus* genes. Both genes were carried on the same plasmid, pNEOGUS15 (7.3 kb), which was constructed by cloning the 1.7-kb *Hin*dIII fragment from pCaMVNEO containing the chimeric *npt*II gene into the *Hin*dIII site of pBI221.1. The resulting construct was introduced into rice protoplasts by electroporation; transformed colonies were selected on medium containing kanamycin or G418. Histochemical analysis showed GUS activity to be 10-fold higher in leaves of transgenic rice plants than in leaves of transgenic tobacco plants carrying the *gus* gene, whereas GUS activity was similar in rice and tobacco roots. In rice, GUS activity confirmed that the CaMV 35S promoter functions in cells of the epidermis, mesophyll, and vascular bundles of leaves, as well as the cortex and vascular cylinder of the root. However, this promoter was only marginally active in the root epidermis.

Most workers have failed to express the frequency of transgenic plant recovery in terms of the number of protoplasts originally plated after plasmid uptake. Calculations using the data of Zhang *et al.* (1988) show that, in their experiments, one transgenic plant was recovered from every 1.0×10^6 protoplasts plated. This result assumes a maximum recovery of kanamycin-resistant colonies following heat-shock treatment of protoplasts prior to plasmid uptake. In practice, 10 times this number of protoplasts may need to be plated to recover one transgenic plant. Other workers have not presented the data to allow comparable calculations of transformation frequency in terms of plant recovery. However, in the absence of such details, some workers claim to have regenerated transgenic rice plants in relatively large numbers from transformed tissues. For example, more than 50% of hygromycin-resistant colonies selected by Shimamoto *et al.* (1989) produced shoots in every experiment. From four different experiments, 6 plants were shown to contain a 1.1-kb fragment that hybridized to the *hph* gene; the presence of multiple bands indicated multiple chromosomal integration sites. The copy number of the integrated gene was between 2 and 10 per diploid plant cell. Cotransformation of protoplasts with pGL2 mixed with pBI221 carrying the *gus* gene did not reduce the frequency of production of hygromycin-resistant colonies. Transgenic plants were recovered that also contained the complete 3.0-kb *gus* gene.

Matsuki *et al.* (1989) also employed hygromycin selection in cotransformation experiments to recover transgenic plants of the Japonica variety 'Yamahoushi' after electroporation-induced uptake of pCH (5.0 kb), carrying the *hph* gene and pUC12-012-GUS (5.8 kb), with the *gus* gene linked to ORF12 (*rol* C) of the Ri plasmid TL-DNA. From several hundred hygromycin-resistant calluses, 26 plants that appeared to be albino were obtained. Two plants expressed GUS activity as shown by fluorimetric analysis. GUS activity directed by the *rol* C promoter in leaves and roots of these transgenic plants was significantly higher than in nontransformed plants. Organ-related activation of the *gus* gene was not observed in rice, in contrast to tobacco, perhaps because of structural differences in vascular development between monocotyledons and dicotyledons. In rice leaves, GUS was localized in vascular tissues, particularly in association with phloem cells. The cellular localization of GUS expression in roots was found to be the same as in leaf sheaths. When comparing their results with those of other workers, Matsuki *et al.* (1989) concluded that tissue-specific expression of the *rol* C promoter was similar in both dicotyledons and monocotyledons.

After the selection of numerous hygromycin-resistant protoplast-derived colonies of the "Indica-type" variety 'Chinsurah Boro II,' Datta *et al.* (1990) regenerated 77 green transgenic plants and 16 albinos. Southern hybridization demonstrated the integration of transforming plasmid DNA into high molecular weight plant DNA, the presence of the 1.033-kb *Bam*HI *hph* gene fragment in transgenic plants, and hybrid fragments between rice DNA and transforming DNA. The resistance of plant tissues to hygromycin B was complemented by enzymatic proof of gene function; specific phosphorylation of the antibiotic was obtained using protein extracts from transgenic plants.

Although most workers have employed antibiotic resistance to select transformed protoplast-derived rice tissues, some workers have omitted this procedure. Zhang and Wu (1988) confirmed the efficiency of DNA uptake into protoplasts of the Japonica rice varieties 'Pi-4' and 'Taipei 309' by counting the number of blue protoplast-derived cells in an *in situ* GUS assay performed 48 hr after PEG-induced uptake of pAI_1GUSN (6.5 kb). Subsequently, 61 of 378 randomly chosen calli showed GUS activity; of 378 regenerated plants of 'Pi-4' (of which 35 were albino), at least 85 were transgenic as determined by

DNA hybridization. The copy number of the GUS gene in transgenic plants ranged from 1 to 10 per cell; the majority of plants carried 1 or 2 copies per cell. Several albino plants showed a relatively high copy number. The ratio of calli that gave rise to green or albino plants was 10:1, indicating that most plantlets had the potential to develop into healthy mature plants. GUS activity in transgenic plants was confirmed using a fluorometric procedure. In two transgenic plants, GUS activity was 5- to 6-fold higher in their roots under anaerobic conditions, which correlated with the presence of the anaerobically stimulated maize *Adh*1 gene, whose promoter region and intron 1 were carried by pAI_1GUSN.

1.5 Inheritance of Foreign Genes in Transgenic Rice Plants

The stable integration of foreign genes into rice germ lines and the continued expression of such genes through subsequent seed generations is essential if the full potential of transformation technology is to be exploited. The production of transgenic plants in a limited number of rice varieties has enabled gene inheritance studies to be initiated. Preliminary results are available through the production of fertile transgenic plants carrying the genes encoding kanamycin resistance (Zhang *et al.*, 1988), hygromycin resistance (Datta *et al.*, 1990), and hygromycin resistance combined with GUS activity (Shimamoto *et al.*, 1989; Terada and Shimamoto, 1990).

Shimamoto *et al.* (1989) selected 4 of their 14 transgenic plants for studies of the transmission of the *hph* gene through seed progeny. Of 209 seeds produced following selfing of the 4 transgenic plants, 179 germinated on medium containing hygromycin at 30 mg/liter to give a resistant:sensitive seedling ratio of 6:1. Survival of the seedlings on this concentration of hygromycin, which normally inhibits growth of nontransformed seedlings, was considered evidence that the active *hph* gene had been transmitted to seed progeny. Additional evidence for the presence of the foreign gene was obtained from DNA hybridization. More detailed analysis of one hygromycin-resistant plant and its seed progeny showed that 10 progeny plants also carried the *gus* gene, and 41 of 57 seeds showed GUS activity in their embryos and endosperms. In a more detailed study of GUS expression, Terada and Shimamoto (1990) analyzed a transgenic plant, that showed GUS activity and two copies of the CaMV35S-*gus* gene, and its selfed seed progeny. Histochemical assay confirmed GUS activity in and around the vascular bundles of leaves, floral organs, and roots. In leaf sheaths and blades, GUS activity was localized in sclerenchymatous cells surrounding the vascular bundles, whereas on the adaxial side of the midrib GUS activity was observed only in the metaphloem cells. In floral organs, *gus* gene expression occurred in the basal part of the ovary and in vascular bundles. Staining of pollen was rare, but staining of root tips was intense. As reported by Shimamoto *et al.* (1989), embryos and endosperm showed staining after incubation with the histochemical reagent X-Gluc (5-bromo-4-chloro-3-indolylglucuronide); the aleurone layer as well as cells in the center of the seed showed activity. A change in localization of GUS activity occurred during seed germination. In germinating embryos, high activity was associated with the outermost layer of the scutellum and leaf or root primordia, in contrast to dormant or imbibed seeds in which staining of the scutellum was not detected. These results indicated differential expression of the CaMV 35S promoter in transgenic rice plants, although, as Terada and Shimamoto (1990) emphasized, there was some degree of uncertainty in the interpretation of GUS expression in tissue sections. The results of experiments involving further selfing and backcrossing of transgenic rice plants produced by Shimamoto *et al.* (1989) should be available soon.

DNA hybridization studies confirmed that the integration pattern of the *hph* gene in 10 seed progeny of plants of the "Indica-type" variety 'Chinsurah Boro II' was identical to that of the parent plant (Datta *et al.*, 1990). In addition, 31 seeds derived from 5 primary transgenic plants were all resistant to hygromycin at 40 μg/ml. The lack of segregation of the integration pattern of the foreign DNA in these offspring indicated that the primary transformed plants were homozygous, probably because the protoplasts used for plasmid uptake were isolated from an embryogenic microspore-derived culture. The authors concluded that further genetic analysis of additional transgenic plants from independent experiments, and determination of the ploidy levels of the cell suspension before transformation, should clarify this point.

Seeds harvested from one of the transgenic R_0 rice plants of variety 'Taipei 309' regenerated from kanamycin-resistant protoplast-derived tissues by Zhang *et al.* (1988) were germinated on medium containing kanamycin at 20 μg/ml. The 16 first generation (R_1) seedlings that remained green in the presence of the antibiotic were raised to maturity in the glasshouse (Davey *et al.*, 1991). DNA dot and Southern blot analysis confirmed the presence of the *npt*II gene in the kanamycin-resistant R_1 plants. Transmission of the *npt*II gene to the R_1 plants in the variety 'Taipei 309' appeared not to exhibit a Mendelian segregation pattern. This could be a disadvantage if transgenic plants are to be incorporated into conventional breeding programs. Although Mendelian segregation of genes has been reported in transgenic dicotyledons after direct DNA uptake into protoplasts (Paszkowski *et al.*, 1984), non-Mendelian segregation of T-DNA has been observed in tobacco (Budar *et al.*, 1986) and *Arabidopsis thaliana* (Feldmann and Marks, 1987). In rice, segregation of introduced foreign genes has yet to be studied through cross-fertilization with pollen from a nontransformed donor. The need for genetic tests to analyze the pattern of foreign gene integration and inheritance has been emphasized for rice (Toriyama *et al.*, 1988) and for other cereals (Gasser and Fraley, 1988; Strange, 1990; Vasil, 1990).

The phenotypic characteristics of transgenic plants may be influenced by the presence of the foreign gene, protoclonal variation, and the methods used for plasmid delivery to protoplasts. Several studies have reported morphological variation in Japonica and Indica rice plants regenerated from protoplast-derived tissues (R_0 plants) and their R_1 seed progeny (Ogura *et al.*, 1987, 1989; Abdullah *et al.*, 1989; Lee *et al.*, 1989). In an attempt to identify any positive or negative phenotypic changes associated with transformation, Davey *et al.* (1991) compared the phenotypic characteristics of R_1 transgenic plants carrying the *npt*II gene with the phenotypic variation already reported in nontransformed seed progeny of protoplast-derived plants of the same cultivar (Abdullah *et al.*, 1989). At maturity, the R_1 transgenic plants produced more tillers, but flowered later than nontransformed R_1 protoclones and seed-derived plants. Flag leaf length:width ratio in the R_1 transgenic plants was similar to that of nontransformed seed-derived plants, but panicles of transformed plants were shorter with fewer primary branches and spikelets than those of nontransformed R_1 protoclones or seed-derived plants. The presence of the *npt*II gene may result in alterations in plant phenotype, particularly a reduction in fertility over nontransformed protoclonal-derived plants. The phenotypic characteristics of the R_2 generation 'Taipei 309' plants, and the inheritance of the *npt*II gene, are being investigated currently in a glasshouse experiment at Nottingham and in a trial at Arizona State University field station.

1.6 Development of Methods Other than Direct Uptake of DNA into Protoplasts for Stable Transformation of Rice

The potential of different methods of transferring genes to cereals was reviewed critically by Potrykus (1990a). Although the only method, to date, for the reproducible transformation of rice involves direct DNA uptake into protoplasts, other methods are being investigated; the merits and limitations of these also have been summarized (Potrykus, 1990b). It may be desirable to develop a transformation system for rice and other cereals that negates the use of protoplasts and eliminates the labor-intensive establishment and maintenance of cell-suspension cultures.

The general recalcitrance of monocotyledons to *Agrobacterium*-mediated gene transfer has precluded the use of this method to deliver genes into rice. Nevertheless, there are claims of rice transformation by Ti and Ri plasmids but, as yet, transgenic plants have not been produced. Baba *et al.* (1986) introduced an octopine wild-type Ti plasmid into rice cells by PEG-induced fusion of rice protoplasts with spheroplasts of *Agrobacterium tumefaciens*. Approximately 1 in 10^4 of the protoplast-derived cell colonies synthesized T-DNA specific opines. Raineri *et al.* (1990) produced transformed rice callus by inoculating mature embryos of the variety 'Nipponbare' with the wide-host-range supervirulent *Agrobacterium* strain A281 (pTi Bo542). In other experiments, embryos of the Japonica variety 'Fujisaka 5' produced roots after infection with a limited-host-range bacterial strain carrying the *npt*II gene. Such roots produced callus when cultured on medium containing kanamycin and also fluoresced when assayed for GUS activity. The lack of plant regeneration from root-derived callus may have resulted from the inhibitory effects of kanamycin in the culture medium used for selection.

Agroinfection has been shown to lead to the systemic spread of maize streak virus in maize plants (Grimsley *et al.*, 1987), showing that viral genomes that are not themselves infective can infect cereal cells when introduced between the T-DNA borders of the *Agrobacterium* Ti plasmid. However, as yet, agroinfection has not resulted in transgenic cereals.

Large volumes of water are taken up during the initial stages of seed germination. In attempting to exploit this phenomenon for transformation, Töpfer *et al.* (1989) soaked isolated dry and viable embryos of the Indica variety 'IR36' for 2 hr in pDWNeo2 containing a dimer of a wheat gemini virus genome in which the *npt*II gene replaced the viral coat protein genome. After 3 days of culture, NPTII activity was detected in the embryos in a transient expression assay. The authors emphasized that the results obtained in this transient assay resulted from DNA uptake and expression in cells of the plant embryos and was not, apparently, due to microbial contamination. In further attempts to exploit imbibition as a method of DNA uptake, Aryan *et al.* (1991) also reported transient gene expression in mature rice embryos after imbibition of dry seeds in solutions of plasmids in which the *gus* gene was linked to the rice glutelin, alcohol dehydrogenase, or CaMV 35S promoters. GUS activity was restricted to embryos, its level depending on the concentration of dimethyl sulfoxide used during imbibition, the inclusion of PEG in the DNA solution, and the duration of soaking. However, stably transformed rice tissues have not been reported, to date, using this technique.

The transfer of foreign DNA into recently pollinated rice florets, known as the pollen tube pathway method, first was reported as a technique for rice transformation by Duan and Chen (1985). Total DNA was transferred from a rice variety with purple leaf and glume pigmentation into a variety with normal pigmentation by cutting the stigmas and applying a drop of DNA solution to the cut surface. DNA was thought to reach the ovules through the open ends of the pollen tubes. Some of the progeny from the DNA-treated florets exhibited pigmentation similar to that of the donor plant. However, there was no molecular evidence to support this claim for rice transformation. Using the same technique,

Luo and Wu (1988) attempted to introduce p35S-*npt*II-*hist*3 into the Japonica rice variety 'Fujisaka 5.' Of 54 seeds resulting from DNA-treated florets, 10 gave positive DNA hybridization signals in a slot-blot assay. Southern blot hybridization, using p35S-*npt*II as probe, suggested integration of the reporter gene. However, the *npt*II gene was not transferred to subsequent generations. Although this method of gene transfer is simple, conclusive molecular evidence for foreign gene integration and expression in rice plants has not been reported.

Biolistics, or the particle gun bombardment method, involving the high velocity microprojectile delivery of DNA or RNA into plant cells as developed by Sanford *et al.* (1987), has received considerable publicity for gene transfer into a number of crop plants. In rice, transient expression of genes delivered by particle bombardment has been achieved in several tissues, including suspension-cultured cells of 'Taipei 309' (Wang *et al.*, 1988) and Lemont (McElroy *et al.*, 1990) varieties, compact calli developed from mature seed scutella (Oard *et al.*, 1990), and embryo sections (Cao *et al.*, 1990). Some of these studies have been directed toward analysis of tissue-specific expression of promoters (Wang *et al.*, 1988). Interestingly, Wang *et al.* (1988) and Oard *et al.* (1990) assessed GUS activity after introduction of the same pAI$_3$GUSN (6.5 kb, carrying the maize *Adh* promoter, the maize *Adh* intron 1, and the *gus* and *nos* genes in pUC8) into recipient cells. Wang and colleagues employed an explosive discharge apparatus, whereas Oard *et al.* (1990) described the use of a relatively inexpensive airgun apparatus and the optimum parameters for gene delivery. Indeed, they provided the first estimate for absolute transformation frequency by biolistics (4.4×10^{-3}).

McElroy *et al.* (1990) used explosive discharge for gene transfer into cultured cells of the 'Lemont' variety. They concluded that foreign gene expression in transformed cells is dependent on the presence of an intact intron sequence in the promoter; in the absence of such a sequence, there is an *in vivo* requirement for intron splicing during the transformation process. In addition, these workers also concluded that the rice *Act*1 promoter may be preferable for gene expression in rice to the maize *Adh*1 promoter used previously in the same laboratory (Zhang and Wu, 1988), since the *Adh*1 promoter is not expressed constitutively in all tissues of transgenic rice plants. Although stably transformed plants arising from particle bombardment have been reported in other crops such as soybean (McCabe *et al.*, 1988) and maize (Fromm *et al.*, 1990; Gordon-Kamm *et al.*, 1990), there are no reports, to date, of transgenic rice plants being produced using this method. If the examples of maize transformation using biolistics are extended to rice, then considerable effort will be required to produce transgenic rice plants using this approach.

Microinjection is an additional approach that has been used to introduce DNA into plant cells. Although laborious, this technique has resulted in transgenic plants of oilseed rape after injection of DNA into microspore-derived proembryos (Neuhaus *et al.*, 1987). Microinjection may have potential for rice transformation if combined with plant regeneration from zygotic proembryos (Potrykus, 1990a). Additionally, this method may hold promise for organelle transfer into rice and other cereals, long-term.

Dekeyser *et al.* (1990) employed electroporation to introduce DNA into intact leaf bases of etiolated seedlings of rice. The essential steps of the technique involved eliminating plant-released nucleases, prolonging the DNA–explant interaction time, and expanding the duration of the electrical pulse. Using the *gus* reporter gene in transient expression assays, these workers demonstrated that the procedure also was applicable to sheath and lamina tissues of etiolated and green rice seedlings. The promoter from a pea light-harvesting chloroplast *a*/*b* binding protein gene displayed both light- and chloroplast-dependent expression in rice; the promoter from the *Arabidopsis* *S*-adenosylmethionine synthase gene was expressed preferentially in cells surrounding the vascular bundles. Future experiments should determine whether stably transformed tissues and transgenic plants of rice and other cereals can be produced by this approach.

1.7 Concluding Remarks

The establishment of reproducible protoplast-to-plant systems for several Japonica rice varieties has allowed transgenic plants to be produced in a number of laboratories. The range of varieties amenable to this technology undoubtedly will increase as further success is achieved in regenerating plants from protoplasts of Indica as well as other Japonica varieties. The chemical and physical conditions to which protoplasts are exposed during direct DNA uptake influence the transformation frequency, as do the cells used as a source of protoplasts. Considerable effort still is required to increase the frequency of transformation during direct DNA uptake, ideally to a level comparable to that obtained in many dicotyledons using *Agrobacterium*-mediated gene delivery. It is probable that biolistic and other methods also will, in due course, be used successfully to produce transgenic rice plants.

To date, most laboratories have focused attention on the uptake of readily assayable reporter genes into rice protoplasts for transient expression studies, and on antibiotic resistance genes in stable transformation experiments. Such transient expression investigations are important in analyzing plasmid constructs and promoter function, whereas the production of transgenic rice plants carrying antibiotic resistance genes should provide basic information relating to gene expression and inheritance. Such studies are fundamental in attempting to predict the effects of introducing agronomically important genes, such as those for virus resistance, seed proteins, and stress tolerance, into rice once the gene cloning has been accomplished and suitable plasmids carrying these traits have been constructed. Anther-specific chimeric ribonuclease genes have been introduced into tobacco and oilseed rape (Mariani *et al.*, 1990); this technology should be applicable to rice for the introduction of nuclear male sterility for hybrid seed production.

The ability to cotransform protoplasts with a mixture of plasmids provides a means of introducing nonselectable traits, as well as selectable markers, into rice. Selectable markers, such as genes for antibiotic resistance, enable transformed protoplast-derived cells to be recovered from nontransformed cells. Protoplasts of such transformed cell lines, if incorporated into somatic hybridization programs, permit the selection of somatic hybrid tissues after protoplast fusion. However, it seems likely that future legislation may prohibit the introduction of antibiotic resistance genes into crop plants such as rice. Thus, attention may need to be focused on alternative selection strategies, such as the use of fluorochromes coupled with flow cytometry, for selection of transformed protoplast-derived cells (Blackhall *et al.*, 1990).

In addition to the introduction of specific genes into rice by transformation, the transfer of genetically undefined characteristics and/or multigenic features from wild rice species and other grasses also offers considerable promise for genetically manipulating the cultivated crop. In this respect, Hayasahi *et al.* (1988) produced somatic hybrid plants between Japonica rice and four wild rice species, *O. officinalis, O. eichingeri, O. brachyantha,* and *Leersia perrieri.* Finch *et al.* (1990) produced putative somatic hybrid cell colonies by fusing cell-suspension-derived protoplasts of 'Taipei 309' with mesophyll protoplasts of the salt-tolerant wild species *Porteresia coarctata.* Another significant development in rice cell culture is the ability to regenerate plants from protoplasts isolated from cell suspensions of the wild rices, such as *O. rufipogon* (Baset *et al.*, 1991). This technique will enable those genetically important species to be incorporated into somatic hybridization programs with *O. sativa,* including the Indica varieties that cannot, as yet, be regenerated into plants from protoplasts.

References

Abdullah, R., Cocking, E. C., and Thompson, J. A. (1986). *Bio/Technology* **4,** 1087–1090.

Abdullah, R., Thompson, J. A., Khush, G. S., Kaushik, R. P., and Cocking, E. C. (1989). *Plant Sci.* **65,** 97–101.

Aryan, A. P., Li, W.-H., and Okita, T. W. (1991). *In* "Proceedings of the Second International Rice Genetics Symposium" p. 597–604. May 14–18, 1990. IRRI, Los Banos, Philippines.

Baba, A., Hasezawa, S., and Syono, K. (1986). *Plant Cell Physiol.* **27,** 463–471.

Baset, A., Finch, R. P., and Cocking, E. C. (1991). *Plant Cell Rep.* **10,** 200–213.

Battraw, M. J., and Hall, T. C. (1990). *Plant Mol. Biol.* **15,** 527–538.

Blackhall, N. W., Finch, R. P., Davey, M. R., and Cocking, E. C. (1990). *In* "Abstracts of the International Conference on Electroporation and Electrofusion," p. 21. Marine Biology Laboratory, Woods Hole, Massachusetts.

Budar, F., Thia-Toong, L., Van Montagu, M., and Hernalsteens, J. P. (1986). *Genetics* **114,** 303–313.
Cao, J., Wang, Y.-C., Klein, T. M., Sanford, J., and Wu, R. (1990). *In* "Plant Gene Transfer." (C. J. Lamb and R. W. Beachy, eds.) pp. 21–33. Wiley, New York.
Cocking, E. C., and Davey, M. R. (1987). *Science* **236,** 1259–1262.
Datta, S. K., Peterhans, A., Datta, K., and Potrykus, I. (1990). *Bio/Technology* **8,** 736–740.
Davey, M. R., Rech, E. L., and Mulligan, B. J. (1989). *Plant Mol. Biol.* **13,** 273–285.
Davey, M. R., Kothari, S. L., Zhang, H., Rech, E. L., Cocking, E. C., and Lynch, P. T. (1991). *J. Exp. Bot.* **42,** 1159–1169.
Dekeyser, R., Claes, B., Marichal, M., Van Montagu, M., and Caplan, A. (1989). *Plant Physiol.* **90,** 217–223.
Dekeyser, R. A., Claes, B., De Rycke, R. M. U., *et al.* (1990). *Plant Cell* **2,** 591–602.
Duan, X., and Chen, S. (1985). *China Agricultural Sci.* **3,** 6–9.
Feldmann, K. A., and Marks, M. D. (1987). *Mol. Gen. Genet.* **208,** 1–9.
Finch, R. P., Lynch, P. T., Jotham, J. P., and Cocking, E. C. (1991). *In* "Biotechnology in Agriculture and Forestry" (Y. P. S. Bajaj, ed.), p. 251–268. Vol. 14. Springer-Verlag, Berlin.
Finch, R. P., Slamet, I. H., and Cocking, E. C. (1990). *J. Plant Physiol.* **136,** 592–598.
Fromm, M. E., Morrish, F., Armstrong, C., Williams, R., Thomas, J., and Klein, T. M. (1990) *Bio/Technology* **8,** 833–839.
Fromm, M. E., Taylor, L. P., and Walbot, V. (1986). *Nature (London)* **319,** 791–793.
Gasser, C. S., and Fraley, R. T. (1988). *Science* **244,** 1293–1299.
Gordon-Kamm, W. J., Spencer, T. M., Mangano, M. L., Adams, T. R., Daines, R. J., Start, W. G., O'Brien, J. V., Chambers, S. A., Adams, W. R., Jr., Willetts, N. G., Rice, T. B., Mackey, C. J., Krueger, R. W., Kausch, A. P., and Lemaux, P. G. (1990). *The Plant Cell* **2,** 603–618.
Grimsley, N. H., Hohn, T., Davies, J. W., and Hohn, B. (1987). *Nature (London)* **325,** 177–179.
Hayashi, Y., Kyozuka, J., and Shimamoto, K. (1988). *Mol. Gen. Genet.* **214,** 6–10.
Hu, C.-Y., Chee, P. P., Chesney, R. H., Zhou, J. H., Miller, P. D., and O'Brien, W. T. (1990). *In* "Progress in Plant Cellular and Molecular Biology" (H. J. J. Nijkamp, L. H. W. Van der Plas, and J. Van Aartrijk, eds.), pp. 195–200. Kluwer Academic Publishers, Dordrecht.
Junker, B., Zimny, J., Lührs, R., and Lörz, H. (1987). *Plant Cell Rep.* **6,** 329–332.
Krens, F. A., Molendijk, L., Wullems, G. J. and Schilperoort, R. A. (1982). *Nature (London)* **296,** 72–74.
Lee, L., Schroll, R. E., Grimes, H. D., and Hodges, T. K. (1989). *Planta* **178,** 325–333.
Luo, Z., and Wu, R. (1988). *Plant Mol. Biol. Rep.* **7,** 69–77.
McCabe, D. E., Swain, W. F., Marinell, B. J., and Christou, P. (1988). *Bio/Technology* **6,** 923–926.
McElroy, D., Zhang, W., Cao, J., and Wu, R. (1990). *The Plant Cell* **2,** 163–171.
Marcotte, W. R., Bayley, C. C., and Quatrano, R. J. (1988). *Nature (London)* **335,** 454–457.
Mariani, C., De Beuckeleer, M., Truettner, J., Leemans, J., and Goldberg, R. B. (1990). *Nature (London)* **347,** 737–741.
Matsuki, R., Onodera, H., Yamauchi, T., and Uchimiya, H. (1989). *Mol. Gen. Genet.* **220,** 12–16.
Morota, H., and Uchimiya, H. (1987). *Jpn. J. Genet.* **62,** 363–368.
Neuhaus, G., Spangenberg, G., Mittelsten-Scheid, O., and Schweiger, H.-G. (1987). *Theor. Appl. Genet.* **7,** 30–36.
Oard, J. H., Paige, D. F., Simmonds, J. A., and Gradziel, T. M. (1990). *Plant Physiol.* **92,** 334–339.
Ogura, H., Kyozuka, J., Hayashi, Y., Koba, T., and Shimamoto, K. (1987). *Theor. Appl. Genet.* **74,** 670–676.
Ogura, H., Kyozuka, J., Hayashi, Y., and Shimamoto, K. (1989). *Jpn. J. Breed* **39,** 47–56.
Ou-lee, T., Turgeon, R., and Wu, R. (1986). *Proc. Natl. Acad. Sci. U.S.A.* **83,** 6815–6819.

Paszkowski, J., Shillito, R. D., Saul, M. W., Mandak, V., Hohn, T., Hohn, B., and Potrykus, I. (1984). *EMBO J.* **3,** 2717–2722.
Peng, J., Lyznik, L. A., Lee, L., and Hodges, T. K. (1990). *Plant Cell Rep.* **9,** 168–172.
Potrykus, I. (1990a). *Bio/Technology* **8,** 535–542.
Potrykus, I. (1990b). *Physiol. Plant.* **79,** 125–134.
Raineri, D. M., Bottino, P., Gordon, M. P., and Nester, E. W. (1990). *Bio/Technology* **8,** 33–38.
Rhodes, C. A., Pierce, D. A., Mettler, I. J., Mascarenhas, D., and Detmer, J. J. (1988). *Science* **240,** 204–207.
Sanford, J. C., Klein, T. M., Wolf, E. D., and Allen, N. (1987). *Part. Sci. Tech.* **5,** 27–37.
Shimamoto, K., Terada, R., Izawa, T., and Fujimoto, H. (1989). *Nature (London)* **338,** 274–276.
Strange, C. (1990). *BioScience* **40,** 5–14.
Terada, R., and Shimamoto, K. (1990). *Mol. Gen. Genet.* **220,** 389–392.
Töpfer, R., Gronenborn, B., Schell, J., and Steinbiss, H. (1989). *Plant Cell* **1,** 133–139.
Toriyama, K., Arimoto, Y., Uchimiya, H., and Hinata, K. (1988). *Bio/Technology* **6,** 1072–1074.
Uchimiya, H., Fushimi, T., Hashimoto, H., Harada, H., Syonok and Sugawara, Y. (1986). *Mol. Gen. Genet.* **204,** 204–207.
Vasil, I. K. (1990). *Bio/Technology* **8,** 296–301.
Wall, R. J., Bolt, D. J., Frels, W. I., Hawk, H. W., King, D., Pursel, W. G., Rexroad Jr., C. E. and Rohan R. M. (1990). *Agbiotech News Informat.* **2,** 391–395.
Wang, Y., Klein, T. M., Fromm, M., Cao, J., Sanford, J. C., and Wu, R. (1988). *Plant Mol. Biol.* **11,** 433–439.
Werr, W., and Lörz, H. (1986). *Plant Cell Rep.* **5,** 89–92.
Yang, H., Zhang, H. M., Davey, M. R., Mulligan, B. J., and Cocking, E. C. (1988). *Plant Cell Rep.* **7,** 421–425.
Zhang, H. M., Yang, Y., Rech, E. L., Golds, T. J., Davis, A. S., Mulligan, B. J., Cocking, E. C. and Davey, M. R. (1988). *Plant Cell Rep.* **7,** 379–384.
Zhang, H., Rech, E. L., Eyles, P. S., Cocking, E. C. and Davey, M. R. (1990). (*In preparation.*)
Zhang, W., and Wu, R. (1988). *Theor. Appl. Genet.* **76,** 835–840.
Zhu, Z., Hughes, K. W., Huang, L., and Li, X. (1990). *Focus* **12,** 41–44.

2

Transgenic Maize

Catherine J. Mackey,
T. Michael Spencer,
Thomas R. Adams,
Albert P. Kausch,
William J. Gordon-Kamm,
Peggy G. Lemaux*, and
Roger W. Krueger**

Department of Discovery Research
DEKALB Plant Genetics
Pfizer Central Research
Groton, Connecticut, 06340

Maize (*Zea mays* L.) is a major agricultural crop in the United States and throughout the world. In 1990, 74.5 million acres of maize were planted in the United States; the value of this crop is estimated at more than $17.8 billion (USDA/National Agricultural Statistics Service). Because of the agronomic importance of maize and the ease of performing crosses with this species, the genetics of maize has become the best characterized of any higher plant (Coe *et al.*, 1988).

Since the first fertile transgenic dicots were produced in 1983 (Barton *et al.*, 1983), agricultural biotechnologists have been challenged to extend this capability to maize. Seven years later this goal was realized (Fromm *et al.*, 1990; Gordon-Kamm *et al.*, 1990). This success followed nearly a decade of attempts

* Present Address: Department of Plant Biology, University of California, Berkeley, California, 94710

** Present Address: Agricultural Research Division—Plant Biology, American Cyanamid Company, Princeton, New Jersey, 08543 thwarted by the lack of a natural gene transfer vector and an inability to manipulate maize developmental biology. In the end, discovery and development of a novel gene transfer strategy were required to conquer this recalcitrant species. Still, in the years to come, this development will mark the beginning of a research challenge more notably than the accomplishment of one.

In this chapter we review the successes and failures encountered in the pursuit of maize transformation, make some predictions regarding future improvements in the technology, and discuss the prospects for transgenic maize as an agricultural crop and experimental organism. For additional reviews relevant to gene transfer in cereals, refer to Potrykus (1989, 1990a,b).

2.1 Historical Perspective

2.1.1 *AGROBACTERIUM*

Once genetic transformation of various dicotyledonous species via *Agrobacterium* was demonstrated, many researchers attempted to apply this strategy to maize. The results of this effort have been discouraging. Although one report (Graves and Goldman, 1986) claimed synthesis of T-DNA (transferred DNA)-encoded opines in maize tissues after inoculation of seedlings with *Agrobacterium,* to date there has been no evidence of T-DNA integration in maize. The fact that transfer of T-DNA from *Agrobacterium* to the nucleus of maize cells can occur was demonstrated by a technique called agroinfection; maize plants, inoculated with strains of *Agrobacterium* carrying tandemly repeated copies of the maize streak virus genome in their T-DNA, become systemically infected with the virus (Grimsley *et al.*, 1987, 1988). The lack of success in obtaining transgenic plants using either this technique or more conventional strategies for *Agrobacterium*-mediated transformation is probably due to fundamental aspects of the biology of maize. Most dicotyledonous plants respond to wounding by cell dedifferentiation and massive cell proliferation followed by cell redifferentiation. In contrast, in maize and most other monocots the damaged cells die and form a protective layer, and the underlying living cells become lignified; cell division is not induced (Kahl, 1982). This response suggests that transgenic maize cultures cannot be recovered from *Agrobacterium*-infected cells because these wound-response cells are developmental dead ends. The lack of success using *Agrobacterium*-mediated transformation has led to investigation of recipient cell systems amenable to physical methods of DNA delivery.

2.1.2 POLLEN

Genetic transformation of mature maize pollen has been pursued over the years, and spurious claims of success have been made; however, to date, no proof of transformation has been obtained. De Wet *et al.* (1985) reported transformation by pollinating an inbred with germinating pollen that had been mixed with either *Tripsacum* or maize genomic DNA. Successful transformation was claimed based on altered phenotype of the resulting seed; however, no molecular proof of transformation was obtained. Sanford *et al.* (1985) pursued a similar strategy using maize genomic DNA and well-defined genetic markers, but observed no transformation events. Using a similar approach, Ohta (1986) claimed high efficiency transformation as determined solely by genetic markers. Booy *et al.* (1989) used a mixture of pollen and plasmid DNA carrying the neomycin phosphotransferase II (*nptII*) gene to fertilize ovules *in vitro,* but no transformation events were observed.

Although mature pollen is an appealing alternative for transformation of maize because it eliminates the problems inherent in cell culture, it presents another set of obstacles. Assuming the barrier presented by the pollen cell wall could be penetrated by physical methods, the transforming DNA still would have to escape the nuclease activity observed in pollen (Matousek and Tupy, 1983). A more formidable barrier may be DNA integration into the genome of mature pollen nuclei, which have completed the S-phase of DNA replication (Alwen *et al.*, 1990). For these reasons, most maize transformation efforts have focused on the use of *in vitro* cell culture to enrich for morphogenically competent cells that are accessible to physical methods of DNA delivery and potentially more receptive to integration of foreign DNA.

2.1.3 PROTOPLAST ELECTROPORATION

Protoplasts have been pursued aggressively as recipients for transformation of maize because of their potential compatibility with DNA delivery methods such as $CaCl_2$ precipitation, polyethylene glycol (PEG)-induced uptake, electroporation, and microinjection. The resistance of most maize cell types to dedifferentiation presents a significant challenge to development of protoplast systems. Unlike dicots, in which division-competent protoplasts can be isolated from a large number of cell types, suspension cultures have been the sole source for division-competent protoplasts in maize. Such protoplasts were isolated first from non-embryogenic cell-suspension cultures of B73 (Potrykus *et al.*, 1979), hybrid C103 × W155 (Polikarpochkina *et al.*, 1979), and Black Mexican Sweet (BMS) (Chourey and Zurawski, 1981). Embryogenic suspension cultures of maize inbred A188 were first reported by Green *et al.* (1983). They used immature embryo-derived friable callus (Type II) for the suspension inoculum. Since then, Type II callus has proven to be a reliable inoculum for initiation of embryogenic suspensions from which division-competent protoplasts can be obtained (Kamo and Hodges, 1986; Vasil and Vasil, 1987). Eventually, regeneration of whole plants (Rhodes *et al.*, 1988a) and regeneration of fertile plants (Prioli and Sondahl, 1989; Shillito *et al.*, 1989; Morocz *et al.*, 1990) from maize protoplasts was achieved, demonstrating the existence of morphogenically competent protoplastable cell types in these suspensions.

The first report of transformed maize cell cultures came when Fromm *et al.* (1986) applied the technique of electroporation to BMS protoplasts. Although the culture was not morphogenically competent, this result demonstrated the existence of cell types that are both capable of yielding division-competent protoplasts and competent for integrative transformation. Subsequently, Rhodes *et al.* (1988b) applied the technique of electroporation to an embryogenic protoplastable cell culture and were able to produce genetically transformed maize plants; however, these plants were not fertile.

2.1.4 SUSPENSION CULTURE BOMBARDMENT

During the mid-1980s, while a great deal of effort was being devoted to development of suspension cultures suitable for protoplasting, Sanford and co-workers were investigating a DNA-delivery method capable of penetrating plant cell walls. Using high velocity microprojectiles for delivering nucleic acids directly into cells, they obviated the need for protoplasts (Klein *et al.*, 1987; Sanford *et al.*, 1987). The potential of this delivery method was realized when transformed maize cell cultures were produced by microprojectile bombardment of cell-suspension cultures (Klein *et al.*, 1989; Spencer *et al.*, 1990). Subsequently, fertile transgenic plants were obtained using this technique (Fromm *et al.*, 1990; Gordon-Kamm *et al.*, 1990; see Section 2.2).

2.1.5 TRANSFORMANT SELECTION

Identification of plant cell transformants has relied on the expression of dominant selectable marker genes assembled from appropriate 5′ and 3′ regulatory sequences and structural genes encoding resistance to antibiotics or herbicides. Fromm *et al.* (1985) were the first to demonstrate functionality of the cauliflower mosaic virus (CaMV) 35S RNA promoter (35S) in maize protoplasts by its ability to direct transient expression of the chloramphenicol acetyltransferase (*cat*) gene. To construct a marker gene for selection of integrative transformants, this promoter was fused to the *nptII* gene. The gene was introduced into non-embryogenic BMS protoplasts via electroporation, and transformants were selected using the antibiotic kanamycin (Fromm *et al.*, 1986). Using a similar vector, selection of transformants from embryogenic cultures was accomplished (Rhodes *et al.*, 1988b).

Genes for herbicide resistance also have proven useful for transformant identification in maize. Fromm *et al.* (1990) used a mutant acetolactate synthase (ALS) gene driven by the 35S promoter to transform embryogenic cell cultures; the herbicide chlorsulfuron was used as the selective agent. Spencer *et al.* (1990)

were the first to demonstrate usefulness of a selectable marker gene containing the 35S promoter and the structural gene encoding phosphinothricin acetyltransferase (PAT) for transformant identification in maize; bialaphos was used to select transformants from bombarded BMS cultures. Subsequently, this marker was used to identify transformed embryogenic cells (Fromm *et al.*, 1990; Gordon-Kamm *et al.*, 1990).

2.2 State of the Art

Two reports have proven maize transformation and passage of the introduced gene(s) to progeny rigorously (Fromm *et al.*, 1990; Gordon-Kamm *et al.*, 1990). The approach described by Gordon-Kamm *et al.* (1990) involved three major components: (1) embryogenic suspension cultures, (2) delivery of DNA via microprojectile bombardment, and (3) a dominant selectable marker system. The suspension cultures were initiated from friable embryogenic Type II callus (Armstrong and Green, 1985) of A188 × B73 and A188 × B84. The microprojectile bombardment process was modified to deliver DNA to a large number of cells while minimizing cell damage (Gordon-Kamm *et al.*, 1990). Modifications included the use of fine-mesh stainless steel screens to disperse the DNA-coated tungsten microprojectiles and alterations in the original DNA precipitation protocol described by Klein *et al.* (1987). This modified bombardment procedure increased the number of cellular foci with transient β-glucuronidase (GUS; Jefferson, 1987) activity from hundreds to thousands per filter. The selectable marker was phosphinothricin acetyltransferase (PAT), which confers resistance to phosphinothricin (PPT), an herbicide that inhibits glutamine synthetase (Tachibana *et al.*, 1986). PAT, encoded by the *bar* gene from *Streptomyces hygroscopicus* (Murakami *et al.*, 1986; Thompson *et al.*, 1987; White *et al.*, 1990), first was shown to be useful as a selectable marker for *Agrobacterium*-mediated transformation of dicots (De Block *et al.*, 1987) and later for electroporated rice protoplasts (Dekeyser *et al.*, 1989) and bombarded BMS maize cells (Spencer *et al.*, 1990).

Bialaphos, a tripeptide antibiotic/herbicide that is cleaved intracellularly to yield the active component PPT, was effective for selection of embryogenic maize cells expressing *bar*. Transformation frequencies of 1–2 independent transformed callus lines per bombarded filter (containing 0.5 ml packed cells) were reported (Gordon-Kamm *et al.*, 1990). No nontransformed callus lines escaped bialaphos selection. Of the 54 independent *bar*-transformed callus lines described, many resulted from cotransformation experiments using two separate plasmids, one containing *bar* and the other encoding GUS. A co-integration frequency of approximately 80% (30/39) and a coexpression frequency of about 20% (7/39) were observed.

Approximately half of the *bar*-transformed callus lines recovered from embryogenic suspension cultures SC82 and SC716 were regenerable (Gordon-Kamm *et al.*, 1990). More than 300 plants (R_0) were regenerated in the absence of selective pressure. Southern blot analyses of genomic DNA from over 50 R_0 plants revealed only one plant that did not have a hybridization pattern identical to the parental callus line (Spencer *et al.*, 1992). The general lack of chimerism for *bar* in transgenic callus lines, as well as in R_0 plants, was evidenced further by germinal transmission and identical hybridization patterns of *bar* in the R_0 parents and their progeny.

Plants regenerated from several transgenic SC82 and SC716 callus lines yielded viable progeny (Gordon-Kamm *et al.*, 1990). Transgenic R_0 plants did not produce sufficient pollen for outcrossing or selfing; hence, they were outcrossed using pollen from nontransformed inbred plants. Kernels developed sporadically on ears as well as on feminized tassels of transgenic R_0 plants. Many kernels developed normally and were germinated in soil. However, some kernels ceased development because of premature plant senescence and required embryo rescue. Most progeny (R_1) populations analyzed have segregated for the introduced genes as expected (Fig. 2.1). Analysis of subsequent generations of transgenic plants has demonstrated that the phenotypic abnormalities observed in the R_0 generation (i.e., reduced fertility, premature senescence) were not heritable.

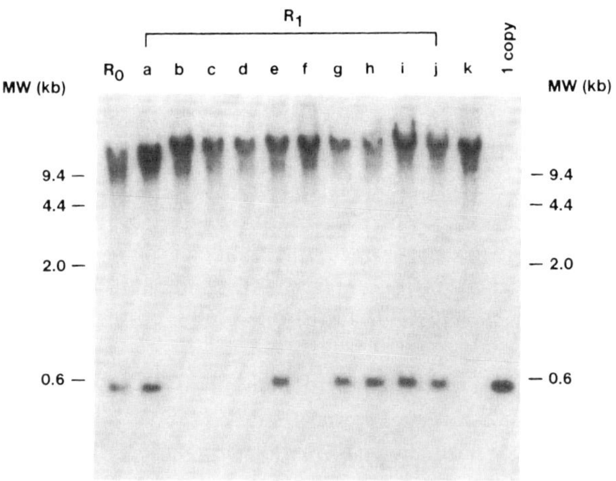

Figure 2.1 Southern blot analysis of a segregating population of progeny (R_1) recovered from plants (R_0) regenerated from a transgenic SC82 callus line. R_0 plants were backcrossed to nontransformed plants. Genomic DNA was digested with *SmaI* and probed with ^{32}P-labeled *bar*. The plasmid used for transformation (pDPG165) contains *bar* on a 559-bp *SmaI* fragment (Gordon-Kamm *et al.*, 1990). Lane designated R_0 contains DNA from a fertile regenerant; lanes labeled a–j contain DNA from R_1 plants; lane k contains DNA from a nontransformed control plant. Lane designated 1 copy contains pDPG165 digested with *SmaI* to approximate 1 copy per genome.

In transgenic lines analyzed thus far, male and female fertility and vigor of most R_1 plants were normal. Mendelian segregation ratios were observed in populations resulting from outcrossing and selfing.

Subsequent to the report by Gordon-Kamm *et al.* (1990), Fromm *et al.* (1990) also described maize transformation using microprojectile bombardment. Their efforts used callus and suspension cultures as transformation targets. They described use of two selectable marker genes, *bar* and a mutant maize ALS gene. Two regenerable *bar*-transformed callus lines were recovered from bombarded callus using PPT selection. These callus lines were confirmed to contain *bar* by polymerase chain reaction (PCR) analysis; however, of the over 150 plants regenerated from this callus in the absence of selection, none were found to contain *bar,* indicating that the callus cultures were chimeric for *bar*. Application of selective pressure during plant regeneration was reported to result in recovery of transgenic plants. Similar results were obtained from bombardment of suspension culture cells with a plasmid containing both a mutant ALS gene and the firefly luciferase gene. Six callus lines were isolated in experiments using chlorsulfuron selection; transformation was confirmed by PCR analysis. Of these callus lines, four were regenerable; however, plants regenerated from two lines lacked the introduced genes, indicating that the callus was chimeric. Only one chlorsulfuron-resistant callus line produced fertile transgenic plants. R_1 plants resulting from outcrosses using nontransformed pollen were sustained by embryo rescue. Luciferase activity was detected in 44% (53/120) of the R_1 plants recovered. The presence of a luciferase-encoding sequence in several luciferase-positive R_1 plants was confirmed by Southern blot analysis. Mutant ALS activity was found to cosegregate with luciferase activity, as determined by *in vitro* enzyme assay and herbicide application.

In addition to the microprojectile bombardment-based approaches just described, Hoechst researchers recently claimed success using a protoplast transformation system (G. Donn, personal communication, 1990). Protoplasts were isolated from a hybrid suspension culture developed by Morocz *et al.* (1990) and transformed using PEG-mediated DNA uptake. Protoplasts were transformed with a synthetically altered PAT coding sequence, originally cloned from *Streptomyces viridochromogenes*. The sequence of this gene was altered to adopt appropriate plant codon usage. Transformed calluses were selected using PPT. Plants were regenerated under selective pressure from over 20 independent transformants; most regenerants proved to be male and female fertile. Transformed progeny were confirmed by *in vitro* enzyme assay, herbicide application, and Southern blot analysis. Based on this report, it now appears that protoplast-mediated transformation of maize is possible.

The successful approaches to maize transformation just described used dominant herbicide resistance markers for selection of transformants from embryogenic cell cultures. Either bialaphos or PPT was used to identify regenerable transgenic callus expressing *bar*. In the experiments described by Gordon-Kamm *et al.* (1990), over 50 independent *bar*-transformed callus lines were selected from bombarded suspension culture cells using bialaphos (1 mg/liter). Plants were regenerated from transgenic callus lines in the absence of selective pressure. In these experiments, only 1 of 14 callus lines analyzed showed evidence of possible chimerism for *bar* (Gordon-Kamm *et al.*, 1990; Spencer *et al.*, 1992). In contrast, experiments reported by Fromm *et al.* (1990) described the use of low levels of PPT (0.05 mg/liter) as a selective agent for callus bombarded with *bar*. Of 2 transformed lines isolated, both initially gave rise to plants that did not contain the gene; however, applying selective pressure during regeneration reportedly resulted in transformed plants. The differing results of the two selection strategies described may be related to target tissue morphology, or to the concentration or the biochemical properties of the selective agent used.

Fromm *et al.* (1990) also used a mutant maize ALS gene and chlorsulfuron selection in maize suspension-culture bombardment experiments. Of 6 chlorsulfuron-resistant callus lines isolated, 2 gave rise to plants; however, they were not transformed. As with some of the *bar*-containing isolates, maintenance of selection pressure during regeneration resulted in plants that contained the ALS gene. Researchers at Pioneer, Johnston, Iowa, also have used a mutant maize ALS gene for maize transformation (D. Tomes, personal communication, 1990).

2.3 Future Improvements

2.3.1 RECIPIENT CELLS

The current technology for maize transformation requires development and maintenance of suspension cultures containing the appropriate recipient or target cells. These cells must be (1) accessible to microprojectile bombardment, (2) competent for integrative transformation, (3) capable of sustained divisions following integrative transformation to allow for development of a transformed clonal sector that can be selected, and (4) morphogenically competent for plant regeneration. A deeper understanding of these cell types and how they are affected by environmental and genetic factors is a prerequisite to improvement of the current technology. Morphological information concerning developmental patterns in cell clusters of embryogenic suspension cultures is scarce. Suspension cultures used for protoplast experiments generally have been described as being finely dispersed and consisting of small groups of cytoplasmically dense cells (Green *et al.*, 1983; Kamo and Hodges, 1986; Vasil and Vasil, 1986; Rhodes *et al.*, 1988a; Prioli and Sondahl, 1989; Shillito *et al.*, 1989). A proportion of the embryogenic cells always becomes nonembryogenic and terminally differentiated, as evidenced by cessation of cell division and rapid elongation (Vasil and Vasil, 1987; Rhodes *et al.*, 1988a). Several reports indicate the existence of clusters with various organizational patterns (Green *et al.*, 1983; Kamo and

Hodges, 1986; Rhodes *et al.*, 1988a). However, there is no morphological information concerning developmental relationships between cell types or cell cluster formation and growth. Such information would be valuable for evaluation of recipient cells for genetic transformation.

As demonstrated by the results of Fromm *et al.* (1990), Type II callus also contains recipient cells that are competent for integrative transformation. Type II callus grows rapidly and develops proembryos and globular embryoids. Its friability and high capacity for regeneration make it an ideal inoculum for suspension-cell cultures for transformation. Histological and ultrastructural examination of Type II callus reveals aggregates of cytoplasmically dense cells that constitute the embryogenic potential of this callus type (Fransz, 1988). These cell aggregates contain 2–8-cell somatic proembryos. This pattern is similar to the growth pattern sustained by some suspension-cell clusters. Still, callus tissue may contain a lower proportion of potential recipient cells than would a suspension culture. In addition, there is a greater potential for differentiation and a higher degree of tissue organization in maize callus than in suspension cultures. This organization may result in a higher proportion of transformed cells that do not continue to divide or that become sectors in organized chimeric tissues that ultimately give rise to chimeric plants.

Other cell types in maize may satisfy the prerequisites for a successful recipient cell system for genetic transformation. Among these possibilities are individual vegetative and atropic *in vitro* apical meristems, pollen- and microspore-derived embryos or sustainable embryogenic callus generated from them, intercalary meristematic cells from leaf bases, or even zygotic proembryos.

The limitations to current applications of transformation by microprojectile bombardment involve particle penetration, ability of bombarded cells to sustain divisions after DNA introduction and to be selected, if necessary, and the ability of transformed cells to generate fertile plants. At the time of this writing, the field of DNA delivery is changing very rapidly, so it is not difficult to imagine that ultimate modification of this technique may allow any morphogenically competent cell to be transformed and recovered to a fertile plant without selection.

2.3.2 DNA DELIVERY

Successful application of microprojectile bombardment requires a balance between the number of cells transformed and the trauma inflicted on the bombarded tissue or cells. Improvements continue to be made in increasing transformation efficiencies while mitigating the damage associated with bombardment. It is well within the range of currently available technology to develop instruments for precise particle acceleration and targeted delivery. Such a system would allow for single particle delivery to individual cells. Physical DNA-delivery methods other than particle acceleration may be necessary to deliver nucleic acid to recipient cells not accessible to particle penetration. Also, introduction of large pieces of DNA or even chromosome fragments may require alternative DNA-delivery strategies.

2.3.3 TRANSFORMANT IDENTIFICATION

Efficient selection of transformed cells from a large population of untransformed cells is a crucial component of the current maize transformation technology. The strategy relies on expression in callus of dominant selectable marker genes. The 5′ expression signals that have been used to drive expression of the selectable marker genes in the successful transformation experiments reported with maize were either the CaMV 35S RNA promoter alone (Gordon-Kamm *et al.*, 1990) or the 35S promoter fused to the first intron of maize alcohol dehydrogenase I (*AdhI;* Fromm *et al.*, 1990). Although these regulatory regions promote sufficient expression to allow selection of resistant calluses, more efficient promoters might result in the isolation of greater numbers of transformants. The herbicide resistance genes *bar* and mutant ALS both have been used successfully as marker genes for maize transformation. Since cell cultures may vary in their response to the herbicides used to select cells transformed with these genes, additional choices of marker gene/selective agent combinations would be desirable.

A number of visual markers have been used in maize, including GUS (Klein *et al.*, 1988, 1989; Lyznik *et al.*, 1989; Fromm *et al.*, 1990; Gordon-Kamm *et al.*, 1990; Spencer *et al.*, 1990), luciferase (Fromm *et al.*, 1990), and genes involved in anthocyanin biosynthesis, such as *A1* and *Bz1* (Klein *et al.*, 1989), and the *R* gene family (Ludwig *et al.*, 1990). However, none of these markers has enabled identification of transformed cells and recovery of transformed plants. Marker systems that rely on visual screening for transformant identification rather than biochemical selection will be important tools in the study of transformation of alternative recipient cell systems. In addition, such tools will be useful for studying the fate of transformed cells and for understanding and optimizing transformation protocols.

2.4 Prospects for Transgenic Maize

The development of maize transformation technology enables us to look to any organism for genes that might improve the quantity and quality of maize harvests. Genes conferring resistance to herbicides have been used as selectable markers for transformation, and herbicide-tolerant maize may be the first genetically engineered product for this species. Genes conferring resistance to two of the most economically important pests of corn in the United States, European corn borer (*Ostrinia nubilalis*) and Western corn rootworm (*Diabrotica vir-*

gifera), are being pursued vigorously. In addition to the potential for foreign gene introduction into maize, the isolation, *in vitro* manipulation, and reintroduction of native maize genes also holds promise. Gene tagging and gene mapping technologies are being developed aggressively for identification and isolation of genes and regulatory sequences in maize. In addition to the global demands for plentiful harvests of nutritionally balanced grain, there is increasing demand for alternative (nonfood) uses of maize, for example, as a source of ethanol or in biodegradable plastics. *In vitro* genetic manipulation should facilitate improvement of maize for these and other alternative uses. Eventually, we may be able to harness the efficient production capability of the maize plant for completely new products.

Maize transformation technology provides a powerful tool, not only for crop improvement but also for basic research. Basic studies of maize biology will be impacted significantly by the ability to make precise genetic alterations and observe the effects of those changes. This technology, in combination with the extensive knowledge base on maize genetics, physiology, and developmental biology, is certain to fortify maize as the experimental organism of choice for studying the biology of higher plants.

2.5 Conclusion

We have entered the 1990s equipped with a powerful tool to assist us in meeting the challenges of this "decade of the environment." Transformation technology will facilitate genetic manipulation of maize to enable high yields under a variety of environmental conditions, diminishing the need to manipulate the environment through use of pesticides, nitrogen fertilizers, or irrigation. Although *in vitro* manipulation of traits such as insect resistance, nitrogen utilization, and drought tolerance is an ambitious objective, we now can redirect some of the resources once focused on development of transformation technology to identification and isolation of genes involved in these and other important agronomic traits. Whether the full agronomic benefits of this technology are realized during this decade may hinge less on technical feasibility than on issues of economics and politics. However, we can be certain that a profound contribution of transgenic maize in the 1990s will be its use to address basic questions in plant biology.

References

Alwen, A., Eller, N., Kastler, M., Benito Moreno, R. M., and Heberle-Bors, E. (1990). *Physiol. Plant.* **79,** 194–196.

Armstrong, C. L., and Green, C. E. (1985). *Planta* **164,** 207–214.

Barton, K. A., Binns, A. N., Matzke, A. J. M., and Chilton, M.-D. (1983). *Cell* **32,** 1033–1043.

Booy, G., Krens, F. A., and Huizing, H. J. (1989). *J. Plant Phys.* **135,** 319–324.
Chourey, P. S., and Zurawski, D. B. (1981). *Theor. Appl. Genet.* **59,** 341–344.
Coe, E. H., Jr., Neuffer, M. G., and Hoisington, D. A. (1988). *In* "Corn and Corn Improvement" (G. F. Sprague, and J. W. Dudley, eds.), pp. 81–258. American Society of Agronomy, Crop Science Society of America, Soil Science Society of America, Madison, Wisconsin.
De Block, M., Botterman, J., Vandewiele, M., Dockx, J., Thoen, C., Gossele, V., Movva, N. R., Thompson, C., Van Montagu, M., and Leemans, J. (1987). *EMBO J.* **6,** 2513–2518.
Dekeyser, R., Claes, B., Marichal, M., Van Montagu, M., and Caplan, A. (1989). *Plant Physiol.* **90,** 217–223.
De Wet, J. M. J., Bergquist, R. R., Harlan, J. R., Brink, D. E., Cohen, C. E., Newell, C. A., and De Wet, A. E. (1985). *In* "Experimental Manipulation of Ovule Tissues" (G. P. Chapman, S. H. Mantell, and R. W. Daniels, eds.), pp. 197–209. Longman, New York.
Fransz, P. F. (1988). Ph.D. Thesis. University of Wageningen, Wageningen, Netherlands.
Fromm, M. E., Taylor, L. P., and Walbot, V. (1985). *Proc. Natl. Acad. Sci. U.S.A.* **82,** 5824–5828.
Fromm, M. E., Taylor, L. P., and Walbot, V. (1986). *Nature (London)* **319,** 791–793.
Fromm, M. E., Morrish, F., Armstrong, C., Williams, R., Thomas, J., and Klein, T. M. (1990). *Bio/Technology* **8,** 833–839.
Gordon-Kamm, W. J., Spencer, T. M., Mangano, M. L., Adams, T. R., Daines, R. J., Start, W. G., O'Brien, J. V., Chambers, S. A., Adams, W. R., Jr., Willetts, N. G., Rice, T. B., Mackey, C. J., Krueger, R. W., Kausch, A. P., and Lemaux, P. G. (1990). *Plant Cell* **2,** 603–618.
Graves, A. C. F., and Goldman, S. L. (1986). *Plant Mol. Biol.* **7,** 43–50.
Green, C. E., Armstrong, C. L., and Anderson, P. C. (1983). *In* "Advances in Gene Technology: Molecular Genetics of Plants and Animals," (K. Downey, R. M. Voellmy, F. Ahmad, and J. Schultz, eds.) pp. 147–157. Academic Press, Orlando, Florida.
Grimsley, N., Hohn, T., Davies, J. W., and Hohn, B. (1987). *Nature (London)* **325,** 177–179.
Grimsley, N. H., Ramos, C., Hein, T., and Hohn, B. (1988). *Bio/Technology* **6,** 185–189.
Jefferson, R. A. (1987). *Plant Mol. Biol. Rep.* **5(4),** 387–405.
Kahl, G. (1982). *In* "Molecular Biology of Plant Tumors" (G. Kahl and J. S. Schell, eds.), pp. 211–267. Academic Press, New York.
Kamo, K. K., and Hodges, T. K. (1986). *Plant Sci.* **45,** 111–117.
Klein, T. M., Wolf, E. D., Wu, R., and Sanford, J. C. (1987). *Nature (London)* **327,** 70–73.
Klein, T. M., Grodziel, T., Fromm, M. E., and Sanford, J. C. (1988). *Bio/Technology* **6,** 559–563.
Klein, T. M., Kornstein, L., Sanford, J. C., and Fromm, M. E. (1989). *Plant Physiol.* **91,** 440–444.
Ludwig, S. E., Bowen, B., Beach, L., and Wessler, S. R. (1990). *Science* **247,** 449–450.
Lyznik, L. A., Ryan, R. D., Ritchie, S. W., and Hodges, T. K. (1989). *Plant Mol. Biol.* **13,** 151–161.
Matousek, J., and Tupy, J. (1983). *Plant Sci. Lett.* **30,** 83–89.
Morocz, S., Donn, G., Nemeth, J., and Dudits, D. (1990). *Theor. Appl. Genet.* **80,** 721–726.
Murakami, T., Anzai, H., Imai, S., Satoh, A., Nagaoka, K., and Thompson, C. J. (1986). *Mol. Gen. Genet.* **205,** 42–50.
Ohta, Y. (1986). *Proc. Natl. Acad. Sci. U.S.A.* **83,** 715–719.
Polikarpochkina, R. T., Gamburg, K. Z., and Kharkin, E. E. (1979). *Z. Pflanzenphysiol.* **95,** 57–67.
Potrykus, I. (1989). *Trends Biotech.* **7,** 269–273.
Potrykus, I. (1990a). *Bio/Technology* **8,** 535–542.
Potrykus, I. (1990b). *Physiol. Plant.* **79,** 125–134.
Potrykus, I., Harms, C. T., and Lorz, H. (1979). *Theor. Appl. Genet.* **54,** 209–214.
Prioli, L. M., and Sondahl, M. R. (1989). *Bio/Technology* **7,** 589–594.
Rhodes, C. A., Lowe, K. S., and Ruby, K. L. (1988a). *Bio/Technology* **6,** 56–60.
Rhodes, C. A., Pierce, D. A., Mettler, I. J., Mascarenhas, D., and Detmer, J. J. (1988b). *Science* **240,** 204–207.

Sanford, J. C., Skubik, K. A., and Reisch, B. I. (1985). *Theor. Appl. Genet.* **69,** 571–574.
Sanford, J. C., Klein, T. M., Wolf, E. D., and Allen, N. (1987). *J. Part. Sci. Technol.* **5,** 27–37.
Shillito, R. D., Carswell, G. K., Johnson, C. M., DiMaio, J. J., and Harms, C. T. (1989). *Bio/Technology* **7,** 581–587.
Spencer, T. M., Gordon-Kamm, W. J., Daines, R. J., Start, W. G., and Lemaux, P. G. (1990). *Theor. Appl. Genet.* **79,** 625–631.
Spencer, T. M., O'Brien, J. V., Start, W. G., Adams, T. R., Gordon-Kamm, W. J., and Lemaux, P. G. (1992). *Plant Mol. Biol.* **18,** 201–210.
Tachibana, K., Watanabe, T., Sekizawa, Y., and Takematsu, R. (1986). *J. Pesticide Soc.* **11,** 27–31.
Thompson, C. J., Movva, N. R., Tizard, R., Crameri, R., Davies, J. E., Lauwereys, M., and Botterman, J. (1987). *EMBO J.* **6,** 2519–2523.
Vasil, V., and Vasil, I. K. (1986). *J. Plant Physiol.* **124,** 399–408.
Vasil, V., and Vasil, I. K. (1987). *Theor. Appl. Genet.* **73,** 793–798.
White, J., Chang, S.-Y. P., Bibb, M. J. and Bibb, M. J. (1990). *Nucleic Acids Res.* **18,** 1062.

3

Transgenic Asparagus

Jean-Pierre Hernalsteens

Laboratorium Genetische Virologie
Vrije Universiteit Brussel
B-1640 Sint-Genesius-Rode, Belgium

Benny Bytebier

Department of Botany
University of Nairobi
Nairobi, Kenya

Marc Van Montagu

Laboratorium voor Genetica
Universiteit Gent
B-9000 Gent, Belgium

Agrobacterium-mediated gene transfer commonly is used for the transformation of dicotyledonous plants. By this technique, a well-defined segment of the Ti plasmid, the T-DNA, is inserted, most often with a low copy number, into the plant genome and is inherited stably. Transgenic plants can be obtained by this approach in any species that regenerates efficiently from tissue explants or cultured tissues. Additionally, new techniques such as seed (Feldmann and Marks, 1987) and meristem (Ulian *et al.*, 1988) transformation may allow its application beyond this range.

Because of their economic importance, monocotyledonous plants are major targets for plant genetic engineering. Although the formation of swellings that, in some cases, contained opines had been reported after inoculation of some monocots with virulent *Agrobacterium* strains (De Cleene and De Ley, 1976; Hooykaas-Van Slogteren *et al.*, 1984; De Cleene, 1985), monocotyledonous plants generally were considered to be insensitive to *Agrobacterium* infection.

In numerous monocots, particularly the members of the family Gramineae, only a very limited number of tissues can proliferate *in vitro*. This inability to reinitiate cell division might be a causal factor for the apparent inability of *Agrobacterium* to transform monocots. We have investigated the interaction of

Agrobacterium tumefaciens with *Asparagus officinalis*, a monocot species of the family Liliaceae that shows good tissue culture properties. Cell divisions can be induced easily in explants from different parts of this plant. Further, *Asparagus* was the first monocot to be regenerated from tissue culture (Wilmar and Hellendoorn, 1968) and protoplasts (Bui Dang Ha and Mackenzie, 1973; Bui Dang Ha *et al.*, 1975).

3.1 Oncogenic Transformation of *Asparagus*

3.1.1 TUMOR INDUCTION BY *AGROBACTERIUM*

The oncogenes of the T-DNA of *Agrobacterium* were used as markers to detect and optimize *Asparagus* transformation. The expression of these genes leads to the formation of tumorous outgrowths on the plant and allows the proliferation of the transformed cells on hormone-free medium that does not support the growth of normal cells.

For *Agrobacterium* transformation, Hernalsteens *et al.* (1984) used young rapidly elongating spears (20 cm long, 2–3 mm in diameter) from 10-month-old seedlings of *Asparagus officinalis* cv. 'Roem van Brunswijk.' These spears were surface sterilized, cut into 3-cm long segments, and planted vertically, 5 mm deep, in hormone-free half-strength Murashige and Skoog (MS; 1962) medium. Immediately afterward, the apical side of the segments was inoculated with *Agrobacterium*. The bacteria were scooped from a fresh culture plate with a spatula and applied directly onto the wound. No reaction was detected in segments that were not infected or inoculated with the nononcogenic strain C58C1 (Van Larebeke *et al.*, 1974), whereas 5–15% of the segments infected with the oncogenic strains C58 (Hamilton and Fall, 1971) and C58C1(pTiB6S3) (Holsters *et al.*, 1978), respectively harboring wild-type nopaline and octopine Ti plasmids, developed tumor tissue. In some segments, large areas of the infected wound surface began to proliferate after 2 wk, whereas in other cases a localized and small tissue outgrowth was obtained only after 1 month.

Spear slices of *A. officinalis* cv. 'Limbras 26' plants cultured in a greenhouse were used for infection also (Prinsen *et al.*, 1990; B. Bytebier, 1985, unpublished results). Initially, all existing spears of these plants were removed to stimulate the synchronous growth of new spears. When approximately 30 cm long, these spears were harvested, surface sterilized, and cut into slices 5–10 mm thick. The slices were incubated on hormone-free medium containing the minerals of Murashige and Skoog (1962) at half-strength and inoculated with 10 μl of a suspension of about 2×10^9 *Agrobacterium* cells per ml. Half the slices were incubated upside down to test the influence of the polarity. After 2 days, the infected slices were washed and transferred onto Linsmaier and Skoog (LS; 1965) medium without phytohormones, and supplemented with 500 mg/liter cefotaximum to eliminate the agrobacteria present in the explants, keeping the infected side up. Transformed tissue became visible after 1 wk (Prinsen *et al.*, 1990). In a typical experiment with strain C58, tumorous proliferation was obtained in 22 of 30 (73%) of the inverted segments, whereas only 1 of 30 (3.3%) of the fragments incubated upright showed tumor formation. The enhanced transformation efficiency of inverted explants is most likely correlated with polar phytohormone transport.

To test the inoculation of *Asparagus* seedlings cultured *in vitro, A. officinalis* cv. 'Roem van Brunswijk' seeds were surface sterilized and sown in glass jars on half-strength MS medium. After 1 month, the shoot tip of the plantlets was removed with a scalpel and the wound was inoculated with bacteria scooped from a fresh culture plate with a spatula. Tumorous proliferation was observed in 11 of 77 (14.3%) of the plantlets inoculated with the nopaline strain C58 and in 4 of 65 (6.1%) of the plantlets infected with the octopine strain C58C1(pTiB6S3). However, the continued culturing of the very small tumors obtained by this procedure was problematic (B. Bytebier, 1985, unpublished results).

Suseelan *et al.* (1987) obtained stem tumors by inoculation of *Asparagus tetragonus* and *Asparagus sprengeri* with an oncogenic *Agrobacterium* strain harboring pTiT37.

Conner *et al.* (1988a,b; A. J. Conner, 1990, personal communication) inoculated 19 ornamental *Asparagus* taxa and 17 genotypes of *A. officinalis* with 10 different *Agrobacterium* strains belonging to the species *A. tumefaciens, A. rhizogenes,* and *A. rubi.* The upper region of young rapidly growing stems of the asparagus plants were pricked with a 1-mm diameter dissecting needle that had been dipped in an *Agrobacterium* colony growing on nutrient medium. Tumors were induced by some strains of all three *Agrobacterium* species on some genotypes of *Asparagus asparagoides, A. densiflorus, A. officinalis,* and *A. setaceus.* Tumors were only observed on *Asparagus myriocladus* after infection by *Agrobacterium tumefaciens* and *A. rhizogenes* strains. Only *A. tumefaciens* induced tumours on *Asparagus laricinus.* Also, there were clear differences in the sensitivities of different genotypes of the same *Asparagus* species. *Asparagus densiflorus* cv. 'Sprengeri' and *A. asparagoides* proved to be excellent hosts for tumor induction, which especially with the *Agrobacterium tumefaciens* strain C58, was very reproducible. On the other hand, the sensitivity of some species, such as *Asparagus myriocladus,* was very low (only one tumor from over 100 inoculations). No tumors were obtained on *Asparagus compactus, A. verticillatus, A. macowanii, A. virgatus, A. scandens, A. sarmentosus, A. racemosus, A. tenuis, A. retrofractus, A. officinalis* var. *pseudoscaber,* and some *A. officinalis* cultivars. These data clearly demonstrate that both the plant and the bacterial genotype are critical for *Asparagus* transformation.

3.1.2 *IN VITRO* CULTURE OF CROWN-GALL TUMOR TISSUE

For further confirmation and analysis, the tumor tissues induced on *Asparagus officinalis* cultivars 'Roem van Brunswijk' (Hernalsteens *et al.*, 1984) and 'Limbras 26' (Prinsen *et al.*, 1990) were excised after 1 month and cultured *in vitro* on hormone-free LS medium, supplemented with 200 mg/liter glutamine and 500 mg/liter cefotaximum. (Glutamine stimulates the growth of *Asparagus* tissue cultures.) Tumorous outgrowths 2–3 mm in diameter or more proliferated routinely on this medium (Hernalsteens *et al.*, 1984). Smaller explants often survived without visible growth; these could be cultured reproducibly by first incubating them on medium with phytohormones (the same medium supplemented with 1 mg/liter 6-benzylaminopurine; BAP and 1 mg/liter 1-naphthaleneacetic acid; NAA) and then transferring them onto hormone-free medium immediately after proliferation began. Prolonged incubation on medium with hormones can lead to the habituation of the nontransformed cells present in the tissue and thereby interfere with the selection of tumor tissue on hormone-free medium. Nontransformed *Asparagus* explants can survive for some time, but do not proliferate, when cultured on hormone-free medium.

Depending on the *Agrobacterium* strain, tumor cultures with different morphology were obtained. Most of the lines induced by the nopaline type *Agrobacterium* C58 and the octopine type *Agrobacterium* C58C1(TiB6S3) grew as unorganized and compact tissues. With C58, teratomatous cultures producing masses of abnormal thickened spears that did not elongate were obtained occasionally (Prinsen *et al.*, 1990; B. Bytebier, 1986, unpublished results).

The tumor cultures described by Hernalsteens *et al.* (1984), Bytebier *et al.* (1987), and Prinsen *et al.* (1990) are completely hormone independent. The oldest of these cultures maintained a constant growth rate for more than 6 yr on hormone-free medium. However, other *Asparagus* tumor lines are only partially hormone independent. A. J. Conner, 1990 (personal communication) selected tumor tissue after cocultivation of regenerating callus with *Agrobacterium tumefaciens* C58. The growth rate of these cultures declined on subculture on hormone-free medium and an occasional subculture on hormone-supplemented media was necessary for the long-term maintenance of these cultures.

3.1.3 OPINE SYNTHESIS IN CROWN-GALL TISSUE

To confirm the transformed status of the *Asparagus* tumors and cultured tumor tissues, the presence of opines was analyzed. Electrophoresis at pH 6.5 was used to demonstrate the presence of nopaline in tumor cultures induced by strain C58. This pH was used because an interfering compound (with the same electrophoretic mobility as nopaline at pH 1.8, which is used normally for nopaline tests) that stains similarly with phenanthrenequinone is present in nontransformed *Asparagus* tissues. Nopaline synthesis was confirmed further (1) by the incorporation of [^{3}H]arginine and electrophoresis at pH 1.8 followed by autoradiography, (2) by the conversion of nopaline into pyronopaline on acid treatment, and (3) by the Ti plasmid-dependent catabolism of the putative nopaline (Hernalsteens *et al.*, 1984). The Sakagushi technique (Shaw *et al.*, 1988) also was applied for the detection of nopaline in *Asparagus* tumors after electrophoresis (Conner *et al.*, 1988b). Agrocinopine A was detected in *Asparagus* tumor cultures induced by the nopaline type *Agrobacterium* strain C58 using the method of Messens *et al.* (1985), which involves incubation of the tissue in medium with $H_3{}^{32}PO_4$ followed by electrophoresis and autoradiography.

The presence of octopine or agropine and mannopine in tumor tissue induced by *Agrobacterium tumefaciens* C58C1(pTiB6S3) was demonstrated by paper electrophoresis at pH 1.8, followed by staining, respectively, with phenanthrenequinone and $AgNO_3$ (Bytebier *et al.*, 1987).

These data confirmed the transformed status of the cultured tissues. Moreover, since the quantities of opines detected were similar to the levels normally found in dicots, this work showed that the regulatory elements of the opine biosynthetic genes are recognized efficiently in *Asparagus* cells.

3.1.4 EXPRESSION OF ONC GENES

In crown gall tumors, indole-3-acetic acid (IAA) is synthesized from tryptophan (Trp) by two T-DNA-encoded enzymes. The gene *1* (*iaaM*) encodes the enzyme tryptophan 2-monooxygenase, converting Trp into indole-3-acetamide (IAM). The product of gene *2* (*iaaH*) is the enzyme indole-3-acetamide hydrolase, converting this intermediate into the active auxin IAA. The product of the T-DNA gene *4* (*ipt*) is the enzyme isopentenyltransferase, catalyzing the first step in cytokinin biosynthesis. The function of the T-DNA oncogenes was reviewed by Inzé *et al.* (1987).

Endogenous phytohormone levels were measured in independent octopine- and nopaline-type tumor cultures and in nontransformed callus of *Asparagus officinalis* cv. 'Limbras 26'. No time-correlated fluctuations in the phytohormone levels were detected. The concentrations of IAM and IAA in the tumor cultures were similar to the levels found in the nontransformed cells. Tryptophan 2-monooxygenase activity was very low in all lines. A rather high and comparable level of conversion of IAM to IAA was detected both in nontransformed tissue and in nopaline-type tumor cells; this level was about two orders of magnitude higher in an octopine-type *Asparagus* tumor culture. Exogenous IAM and IAA was toxic for both transformed and nontransformed *Asparagus* lines, suggesting that a high expression level of the T-DNA-encoded auxin biosynthetic system would, most likely, be lethal for *Asparagus* tissues. Hypermethylation of the corresponding region of the T-DNA was observed and might be the cause of the low expression level of the auxin biosynthetic genes in *Asparagus*. Endogenous cytokinin levels were higher in both octopine- and nopaline-type tumor cultures than in nontransformed callus. The highest levels were observed in fast-growing cultures. In conclusion, the hormone-independent proliferation of *Asparagus* crown gall tissue cultures seems only to be dependent on increased cytokinin levels induced by gene *4* activity (Prinsen *et al.*, 1990).

3.1.5 *T-DNA* ORGANIZATION IN CROWN-GALL TISSUE

Total genomic DNA, with a size larger than 50 kb, was extracted from *in vitro* cultured axenic *Asparagus officinalis* tumor tissues by the procedure of Dellaporta *et al.* (1983). By DNA hybridization with ^{32}P-labeled probes (Southern, 1975), fragments covering the whole T-DNA over a length of 23 kb, without detectable deletions, were detected in a nopaline-type tumor culture induced by C58 on *A. officinalis* cv. 'Roem van Brunswijk.' Hybridization with border segments showed that at least 4 copies of the T-DNA were integrated in this line (Bytebier *et al.*, 1987).

In an octopine-type *Asparagus* tumor line synthesizing octopine (determined by T_L-DNA) as well as agropine and mannopine (determined by T_R-DNA), independent integrations of the entire T_L- and T_R-DNA were detected (Bytebier *et al.*, 1987).

3.2 Isolation of Transgenic Plants

3.2.1 INFECTION AND SELECTION

The experiments with oncogenic strains discussed previously demonstrated that *Agrobacterium* can transfer and integrate its T-DNA in *Asparagus*, allowing the use of nononcogenic *Agrobacterium* strains to obtain transgenic plants. The infection procedures that had been optimized previously using oncogenic *Agrobacterium* strains were applied with this aim.

Bytebier *et al.* (1987) obtained transgenic *Asparagus* plants using the *Agrobacterium* strain C58C1pGV3850::1103neo(dim). This strain harbors a co-integration-type vector devoid of all oncogenic sequences and carries the nopaline synthase (nopaline dehydrogenase) gene and a chimeric gene for kanamycin resistance between T-DNA border sequences (Czernilofsky *et al.*, 1986). Rapidly elongating spears (20 cm long, 2–3 mm in diameter) from 10-month-old seedlings of *Asparagus officinalis* cv. 'Roem van Brunswijk' were surface sterilized and cut into 3-cm long segments. These were inverted and planted 5 mm deep in half-strength MS medium. The bacteria were taken from a fresh culture plate with a spatula and applied directly onto the upper ends of the segments. After 1 month of incubation, a 2-mm thick slice of infected tissue was excised and cultured twice for 1 month on nonselective medium (LS medium supplemented with 200 mg/liter glutamine, 500 mg/liter cefotaximum, 1 mg/liter BAP, and 1 mg/liter NAA). The resulting callus was transferred onto the same medium supplemented with 50 mg/liter kanamycin sulfate. From this procedure, kanamycin-resistant callus was recovered from about 15% of the infected segments. The recovery of transformed lines was inefficient when the selection was initiated earlier. The three kanamycin-resistant lines tested contained large amounts of nopaline.

Alternatively, *in vitro* cultured seedlings were decapitated and infected, as described earlier, with *A. tumefaciens* C58C1pGV3850::1103neo(dim). After 1 month, a 1.5-cm long infected spear segment was excised and incubated on MS medium with 1 mg/liter BAP, 0.1 mg/liter NAA, and 500 mg/liter cefotaximum. The calluses formed at the infected sides were subcultured 1 month later onto the same medium supplemented with 50 mg/liter kanamycin. Green kanamycin-resistant transformed callus containing nopaline was obtained from 1 of 8 infected explants (B. Bytebier, 1985, unpublished results).

Similarly, Conner *et al.* (1988a,b) cocultivated tissue from *in vitro* cultures of *Asparagus officinalis* cv. 'CRD157' with the *Agrobacterium* strain C58C1 pGV3850::1103neo (Hain *et al.*, 1985). The explants were dipped for about 30 sec in a bacterial suspension, blotted on filter paper, and incubated on callus-inducing medium (MS medium supplemented with 200 mg/liter glutamine, 1 mg/liter kinetin, and 1 mg/liter NAA). After 2–3 days, the explants were subcultured onto the same medium supplemented with 250 mg/liter cefotaximum. Kanamycin selection was initiated by transfer onto culture medium containing 100 mg/liter kanamycin after 10 days. Kanamycin-resistant callus was obtained from 3 of 125 explants after 7 months.

3.2.2 PLANT REGENERATION

Plant regeneration was induced by transferring bacteria-free kanamycin-resistant callus onto shoot-induction medium (LS medium without antibiotics and supplemented with 40 mg/liter adenine, 4 mg/liter BAP, and 1 mg/liter NAA). Since single *Asparagus* shoots cannot form roots, the resulting shoots were transferred first onto LS medium with 0.1 mg/liter BAP for shoot proliferation. Rooting was obtained routinely by subdividing the clumps of shoots obtained after this treatment into fragments containing 3–10 spears and incubating these on LS medium with 0.5 mg/liter IAA. This regeneration procedure worked reasonably well for the cultivar 'Roem van Brunswijk;' in the initial experiment, transgenic plants were obtained from 2 of 3 transformed lines (Bytebier *et al.*, 1987). For efficient regeneration, the period of undifferentiated growth should be kept as short as possible. Regeneration is influenced also by the genotype, for example, transformed lines of the cultivar 'Limbras 26' do not regenerate efficiently by this procedure (B. Bytebier, 1986, unpublished results).

Conner *et al.* (1988b) obtained shoots from transformed callus after transfer onto MS medium supplemented with 200 mg/liter glutamine and 1 mg/liter kinetin. These shoots can grow on medium containing kanamycin and produce nopaline. Rooting of these clones was achieved on MS medium with 6% sucrose supplemented with 200 mg/liter glutamine and 1 mg/ liter NAA.

3.2.3 *IN VIVO* CULTURE OF TRANSGENIC PLANTS

Careful acclimatization of the transgenic plants was essential for their survival. Before transfer to the greenhouse, Bytebier *et al.* (1987) transferred the rooted plantlets into glass jars containing a 3-cm layer of sterile perlite, moistened with half-strength minerals of MS medium and covered with Reynolds plastic film. The film was punctured daily to expose the plants to their surroundings gradually. Finally, the plants were transferred to plastic flowerpots with perlite and watered daily with a 200 mg/liter solution of Peters 20+20+20 fertilizer. In the greenhouse, the transgenic plants were morphologically indistinguishable from plants regenerated from nontransformed callus or grown from seeds. The presence of nopaline was demonstrated by electrophoresis at pH 6.5, followed by phenanthrenequinone staining, of crude extracts of these plants.

The transgenic plants obtained by Conner *et al.* (1988b) from a single transformed clone were morphologically similar to untransformed plants when cultured under greenhouse conditions. In the field however, these plants grew slowly and displayed a highly branched and bushy phenotype, most likely due to somaclonal variation (A. J. Conner, 1990, personal communication).

Male plants from this transgenic clone were crossed with a nontransformed female plant. Unexpectedly, expression of the T-DNA markers, kanamycin resistance, and nopaline synthesis was not observed in any of the 95 F_1 seedlings tested (A. J. Conner, 1990, personal communication).

3.2.4 *T-DNA* STRUCTURE IN TRANSFORMED CALLUS AND TRANSGENIC PLANTS

The structure of the T-DNA was determined in three independent callus lines transformed by *Agrobacterium* C58C1pGV3850::1103neo(dim). Southern hybridization showed that all fragments between the left and the right T-DNA borders are present in the transgenic plants. Hybridization with a right border probe detected several border fragments, demonstrating that more than one copy of the T-DNA is integrated in the *Asparagus* genome. No differences were found among the T-DNA structure in transformed callus and in three transgenic plants regenerated from this line (Bytebier *et al.*, 1987).

Conner *et al.* (1988a,b) proved the presence of the T-DNA in *Asparagus officinalis* cv. 'CRD157' callus transformed by *Agrobacterium* strain C58C1 pGV3850::1103neo (Hain *et al.*, 1985) by Southern hybridization using an *nptII*-specific probe.

3.3 Factors Influencing Efficiency of Transformation

The studies on *Asparagus* transformation have revealed a number of factors critical for transformation efficiency.

A very low transformation efficiency (only 2 transformed calluses from a total of 676 infected stem segments) was observed after cocultivation of *Asparagus* stem fragments in liquid medium followed by callus induction and selection on kanamycin-containing solid medium (B. Bytebier, 1987, unpublished results). This result may indicate that a very large number of bacteria is required at the wound surface for efficient transformation.

Suspensions of isolated viable cells can be obtained by mechanical homogenization of *Asparagus* cladodes. After plating, the cells develop efficiently into callus (Jullien, 1974). We attempted to transform both freshly isolated cell suspensions and dividing cells with several *Agrobacterium* strains. Acetosyringone induction of *vir* gene expression (Sheikholeslam and Weeks, 1987) was included in some experiments. The hormone independence conferred by the T-DNA oncogenes and the kanamycin resistance conferred by a chimeric gene, both efficiently expressed after infection of stem segments, were tested as selectable markers, not one transformed clone was observed in several large scale experiments (J.-P. Hernalsteens and F. Salomon, 1987, unpublished results).

The efficiency of tumor induction on *Asparagus* stem segments is reproducibly higher for the *Agrobacterium* strains A281 (Montoya *et al.*, 1977) harboring the hypervirulent agropine-type Ti plasmid pTiBo542 (Hood *et al.*, 1984) than for octopine and nopaline strains. The *Agrobacterium* strain EHA101, which contains a derivative of pTiBo542 deleted of its T-DNA (Hood *et al.*, 1986), is particularly useful for the transformation of *Asparagus* using binary vectors harboring the T-DNA (B. Bytebier, 1989, unpublished results).

These data indicate that the choice of plant material, bacterial strain, and conditions of infection are all crucial factors for *Agrobacterium*-mediated transformation of *Asparagus* and possibly of other monocots as well.

3.4 Regulated Expression of Chimeric Genes in *Asparagus*

The possibility of obtaining transformed *Asparagus* cell lines and transgenic plants allows the use of this species for the study of gene expression in monocots. The chloramphenicol acetyltransferase (*cat*) of Tn*9* (Herrera-Estrella *et al.*, 1983) and the *Escherichia coli gus* or *uidA* gene encoding β-glucuronidase (Jefferson *et al.*, 1987) were used as reporter genes to test the activity of regulated promoters from dicots in *Asparagus*. The coding sequences of these genes were cloned in the T-DNA under the control of the 5′ regulatory sequences of a pea (*Pisum sativum*) chlorophyll *a*/*b* binding protein gene (Simpson *et al.*, 1985) or of an *Arabidopsis thaliana* ribulose-1,5-bisphosphate carboxylase small subunit gene (Krebbers *et al.*, 1988). Transformed calluses, and in some cases transgenic plants, harboring these constructs were isolated. This study revealed that both promoters mediated light inducible gene expression in *Asparagus*. Organ-specific expression was observed also: a chimeric gene consisting of the *cat* coding sequence under the control of the *Arabidopsis* small subunit promoter is expressed at a higher level in cladodes than in stem and root tissues of a transgenic *Asparagus officinalis* plant (B. Bytebier, 1989, unpublished results).

3.5 Discussion

The work on *Asparagus* has shown that *Agrobacterium* strains with octopine-, nopaline-, and agropine-type Ti plasmids can induce tumors in this plant. Cells from these tumors can be cultured on hormone-free media, on which nontransformed *Asparagus* cells are unable to proliferate, and synthesize all opines that are present in tumors induced on dicots by the same *Agrobacterium* strains. The *Asparagus* crown gall cells contain the same T-DNA as the corresponding lines of dicot origin.

Transformed tissues were obtained after infection of *Asparagus* stem explants with nononcogenic *Agrobacterium* strains harboring Ti plasmid vectors with chimeric marker genes. Transgenic plants were regenerated from these cultures.

Important problems for *Asparagus* transformation are the low frequency of the process, as well as the surprisingly large differences in the transformation and regeneration frequencies obtained in different experiments. These fluctuations were observed both in our studies and by A. J. Conner, 1989 (personal communication). This inconsistency is probably somehow linked to unapparent physiological differences in the plant material used. These factors must be studied in more detail before the *Agrobacterium*-mediated transformation of *Asparagus* can become routine.

Reproducible high frequency transformation is probably not essential for gene transfer to improve the horticultural properties of *Asparagus* lines, since a limited number of primary transformants may be sufficient for this aim. Therefore the transfer of genes determining traits such as insect, virus, and herbicide resistance already could be accomplished.

A serious problem for *Asparagus* transformation is the slow growth and development of *Asparagus* tissue cultures and transgenic plants that is probably inherent to this plant species. These factors will limit the use of *Asparagus* as a model system for molecular genetic studies.

In addition to the work on *Asparagus* reviewed in this chapter, recent studies have shown that *Agrobacterium* can interact with several other monocots. Transformed lines were obtained in *Dioscorea* (Schäfer *et al.*, 1987) and rice (Raineri *et al.*, 1989); in both plants, the presence of the T-DNA was demonstrated by DNA hybridization. The T-DNA present in the *Dioscorea* culture was shown to correspond to the T-DNA found in dicots.

These data demonstrate convincingly that *Agrobacterium* can transfer and integrate its T-DNA in a wide range of monocots. The real challenge now is to optimize the conditions for transformation, selection, and regeneration to obtain transgenic plants at a workable frequency.

Acknowledgments

We thank T. Conner for generous communication of unpublished data and comments on *Asparagus* transformation. Critical reading of the manuscript by D. André, C. Bowler, and H. De Greve was most helpful. J.-P. Hernalsteens is a Research Associate of the National Fund for Scientific Research (Belgium). B. Bytebier is indebted to the Instituut tot Aanmoediging van het Wetenschappelijk Onderzoek in Nijverheid en Landbouw (Belgium) for a fellowship. The research in the laboratory of the authors at the Vrije Universiteit Brussel was supported by grants from the Fonds voor Kollectief Fundamenteel Onderzoek (F.K.F.O. 2.007.77), the Belgian National Fund for Scientific Research (Krediet aan Navorsers, Dienstjaar 1985–1986), and the services of the Belgian Prime Minister (IUAP #1).

References

Bui Dang Ha, D., and Mackenzie, I. A. (1973). *Protoplasma* **78,** 215–221.

Bui Dang Ha, D., Norreel, B., and Masset, A. (1975). *J. Exp. Botany* **26,** 263–270.

Bytebier, B., Deboeck, F., De Greve, H., Van Montagu, M., and Hernalsteens, J.-P. (1987). *Proc. Natl. Acad. Sci. U.S.A.* **84,** 5345–5349.

Conner, A. J., Williams, M. K., Deroles, S. C., and Gardner, R. C. (1988a). *In* "Proceedings of the Ninth Australian Plant Breeding Conference" (K. S. McWhirter, R. W. Downes, and B. J. Read, eds.), pp. 131–132. Agricultural Research Institute, Wagga Wagga, Australia.

Conner, A. J., Williams, M. K., Lancaster, J. E., Shaw, M. L., Deroles, S. C., and Gardener, R. C.

(1988b). *In* "Abstracts of the Second International Congress of Plant Molecular Biology." International Society of Plant Molecular Biology, Jerusalem.
Czernilofsky, A., Hain, R., Herrera-Estrella, L., Lörz, H., Goyvaerts, E., Baker, B., and Schell, J. (1986). *DNA* **5,** 101–113.
De Cleene, M. (1985). *Phytopath. Z.* **113,** 81–89.
De Cleene, M., and De Ley, J. (1976). *J. Bot. Rev.* **42,** 289–466.
Dellaporta, S. L., Wood, J., and Hicks, J. B. (1983). *Plant Mol. Biol. Rep.* **1,** 19–21.
Feldmann, K. A., and Marks, M. D. (1987). *Mol. Gen. Genet.* **208,** 1–9.
Hain, R., Stabel, P., Czernilofsky, A. P., Steinbiss, H. H., Herrera-Estrella, L., and Schell, J. (1985). *Mol. Gen. Genet.* **199,** 161–168.
Hamilton, R. H., and Fall, M. Z. (1971). *Experientia* **27,** 229–230.
Hernalsteens, J.-P., Thia-Toong, L., Schell, J., and Van Montagu, M. (1984). *EMBO J.* **3,** 3039–3041.
Herrera-Estrella, L., Depicker, A., Van Montagu, M., and Schell, J. (1983). *Nature (London)* **303,** 209–213.
Holsters, M., De Waele, D., Depicker, A., Messens, E., Van Montagu, M., and Schell, J. (1978). *Mol. Gen. Genet.* **163,** 181–187.
Hood, E., Jen, G., Kayes, L., Kramer, J., Fraley, R. T., and Chilton, M.-D. (1984). *Bio/Technology,* **2,** 702–709.
Hood, E., Helmer, G. L., Fraley, R. T., and Chilton, M.-D. (1986). *J. Bacteriol.* **168,** 1291–1301.
Hooykaas-Van Slogteren, G. M. S., Hooykaas, P. J. J., and Schilperoort, R. A. (1984). *Nature (London)* **311,** 763–764.
Inzé, D., Follin, A., Van Onckelen, H., Rüdelsheim, P., Schell, J., and Van Montagu, M. (1987). *In* "Molecular Biology of Plant Growth Control" (J. A. Fox and M. Jacobs, eds.), pp. 181–196. Liss, New York.
Jefferson, R. A., Kavanagh, T. A., and Bevan, M. W. (1987). *EMBO J.* **6,** 3901–3907.
Jullien, M. (1974). *C. R. Hebd. Seances Acad. Sci., Ser. D* **279,** 747–750.
Krebbers, E., Suerinck, J., Herdies, L., Cashmore, A. R., and Timko, M. P. (1988). *Plant Mol. Biol.* **11,** 745–759.
Linsmaier, E. M., and Skoog, F. (1965). *Physiol. Plant* **18,** 100–127.
Messens, E., Lenaerts, A., Hedges, R. W., and Van Montagu, M. (1985). *EMBO J.* **4,** 571–577.
Montoya, A. L., Chilton, M.-D., Gordon, M. P., Sciacky, D., and Nester, E. W. (1977). *J. Bacteriol.* **129,** 101–107.
Murashige, T., and Skoog, F. (1962). *Plant Physiol.* **15,** 473–497.
Prinsen, E., Bytebier, B., Hernalsteens, J.-P., De Greef, J., and Van Onckelen, H. (1990). *Plant Cell Physiol.* **31,** 69–75.
Raineri, D. M., Bottino, P., Gordon, M. P., and Nester, E. W. (1989). *Bio/Technology* **8,** 33–38.
Schäfer, W., Görz, A., and Kahl, G. (1987). *Nature (London)* **327,** 529–532.
Shaw, M. L., Conner, A. J., Lancaster, J. E., and Williams, M. K. (1988). *Plant Mol. Biol. Rep.* **6,** 155–164.
Sheikholeslam, S. N., and Weeks, D. P. (1987). *Plant Mol. Biol.* **8,** 291–298.
Simpson, J., Timko, M. P., Cashmore, A. R., Schell, J., Van Montagu, M., and Herrera-Estrella, L. (1985). *EMBO J.* **4,** 2723–2729.
Southern, E. M. (1975). *J. Mol. Biol.* **98,** 503–517.
Suseelan, K. N., Bhagwat, A., Mathews, H., and Bhatia, C. R. (1987). *Curr. Sci.* **56,** 888–889.
Ulian, E. C., Smith, R. H., Gould, J. H., and McKnight, T. D. (1988). *In Vitro Cell. Dev. Biol.* **24,** 951–954.
Van Larebeke, N., Engler, G., Holsters, M., Van Der Elsacker, S., Zaenen, I., Schilperoort, R. A., and Schell, J. (1974). *Nature (London)* **252,** 169–170.
Wilmar, C., and Hellendoorn, M. (1968). *Nature (London)* **217,** 369–370.

Part II

Transgenic Dicotyledonous Plants

4

Transgenic Plants from Solanaceae

C. M. Houck and D. Facciotti
Calgene, Inc.
Davis, California, 95616

R. M. Goodman
Russell Laboratories
University of Wisconsin
Madison, Wisconsin, 53705

The family Solanaceae includes about 90 genera and 2000–3000 species (Bateman, 1985). The family contains a number of economically valuable vegetables and fruits such as potatoes (*Solanum tuberosum*), tomatoes (*Lycopersicon esculentum*), eggplant (*Solanum melongena*), and peppers (*Capsicum* species). Many *Capsicum* varieties are used as flavorings or spices in the form of chilies, pimentos, paprika, cayenne pepper, curries, and green and red bell peppers. Other Solanaceae such as *Petunia* and some species of *Datura* and *Nicotiana* are cultivated as ornamentals. Tobacco (*Nicotiana tabacum*) is also an economically important solanaceous crop. A number of other solanaceous plants contain alkaloids that are extracted for use in the pharmaceutical industry.

Many solanaceous species are fairly easy to manipulate in tissue culture; plants can be regenerated from a number of explant tissues, including protoplasts. Because of the ease of manipulation, much of the early development of plant transformation and regeneration systems was done using solanaceous plants. In particular, *Nicotiana* species have been used extensively as model systems for transformation. Transgenic tobacco plants have been obtained by *Agrobacterium* cocultivation of tissue explants (Spielmann and Simpson, 1986) and protoplasts (De Block *et al.*, 1984; Horsch *et al.*, 1984), as well as by electroporation (Riggs and Bates, 1986; Lurquin and Paszty, 1988), microprojectile bombardment (Klein *et al.*, 1988; Tomes *et al.*, 1990), direct DNA uptake (Hain *et al.*, 1985), and microinjection (Crossway *et al.*, 1986). A further advantage of *Nicotiana* as a model system is the fact that several different species with different ploidy levels have been transformed and regenerated (Andre *et al.*, 1986; Kiernan *et al.*, 1989).

We will discuss some of the basic technology in plant transformation and gene expression in transgenic plants that has been developed using solanaceous systems and give a brief status report on the progress that has been made in transforming solanaceous species. We also will review some of the many studies of tissue-specific regulation of plant genes that have been done using transgenic Solanaceae, frequently because the homologous species was difficult to transform or regenerate.

A number of important phenotypes have been obtained in transgenic Solanaceae. Many of these phenotypes improve the agronomic characteristics of the plant by conferring herbicide tolerance, insect resistance, or disease resistance. In other cases, the transgenic plants are expected to have improved quality traits in the harvested product, for example, fruit, tubers, or flowers. It has been demonstrated that male sterile transgenic plants can be produced; this advance may make it possible to improve the efficiency of hybrid production. In other cases, transgenic mutants have been described that may provide insights into the developmental processes in the plant or the function of a particular gene.

Finally, we will discuss the current status of field testing of transgenic Solanaceae and review the development of regulations governing the release of transgenic crop plants.

4.1 Basic Technology Development

4.1.1 VECTORS AND GENE TRANSFER SYSTEMS

Gene transfer technology in higher plants was developed almost entirely using solanaceous species, especially tobacco and petunia. Early research on transformation followed two approaches. One was modeled on the direct DNA insertion and selection techniques that had been used successfully in animal cells, notably microinjection of DNA (Diakumakos *et al.*, 1970), uptake of DNA–polyamine complexes (Carroll, 1972), calcium phosphate coprecipitation (Graham and Van der Eb, 1971), uptake of liposome-coated DNA (Poste *et al.*, 1976), and electroporation of nucleic acids (Wong and Neumann, 1982). The other approach was based on *Agrobacterium tumefaciens* (Chilton *et al.*, 1977) and, to some extent, *Agrobacterium rhizogenes* (White *et al.*, 1982).

During transformation of animal cells, the DNA enters the cell directly through the plasmalemma. After reaching the nucleus, or during cell division in the absence of the nuclear membrane, the DNA can be integrated stably into the cell genome at a low frequency (Toneguzzo *et al.*, 1986). The closest plant equivalent to the animal cell, with respect to direct access to the plasmalemma, is a protoplast produced by mechanical or enzymatic removal of the cell wall (Zaitlin, 1959; Cocking, 1960, 1972). A number of published reports indicated that protoplasts were able to take up macromolecules (Aoki and Takebe, 1969; Hoffman and Hess, 1972; Ohyama *et al.*, 1972), organelles, nuclei (Potrykus and Hoffmann, 1973), and even cyanobacteria (Bradley and Leith, 1979). Much early work in studies on plant viruses followed the demonstration of methods for preparing plant protoplasts in large numbers and their efficient inoculation and infection with plant viruses (Takebe, 1975). It was, therefore, obvious to try direct DNA uptake methods. Protoplasts also offered the advantage of transformation and regeneration of entire plants from individual plant cells (Steward *et al.*, 1964, 1969). Selection at the single-cell level also had the potential to reduce cross-feeding effects that might occur with multicellular tissue explants.

In the early 1980s, protoplast isolation and culture protocols allowing the regeneration of cell walls had been developed in a few species. Under specific *in vitro* culture conditions, the regenerated cells of some of these species were capable of forming whole plants (Gamborg, 1981). Among the most efficient protoplast cultures available were those of *Nicotiana* and *Petunia,* two solanaceous species. Carrot protoplast cultures were also efficient, with respect to cell and plant regeneration, and were used sometimes in transformation experiments (Ohgawara *et al.*, 1983). However, the Solanaceae were preferred because they are annuals (carrots are biennials), which simplified studies of stability of transformation in the progeny.

As early as 1979, the first reports of transformation of tobacco protoplasts by cocultivation with *Agrobacterium tumefaciens* appeared (Marton *et al.*, 1979). Cell lines derived from transformed protoplasts expressed the oncogenes carried by the Ti plasmid and were selected for their ability to grow in a phytohormone-free medium. The first report of protoplast transformation using direct DNA uptake is the work of Davey *et al.* (1980). Ti plasmid DNA isolated from *Agrobacterium* was incubated with *Petunia* protoplasts. *Petunia* was chosen over tobacco because of stricter hormone needs for its growth. As in the work of Marton *et al.* (1979) using *Agrobacterium* cocultivation, the transformed cell lines were selected for their ability to grow in hormone-free medium. Over a period of 2 yr, hormone-independent colonies were recovered at a very low frequency of $1/10^5$ protoplasts plated. Opines, chemicals found only in tumors induced by *Agrobacterium,* were also detected in these colonies (Krens *et al.*, 1982). Although very encouraging, these results were received with caution because of the novelty of the approach, the low frequency of transformation, the time necessary for the recovery of transformed colonies, and, most of all, the absence of direct evidence for the presence of the transfered DNA.

At the same time, considerable effort was devoted to attempts to deliver nucleic acids more efficiently using liposomes, artificial phospholipid vesicles containing encapsulated macromolecules. Liposomes were able to fuse with protoplasts, resulting in delivery of the encapsulated nucleic acids. Most of the research used liposomes carrying viruses or viral RNA (Nagata *et al.*, 1981; Fraley *et al.*, 1982; Sheehy and Lurquin, 1983). Successful delivery resulted in viral infection. The response was amplified by the rapid multiplication of the virus and the relatively abundant production of easily detectable viral proteins. However, liposome transformation was of limited value. The preparation of liposomes was not simple and their quality varied considerably from preparation to preparation and from laboratory to laboratory. Further, encapsulation of DNA was shown to be very size dependent (Mannino *et al.*, 1979). The delivery of Ti plasmid was attempted also using *Agrobacterium* spheroplasts prepared by digesting the bacterial cell wall with lysozyme. The spheroplasts were fused with protoplasts to deliver the Ti plasmid and the transformed cells were selected for hormone-independent growth and assayed for opine production (Hain *et al.*, 1984; Millman and Lurquin, 1985).

Studies with liposomes and virus or with *Agrobacterium* spheroplasts suffered from the lack of good detectable or selectable marker genes capable of being expressed in plants, thus allowing proper evaluation and improvement of the gene delivery system. Selection in animal tissue culture had been developed using methotrexate, with resistance conferred by a gene encoding a dihydrofolate reductase that was insensitive to methotrexate, or genticin 418 (G418), with resistance conferred by expression of a bacterial neomycin phosphotransferase gene that detoxified aminoglycoside antibiotics (such as G418, kanamycin, or neomycin). Although methotrexate was not found to be a useful selectable marker in plant cells, kanamycin is the most widely used selectable marker for plant cell transformation today. Many different promoters have been used to express the bacterial neomycin phosphotransferase (*nptII*) gene (see subsequent text), resulting in plant cells resistant to the antibiotic. Cells originating from protoplasts could be selected for resistance to kanamycin, making it possible to compare directly the efficiency of the different delivery systems: direct DNA uptake, liposomes and spheroplasts, or *Agrobacterium*. Despite interesting successful experiments with spheroplasts from *Escherichia coli* (Hain *et al.*, 1984), it was clear that *Agrobacterium* was the most efficient system.

With the development of efficient selectable markers, transformation systems were no longer limited to those that could be regenerated from protoplasts. Other tissues, such as stem segments or leaf explants, were targeted and resulted in transformed cell lines (Horsch *et al.*, 1985). The efficient use of *Agrobacterium* as a transformation vector required the suppression of the oncogenes. Infection of plant cells with wild-type *Agrobacterium* resulted in the formation of crown gall tumors that precluded the regeneration of plants. Although there were early reports of regeneration of normal fertile plants that had deleted portions of the *Agrobacterium* T-DNA (Otten *et al.*, 1981; Yang and Simpson, 1981), it was not clear whether the oncogenes were necessary for high frequency gene transfer or gene expression. However, analysis of strains with transposon insertions in the oncogenes showed that the strains were still capable of delivering the DNA into the host cell genome and "disarmed" *Agrobacterium* strains with deleted oncogenes were developed rapidly. Two types of systems were used. In one, the Ti plasmid itself was engineered by replacing the oncogenes with the marker gene and any other gene of interest (Zambryski *et al.*, 1983). In the second system, the transferred region (T-region) was maintained on a second plasmid that functioned in *trans* in the *Agrobacterium* strain (the "binary" system) (Hoekema *et al.*, 1983). This system is the one most commonly used today for plant transformation.

That *Agrobacterium rhizogenes* has not been used more widely in plant transformation is somewhat surprising. In the early 1980s, it was reported that plants could be regenerated from hairy roots induced by *A. rhizogenes* (Spano and Costantino, 1982; Tepfer, 1982). The plants had abnormal morphology, but they were fertile, which was sufficient for the study of transferred genes at the whole plant level and in transformed progeny (Comai *et al.*, 1985). The Ri plasmids, however, were not as well characterized as the Ti plasmids, which limited their usefulness.

Although *Agrobacterium* transformation is now the standard for transformation of solanaceous species, its use is limited in many other species. The most important crops among cereals and legumes are difficult or impossible to transform using *Agrobacterium,* giving impetus to attempts to develop new gene transfer technology. Since cell culture and regeneration in these other crops can be difficult, they are not well suited for the development of a technology that requires considerable trial and error. Solanaceae, especially *N. tabacum,* thus became the model system for these efforts. The development of electroporation systems (Riggs and Bates, 1986) and improved methods of direct DNA uptake (Paszkowski *et al.*, 1984; Hain *et al.*, 1985; Uchimiya *et al.*, 1986) were done using tobacco protoplasts. Even microinjection methods intended for use in pollen and other tissues inaccessible to *Agrobacterium* were developed in tobacco (Crossway *et al.*, 1986).

More recently, tobacco was used in the development of the particle gun, a system in which a microscopic particle of heavy metal (gold, tungsten, platinum) is coated with DNA and literally shot into the target cell. The first reports of transgenic plants using this system were from tobacco leaf tissues bombarded with a selectable marker gene (Klein *et al.*, 1988; Tomes *et al.*, 1990). The particle gun has since been used successfully for the production of transgenic soybean (McCabe *et al.*, 1988) and corn (Gordon-Kamm *et al.*, 1990) plants. This system also has been used in conjunction with selectable markers designed to function in plastids to produce tobacco plants containing transformed chloroplasts (Svab *et al.*, 1990).

4.1.1.1 Status Report of Transgenic Solanaceous Plants

Cocultivation of explants with disarmed *Agrobacterium tumefaciens* (or *A. rhizogenes*) is today the preferred and most efficient method of transformation in Solanaceae. In some species, however, especially *N. tabacum, L. esculentum,* and *P. hybrida,* other approaches were developed for protoplast transformation. These methods included cocultivation of protoplasts with *Agrobacterium,* direct DNA uptake, liposome-mediated uptake, protoplast–spheroplast fusion, electroporation, and microinjection. Table 4.1 reviews the solanaceous species that have been transformed and regenerated. It does not include all the reports of transformation of Solanaceae but provides at least one example of each approach. In contrast to most other solanaceous species that have been tested, *Capsicum annum* has not yet been transformed and regenerated. To date, only transformed shoot-like structures have been produced (Liu *et al.*, 1990).

Hairy roots from at least 28 solanaceous species have been induced by *Agrobacterium rhizogenes,* grown *in vitro* (organ culture) and used in studies on tropane alkaloid biosynthesis. These include species of *Hyoscyamus* (Jaziry *et al.*, 1988; Knopp *et al.*, 1988; Signs and Flores, 1989), *Scopolia* (Knopp *et al.*, 1988; Zhang, 1988), *Duboysia* (Deno *et al.*, 1987; Knopp *et al.*, 1988; Mano *et al.*, 1989), *Datura* (Jaziry *et al.*, 1988; Knopp *et al.*, 1988; Kephalas *et al.*, 1989), and *Atropa* (Jung and Tepfer, 1987; Knopp *et al.*, 1988).

4.1.2 GENE EXPRESSION AND REGULATION

Because tobacco was the first plant to be transformed and regenerated (Otten *et al.*, 1981; Yang and Simpson, 1981; Spano and Costantino, 1982; Barton *et al.*, 1983), much of the early work on expression of foreign genes in transgenic plants was done using Solanaceae. T-DNA genes encoding opine synthases were among the first to be used. For example, the nopaline synthase (nopaline dehydrogenase) promoter fused to neomycin phosphotransferase, chloramphenicol acetyltransferase, dihydrofolate reductase, hygromycin phosphotransferase, or other selectable markers was expressed in tobacco cells (Bevan *et al.*, 1983; Herrera-Estrella *et al.*, 1983a,b; Van Den Elzen *et al.*, 1985; Waldron *et al.*, 1985), petunia cells (Fraley *et al.*, 1983), and later in transgenic plants (De Block *et al.*, 1984; Hain *et al.*, 1985). Although this promoter was, at first, thought to produce constitutive expression, later work showed that the expression was higher in certain organs and in certain developmental stages (An *et al.*, 1988). The octopine synthase (octopine dehydrogenase; Comai *et al.*, 1985; Waldron *et al.*, 1985; Thompson *et al.*, 1987; Houck *et al.*, 1990) and mannopine synthase (mannopine dehydrogenase; Velten *et al.*, 1984; Comai *et al.*, 1985; McBride and Summerfelt, 1990) promoters also have been used widely.

TABLE 4.1
Transgenic Solanaceous Species

Genus and species	Explant cocultivation	Protoplast transformation
Nicotiana		
alata	Ebert and Clark (1990)	
bigelowii	Schoelz *et al.* (1990)	
clevelandii	Tavazza *et al.* (1988)	
edwardsonii	Kiernan *et al.* (1989)	
glauca	Amasino *et al.* (1983); Ono *et al.* (1986)	
hesperis	Hamill *et al.* (1986)	
plumbaginifolia	Trinh *et al.* (1987)	Liposome (Coboche and Deshayes, 1984)
rustica	Hamill and Rhodes (1988)	
tabacum	Zambryski *et al.* (1983); Horsch *et al.* (1984); De Block *et al.* (1984); Sengupta Gopalan *et al.* (1985)	Cocultivation (Memelink *et al.*, 1983; Nagy *et al.*, 1985); direct DNA uptake (Paszkowski *et al.*, 1984; Hain *et al.*, 1985; Uchimiya *et al.*, 1986); microinjection (Crossway *et al.*, 1986); electroporation (Riggs and Bates, 1986)
Petunia hybrida	Beachy *et al.* (1985); Jones *et al.* (1985)	Cocultivation (Nagy *et al.*, 1985)
Lycopersicon		
esculentum	McCormick *et al.* (1986); Shahin *et al.* (1986); Chyi *et al.* (1986); Fillatti *et al.* (1987)	Cocultivation (De Block *et al.*, 1987); electroporation (Bellini *et al.*, 1989)
peruvianum	Morgan *et al.* (1987)	Electroporation (Bellini *et al.*, (1989)
Solanum		
tuberosum	Shahin and Simpson (1986); Ooms *et al.* (1987); Twell and Ooms (1987); De Block (1988); Hanisch *et al.* (1988); Sheerman and Bevan (1988); Stiekema *et al.* (1988); Visser *et al.* (1989)	
dulcamara	Lee and Davey (1988)	
nigrum	Johnson *et al.* (1990)	
melongena	Filippone and Lurquin (1989); Guri and Sink (1988); Rotino and Gleddie (1990)	
Atropa belladonna	Ondrej and Vlasak (1987)	
Datura innoxia	Kiernan *et al.* (1989)	
Hyoscyamus muticus	Oksman-Caldentey *et al.* (1989)	

Viral promoters (Kridl and Goodman, 1986) such as the 35S (Odell *et al.*, 1985; Ow *et al.*, 1986; Shah *et al.*, 1986) and 19S (Paszkowski *et al.*, 1984; Balazs *et al.*, 1985; Uchimiya *et al.*, 1986) promoters of cauliflower mosaic virus (CaMV) and the related 34S promoter of the figwort mosaic virus (Sanger *et al.*, 1990) are of special significance because they give high levels of expression in many cell and tissue types. Recently, variations of these promoters designed to enhance expression levels even further have been used successfully. These include duplications of the enhancer region of the 35S promoter (Kay *et al.*, 1987) and hybrid mas-35S promoters (Comai *et al.*, 1990). Dissection of the 35S caulimovirus promoters has been done and several functional regions have been identified (Odell *et al.*, 1988; Benfey *et al.*, 1990a,b).

4.1.2.1 Environmental Responses: Light, Heat Shock, and Wounding

The best venue in which to study the sequences responsible for inducible, developmentally regulated, and/or tissue-specific expression of plant genes is regenerated transgenic plants. The effects of flanking sequences, coding regions, and introns can be probed independently by comparing different promoter–gene constructions. A number of genes that are responsive to environmental influences such as light, temperature, and wounding have been examined in transgenic Solanaceae.

For example, light-inducible expression of ribulose-1,5-bisphosphate carboxylase small subunit (*rbcS*) genes from pea have been studied in transgenic petunia (Nagy *et al.*, 1985) and tobacco (Simpson *et al.*, 1986). DNA binding to a nuclear factor (GT-1) has been correlated with light-specific expression in transgenic tobacco (Lam and Chua, 1990). The expression of *rbcS8B* from *N. plumbaginifolia* has been studied in transgenic *N. plumbaginifolia* and *N. tabacum* (Poulsen *et al.*, 1986). Promoter deletions have allowed the separation of sequences responsible for light regulation from those necessary for organ specificity (Poulsen and Chua, 1988). Comparisons of expression levels of different petunia *rbcS* genes and gene chimeras have been performed in transgenic tobacco (Dean *et al.*, 1989). A short sequence that is sufficient to confer light regulation has been mapped also from an *Arabidopsis rbcS* gene in transgenic tobacco (Donald and Cashmore, 1990). However, a wheat *rbcS* gene is not expressed in transgenic tobacco under control of the wheat promoter. When the wheat gene is expressed in tobacco under the control of the CaMV 35S promoter, RNA splicing of the introns is inefficient and alternative poly(A) sites are used (Keith and Chua, 1986).

Promoters from other light-responsive genes such as the chlorophyll *a/b* binding proteins from pea (Simpson *et al.*, 1986), petunia (Gidoni *et al.*, 1988), and wheat (Nagy *et al.*, 1986), and the *Antirrhinum* chalcone synthase (Kaulen *et al.*, 1986) have been shown to be light regulated in transgenic tobacco.

In response to high temperature stress, most plants synthesize large quantities of heat-shock proteins that have highly conserved carboxy-terminal domains (Vierling *et al.*, 1989). Heat-shock promoters from soybean and maize have been shown to be heat inducible in transgenic tobacco (Baumann *et al.*, 1987; Medford *et al.*, 1989; Schoeffl *et al.*, 1989; Ainley and Key, 1990). The soybean heat-shock promoters contain sequences homologous to a heat-shock consensus element from *Drosophila,* but it is nonetheless interesting to note that the *Drosophila Hsp70* promoter fused to a reporter gene was capable of producing heat-regulated expression in roots, stem, and leaves (but not pollen) of transgenic tobacco (Spena and Schell, 1987).

Plants respond to wounding by expressing a variety of genes thought to be involved in specific defense mechanisms such as resistance to insect or pathogen attack. Several wound-inducible genes have been isolated from solanaceous plants, including serine protease inhibitor genes from potato (Sanchez-Serrano *et al.*, 1986) and tomato (Graham *et al.*, 1986) and a carboxypeptidase inhibitor gene from tomato (Martineau *et al.*, 1991). Other wound-inducible genes have been isolated that do not yet have well-characterized enzymatic activities (Stanford *et al.*, 1989). Several studies have shown that wound-induced genes are also responsive to other stimuli such as oligosaccharides (Sanchez-Serrano *et al.*, 1987), ethylene (Margossian *et al.*, 1988), and abscissic acid (Pena-Cortes *et al.*, 1989). In one case, the wound responsiveness of the protease inhibitor II (*PI-II*) gene from potato was shown to be induced by cytokinin or ethylene and repressed by auxin (Kernan and Thornburg, 1989).

Wound inducibility appears to be maintained in transgenic plants containing wound-inducible promoter constructs. For example, potato *PI-II* genes and constructs fusing the *PI-II* promoter to marker genes are wound inducible in transgenic potato (Keil *et al.*, 1989) and tobacco (Sanchez-Serrano *et al.*, 1987; Thornburg *et al.*, 1987; Johnson and Ryan, 1990). The tomato *PI-I* gene is also wound inducible in transgenic nightshade (Johnson *et al.*, 1990). Wound-induced expression driven by other potato wound-inducible genes of unknown function is maintained also in transgenic potato (Stanford *et al.*, 1990) and tobacco (Logemann *et al.*, 1989; Seibertz *et al.*, 1989). In addition, a *Phaseolus* promoter isolated from a wound-inducible glycine-rich protein is capable of producing wound-inducible expression of β-glucuronidase in transgenic tobacco (B. Keller *et al.*, 1989).

An interesting case is that of the phenylalanine ammonia-lyase gene from bean, which has been studied in transgenic tobacco. In the transgenic plants, the gene is expressed in a complex developmental pattern similar to the endogenous pattern in bean, but the expression is influenced also by wounding and by light (Liang *et al.*, 1989).

A short region from the maize alcohol dehydrogenase promoter is capable of anaerobic induction in transgenic tobacco, but expression is obtained only when enhancer regions of the *ocs* promoter or the CaMV promoter are used upstream (Ellis *et al.*, 1987). This result contrasts with that for a maize sucrose synthase promoter, which does not require an additional enhancer to be induced anaerobically in roots of transgenic tobacco plants. The maize sucrose synthase promoter also regulates the expression of a marker gene in phloem cells and in the endosperm of the transgenic seed (Yang and Russell, 1990).

4.1.2.2 Developmentally Regulated Expression

a. Organ or Tissue-Specific Regulation (Seeds, Fruits, Tubers, Floral Tissues). Much early work on tissue-specific expression was done using the abundant seed-storage proteins of legumes. However, because legumes are difficult to transform and regenerate, studies of expression of these genes in transgenic plants generally were done first in solanaceous plants. For example, soybean β-conglycinin was shown to be expressed in the seeds of transgenic petunia (Beachy *et al.*, 1985; Chen *et al.*, 1986; Naito *et al.*, 1988) and tobacco (Bray *et al.*, 1987). Bean phaseolin (Greenwood and Chrispeels, 1985; Sengupta Gopalan *et al.*, 1985), legumin (Baeumlein *et al.*, 1987), and phytohemagglutinin (Voelker *et al.*, 1987), and pea legumin (Ellis *et al.*, 1988; Shirsat *et al.*, 1989), vicillin (Higgins *et al.*, 1988), and convicillin (Newbigin *et al.*, 1990) all were expressed in the seeds of transgenic tobacco. In general, the pattern of expression of the seed protein was maintained in the transgenic plants, but the level of expression of the mRNA and protein was much lower than in the endogenous species. Many of these seed-specific promoters also have been characterized using reporter genes such as chloramphenicol acetyltransferase (CAT) or β-glucuronidase (GUS).

Seed proteins from nonlegumes, for example, an *Arabidopsis* 2S albumin (De Clercq *et al.*, 1990), and even monocot proteins such as wheat glutenins (Colot *et al.*, 1987; Robert *et al.*, 1989b) also are expressed in seeds of transgenic tobacco. One interesting exception is the case of a maize zein gene which was transcribed not only in seeds but in other parts of transgenic petunia plants (Ueng *et al.*, 1988). In another study, however, a zein promoter–GUS fusion resulted in seed-specific expression in transgenic tobacco (Schernthaner *et al.*, 1988). In addition, a sunflower helianthinin–35S–GUS fusion (Jordano *et al.*, 1989), a rice glutenin–CAT fusion (Leisy *et al.*, 1990), and a barley β-hordein–CAT fusion (Marris *et al.*, 1988) have been expressed in transgenic tobacco seeds.

Fruit-specific regulation of genes has been studied primarily in tomato. Most of the work has focused on ripening-related genes (Mansson *et al.*, 1985; Slater *et al.*, 1985; Bathgate *et al.*, 1986; Grierson *et al.*, 1986; Holdsworth *et al.*, 1987; Ray *et al.*, 1987) or ethylene-regulated expression (Lincoln *et al.*, 1987; Maunders *et al.*, 1987; Deikman and Fischer, 1988; Lincoln and Fischer, 1988;

Margossian *et al.*, 1988), although other genes have been characterized that are expressed at earlier stages of fruit development (Pear *et al.*, 1989). Several fruit-specific gene constructs have been reintroduced into transgenic tomato plants; for example, the polygalacturonase (PG) promoter fused to CAT (Bird *et al.*, 1988), the ethylene-regulated *E8* gene fused to PG (Giovannoni *et al.*, 1989), and the fruit-specific gene *2A11* fused to GUS (VanHaaren and Houck, 1991). Each of these constructs is expressed at the correct stage of development in transgenic plants but the level of expression of the introduced gene can vary.

Expression in potato tubers has been achieved using promoters from patatin genes. Patatins are abundant glycoproteins, found in tubers, that have a lipid acylhydrolase activity (Jefferson *et al.*, 1990). Class I patatin genes are expressed primarily in potato tubers, but can be induced by sucrose in other parts of the plant. Marker genes expressed in transgenic potato and tobacco under the control of class I patatin promoters maintain the sucrose inducibility (Rocha Sosa *et al.*, 1989; Wenzler *et al.*, 1989a,b; Jefferson *et al.*, 1990). Class II patatin genes are not sucrose inducible; transgenic potato plants containing Class II patatin promoters express marker genes in tubers and root tips (Koester *et al.*, 1989). The *PI-II* genes of potato mentioned earlier also are expressed constitutively in stolons and tubers, while maintaining wound inducibility in leaves and stems of transgenic plants (Keil *et al.*, 1989).

Genes expressed specifically in reproductive tissues of solanaceous plants have been characterized and shown to maintain their specificity in transgenic plants. Interesting examples are the chalcone isomerase genes of petunia. The coding region of one gene is regulated by two distinct promoters that can function independently, one promoting expression in limb, tube, seed, sepal, leaf, and stem tissues and the other expressed in the pollen of transgenic plants. A second chalcone isomerase gene is expressed anther-specifically in pollen and tapetal cells of transgenic petunia plants (Van Tunen *et al.*, 1990). *Brassica* self-incompatibility proteins are expressed appropriately in the transmitting tissue of styles and stigma of transgenic tobacco plants (Kandasamy *et al.*, 1990). A potato gene that is wound inducible in leaves was shown to be expressed in anthers and pollen of unwounded transgenic tobacco plants (Seibertz *et al.*, 1989). Other tobacco genes of unknown function have been characterized that are expressed in developing anthers, in some cases specifically in tapetal cells (Goldberg, 1988). Promoters from these genes have been fused to reporter genes but, more importantly, also have been fused to RNase genes causing degeneration of the tapetal cells and male sterility in transgenic tobacco and *Brassica* plants (Mariani *et al.*, 1990).

b. Hormone-Regulated Expression. Many studies have demonstrated the influence of phytohormones on plant gene expression. Complex interactions between hormones and other types of stimuli often exist. Examples of wound-inducible and ripening-related genes that are also responsive to various hormones have been discussed earlier. A further example is a family of auxin-induced genes isolated from soybean that maintain their auxin responsiveness in transgenic petunia (Wright *et al.*, 1989). An interesting study of genes induced by cytokinin was carried out using transformed shoots containing an isopentenyltransferase gene from the *Agrobacterium* T-DNA. The shoots had increased levels of five cDNAs, four of which (extensin, chitinase, PR1, and a PR1-like sequence) had been characterized previously as plant defense-related sequences (Memelink *et al.*, 1987). A defense-related bean chitinase gene was shown to be ethylene inducible in transgenic tobacco; the ethylene-responsive regions were mapped by deletion analysis (Broglie *et al.*, 1989).

4.1.2.3 Antisense and Cosuppression

The recent discovery that genes can be down-regulated by expression of antisense RNA in transgenic plants is one of the most important technical innovations of the last few years. Early experiments used introduced marker genes to demonstrate that the technique was feasible. For example, nopaline synthase expression was decreased by 85% in tobacco plants containing the nopaline synthase gene and antisense regions of the nopaline synthase gene driven by the petunia chlorophyll *a*/*b* binding protein promoter (Sandler *et al.*, 1988). Transgenic tobacco plants expressing CAT were transformed with antisense CAT sequences, resulting in plants containing no detectable CAT activity (Delauney *et al.*, 1988). Progeny of plants in which the sense and antisense sequences segregated independently were found to have restored CAT activity when the antisense sequences were not present (Cheon *et al.*, 1990). Retransformation of tobacco plants expressing GUS activity with antisense GUS sequences resulted in 90% reduction of GUS activity (Robert *et al.*, 1989a).

A much more important observation was that the expression of endogenous plant genes could be reduced by antisense sequences. One example that is of commercial importance is the reduction of PG activity in transgenic tomato fruit (Sheehy *et al.*, 1988; Smith *et al.*, 1988). Plants with greater than 90% reduction in PG activity do not soften or rot as quickly as control fruit, but otherwise ripen normally (Kramer *et al.*, 1989; Smith *et al.*, 1990). Antisense expression of a another ripening-related gene, *pTOM13*, in transgenic tomato reduced ethylene levels in fruit (Hamilton *et al.*, 1990). In an elegant study of coordinate regulation, the level of mRNA and protein was reduced in transgenic tobacco by expression of antisense *rbcS*. This reduction caused a corresponding reduction in the carboxylase large subunit (*rbcL*) protein, but not in the *rbcL* mRNA, demonstrating that *rbcL* is regulated by translational and posttranslational factors and is sensitive to *rbcS* levels (Rodermel *et al.*, 1988).

Expression of antisense viral sequences has been shown to give some protection against tobacco mosaic virus (TMV; Powell *et al.*, 1989) and cucumber mosaic virus (CMV; Rezaian *et al.*, 1988) in transgenic tobacco. In the case of TMV, expression of RNA sequences complementary to the coat protein sequences did not provide protection, but the addition of the sequences complementary to the tRNA-like 3′ region did result in reduced symptoms, although the level of protection was less than that provided by constitutive expression of the coat protein (Powell *et al.*, 1989). Expression of antisense CMV resulted in some protection against viral infection, but the level of protection did not correlate with the level of antisense RNA.

Chalcone synthase levels have been reduced in transgenic petunia plants using antisense constructions with either the 35S promoter or the chalcone synthase promoter. Sequences complementary to the full length mRNA or to the 3′ end of the mRNA created reduced chalcone synthase levels, but sequences complementary to the 5′ end of the mRNA had no effect (Van Der Krol *et al.*, 1988, 1990c). There was some variability in the flower color phenotype in some of the transgenic plants (Van Der Krol *et al.*, 1990b). An extremely interesting phenomenon, termed cosuppression, was discovered in the course of manipulating these flower pigment genes. The observation was that expressing genes in the sense, as well as the antisense, orientation resulted in decreased levels of expression of the encoded protein and a resulting lack of pigment in the petals of the transgenic plants. Cosuppression was observed using either the chalcone synthase gene or the dihydroflavanol reductase genes (Van Der Krol *et al.*, 1990a); up to 42% of the transformants showed either all white or a pattern of white and violet sectors on the petals (Napoli *et al.*, 1990). The means by which reintroduction of a gene causes reduction of expression levels and alteration of the pattern of expression is not yet understood.

4.1.2.4 Transposons

No active transposon systems have been reported to occur in solanaceous species, although a transposon-like sequence has been reported in the promoter region of an inactive class II patatin gene isolated from potato (Koester *et al.*, 1990). The insertion was characterized by the presence of an 11-bp inverted repeat sequence at the termini and an 8-bp duplication at the insertion site (by comparison with other class II patatin genes).

Several groups have been successful in transferring functional transposon systems to transgenic Solanaceae. The most widely used elements are from the maize *Ac*/*Ds* system. The transposase encoded by the *Ac* element causes transposition of *Ac* or *Ds* elements in transgenic tobacco (Baker *et al.*, 1986), tomato (Yoder *et al.*, 1988), and petunia (Haring *et al.*, 1989). Transposition has been assayed in the initial transformants and in the progeny (Belzile *et al.*, 1989; Taylor *et al.*, 1989) using selectable markers (Baker *et al.*, 1987; Jones *et al.*, 1989; Masterson *et al.*, 1989), histochemical staining (Finnegan *et al.*, 1989), and Southern analysis of DNA (Baker *et al.*, 1986; Yoder *et al.*, 1988). Plants expressing the *Ac* transposase can be crossed with plants containing *Ds* elements to give rise to progeny that have undergone transposition (Hehl and Baker, 1989; Lassner *et al.*, 1989).

Other maize elements such as the *En-Spm* elements have been shown to be functional in transgenic tobacco (Masson and Fedoroff, 1989; Pereira and Saedler, 1989) and potato (Frey *et al.*, 1989). The *Tam3* transposable element from *Antirrhinum* is also functional in transgenic tobacco (Martin *et al.*, 1989) and petunia (Haring *et al.*, 1989). *Tam3* sequences in tobacco transformants are rapidly methylated; it is thought that the methylation may suppress the transposition (Martin *et al.*, 1989).

Attempts to cause transposition using the *Drosophila P*-element in transgenic tobacco were not successful, apparently because of incorrect processing of the transposase RNA transcript in tobacco (Martinez Zapater *et al.*, 1988).

4.2 Phenotypes of Transgenic Plants

A wide range of phenotypes has now been obtained in transgenic Solanaceae as a result of the expression of cloned genes. We have divided these phenotypes into agronomic traits, developmental traits, and quality traits. Many of these traits, especially those that confer agronomic benefits on the plants, have been reviewed elsewhere in this volume and only a brief discussion is given here. We expect that many of these traits will be commercialized in the future.

4.2.1 AGRONOMIC TRAITS

4.2.1.1 Herbicide Resistance

Transformed plants tolerant to a variety of herbicides have been developed using several different strategies. For example, tolerance to the herbicide glyphosate, which inhibits the enzyme 5-enolpyruvylshikimate-3-phosphate synthase (EPSP synthase), has been achieved in transgenic tobacco (Comai *et al.*, 1985) and tomato (Fillatti *et al.*, 1987) plants by expressing a mutant *Salmonella* gene that encodes an EPSP synthase less sensitive to glyphosate inhibition that maintains its enzymatic activity. Similar work has been done using an *E. coli* EPSP synthase in transgenic petunia (Della Cioppa *et al.*, 1987). Some tolerance also has been achieved by overexpression of a petunia EPSP synthase in transgenic petunia (Shah *et al.*, 1986).

An herbicide degradation strategy has been used to confer resistance to the herbicide bromoxynil (a photosystem II inhibitor) in transgenic tobacco using a nitrilase enzyme isolated from a *Klebsiella* species that specifically degrades bromoxynil (Stalker *et al.*, 1988).

Resistance to phosphinothricin (an inhibitor of glutamine synthase) has been achieved in transgenic tobacco, potato, and tomato by expression of a bacterial gene encoding a phosphinothricin acetyltransferase (De Block *et al.*, 1987; Wohllenben *et al.*, 1988) and in transgenic tobacco by overexpression of an alfalfa glutamine synthase (Eckes *et al.*, 1989).

Mutant acetolactate synthase (ALS) genes have been isolated that confer resistance to sulfonylureas and imidazolanones. For example, a sulfonylurea-resistant ALS from *Arabidopsis* confers resistance to transgenic tobacco (Haughn *et al.*, 1988; Charest *et al.*, 1990); the sequences of two herbicide-resistant ALS genes from tobacco have been reported (Lee *et al.*, 1988). The different ALS mutants vary in their ability to confer resistance to other sulfonylureas and imidazolanones (Gabard *et al.*, 1989). Resistance to the herbicide 2,4-D has been achieved in transgenic tobacco using a 2,4-D monooxygenase from *Alcaligenes* (Lyon *et al.*, 1989); resistance to atrazine has been conferred on transgenic tobacco by expressing the chloroplast-encoded photosynthetic QB protein (*psbA*) from resistant *Amaranthus* using chloroplast targetting signals (Cheung *et al.*, 1988).

4.2.1.2 Insect Tolerance

Tolerance to insect predation is an extremely important agronomic trait that has been approached using a variety of methods. By far the most successful strategy used to date is expression of the *Bacillus thuringiensis* δ-endotoxin (or a truncated version of it) in transgenic plants. For example, expression of a truncated *B. thuringiensis* toxin results in tobacco that is toxic to *Manduca sexta* larvae (Vaeck *et al.*, 1987a,b); the trait is transmitted to the progeny (Barton *et al.*, 1987). A slightly different version of the *B. thuringiensis* toxin confers resistance to *M. sexta, Heliothis zea,* and *Keiferia lycopersicella* in transgenic tomato in the greenhouse (Fischhoff *et al.*, 1987) and in the field (Delannay *et al.*, 1989). The major technical limitation to the *B. thuringiensis* strategy is the fact that individual *B. thuringiensis* toxins are efficacious against only a limited spectrum of insects.

Other insect resistance strategies have been devised based on the natural defense mechanisms of various plant species. These include the use of protease inhibitors such as the cowpea trypsin inhibitor, which has been shown to confer some tolerance to *Heliothis virescens* on transgenic tobacco (Hilder *et al.*, 1987, 1989). An α-amylase inhibitor (lectin-like protein) from bean seeds has been expressed in transgenic tobacco. This protein inhibits the activity of insect and mammalian amylases (Altabella and Chrispeels, 1990) and may be useful in insect control.

A number of solanaceous plants are sources of protease inhibitors (see

Section 4.1.2.1) that may be useful insect control agents [for review, see Ryan (1973, 1989).] A recent report indicates that expression of potato PI-II in transgenic tobacco plants does confer resistance to *M. sexta* larvae (Johnson *et al.*, 1989).

4.2.1.3 Disease Resistance

Many groups have reported on transgenic Solanaceae that have reduced susceptibility to viral infection as a result of antisense expression of viral components (discussed earlier) or expression of viral coat proteins. For example, TMV coat protein expressed in transgenic tobacco or tomato plants delays symptom development when plants are challenged with TMV (Abel *et al.*, 1986; Li *et al.*, 1989), although high temperature treatment reduces coat protein levels and reduces resistance (Nejidat and Beachy, 1989). Expression of alfalfa mosaic virus (AMV) coat protein in transgenic tobacco and tomato delays AMV symptoms (Tumer *et al.*, 1987). In addition, there are reports that the coat protein of one virus can provide protection against attack by another virus. For example, TMV coat protein expressed in transgenic tobacco plants gives a slight delay in symptom development when plants are inoculated with potato virus X (PVX), potato virus Y (PVY), CMV, or AMV; AMV coat protein expressed in transgenic tobacco plants gives a slight delay of symptoms when challenged with PVX and CMV (Anderson *et al.*, 1989). Expression of TMV coat protein also delays or reduces the effect of tomato mosaic virus, tobacco mild green mosaic virus, pepper mild mottle virus, and *Odontolassum* ringspot virus, although no effect was observed on ribgrass mosaic virus symptoms (Stark and Beachy, 1989; Nejidat and Beachy, 1990).

Transgenic plants (Rogers *et al.*, 1986; Sunter *et al.*, 1987; Hanley-Bowdoin *et al.*, 1989, 1990) have been instrumental in studies on the replication mechanisms of the geminiviruses (Harrison, 1985), showing that genome component A of tomato golden mosaic virus is capable of replication and encapsidation, but not of cell-to-cell movement, in the absence of genome component B and that genome component B is dependent on genome component A for its replication and encapsidation. This work has implications for a better understanding of disease mechanisms, as well as for DNA replication in and vector development for plants.

Recently, tobacco plants resistant to bacterial infection by *Pseudomonas syringae* have been reported. These plants express an acetyltransferase that detoxifies the pathogenic tabtoxin produced by the bacteria (Anzai *et al.*, 1989).

More general approaches to disease resistance are being explored currently by a number of groups studying the plant response to pathogen attack. This has led to such strategies as expression of a bacterial chitinase gene in transgenic tobacco (Jones *et al.*, 1988).

4.2.2 DEVELOPMENTAL TRAITS

Advances in plant transformation technology make it relatively easy to express foreign genes in transgenic plants. The literature provides numerous descriptions of interesting morphological changes in transgenic plants caused by foreign gene expression or ectopic expression of endogenous genes. This approach will continue to expand our understanding of plant developmental processes. In this section, we describe transgenic phenotypes that have given rise to changes in overall plant development or plant architecture.

One example of an altered morphology in transgenic plants was caused by expression of an oat phytochrome under the control of the CaMV 35S promoter in transgenic tobacco and tomato. In tomato, expression of oat phytochrome results in plants that are dwarfed, with dark green foliage and fruits, and shortened hypocotyls with elevated anthocyanin levels (Boylan and Quail, 1989). In tobacco (either *N. tabacum* or *N. plumbagenifolia*), transgenic plants producing oat phytochrome are semi-dwarf with dark green leaves, increased tillering, and decreased apical dominance (J. Keller *et al.*, 1989). Interestingly, rice phytochrome is biologically active in transgenic tobacco plants, but produces no altered morphology (Kay *et al.*, 1989).

Other types of morphological changes have been produced in transgenic plants when genes involved in the synthesis of plant hormones are expressed constitutively. For example, when different regions of the *Agrobacterium* T-DNA are transferred to tobacco, plants with various hormone-related phenotypes have been produced (Krens *et al.*, 1985). T-DNA hormone genes, particularly the isopentenyltransferase, have been expressed in transgenic tobacco plants under the control of the maize *Hsp70* promoter (Medford *et al.*, 1989) and in transformed regenerated shoots under the control of the 35S and *nos* promoters (Smigocki and Owens, 1988, 1989).

4.2.3 QUALITY TRAITS

We refer to quality traits as those phenotypes that result in an improved plant product, generally in the harvested part of the plant. In tomato, fruit-quality traits such as solids content long have been an important goal for plant breeders and for genetic engineering. Significant progress has been made in altering another fruit quality trait: firmness. This change has been achieved by antisense regulation of tomato fruit PG (Sheehy *et al.*, 1988; Smith *et al.*, 1988, 1990; Giovannoni *et al.*, 1989; Kramer *et al.*, 1989), which results in fruit that is less perishable and possibly less susceptible to fungal diseases. Manipulation of ethylene levels in fruit also has the potential to improve fruit quality (Holdsworth *et al.*, 1988; Hamilton *et al.*, 1990).

Manipulation of flower color is a means of increasing the value of ornamental species. For example, a novel flower color was produced in transgenic petunia that expressed a maize dihydroflavonol reductase gene (Meyer *et al.*, 1987). In addition, antisense and cosuppression (see previous text) of chalcone synthase in transgenic petunia produced new patterns of pigment expression in the flowers (Napoli *et al.*, 1990; Van Der Krol *et al.*, 1988, 1990a,b,c).

Other efforts have been directed toward improving the amino acid composition of potato tuber protein by increasing the fraction of essential amino acids in the potatoes (Yang *et al.*, 1989).

Manipulation of seed-storage proteins with the goal of improving protein quality in a seed crop has been a long-term goal of several groups. Thus, for example, soybean β-conglycinin has been expressed in transgenic petunia seeds (Beachy *et al.*, 1985; Naito *et al.*, 1988), pea vicillins have been expressed in transgenic tobacco seeds (Higgins *et al.*, 1988), and zeins have been expressed in transgenic tobacco (Schernthaner *et al.*, 1988). However, since seeds of solanaceous species generally are not eaten, these are examples of test systems that are being explored to determine the feasibility of changing the protein composition of seed-storage proteins in other crops.

4.3 Field Testing and Regulations

4.3.1 FIELD TESTING WITH SOLANACEAE

In the United States, permits are required currently for field testing of transgenic plants and are issued by the Animal and Plant Health Inspection Service (APHIS) of the U.S. Department of Agriculture (USDA) under a federal regulation called 7 CFR 340. Approvals for the first field trials of transgenic plants were issued in late 1987 and early 1988 to three companies: Calgene, Monsanto, and Du Pont, whose scientists had developed transgenic tobacco and tomato expressing genes for herbicide resistance (Calgene and Du Pont) and for insect and virus resistance (Monsanto).

Solanaceous crops were the first to be tested in transgenic field trials because they were the first to be transformed experimentally. However, other characteristics of certain species made them particularly well suited for the first field tests of this new technology. Tobacco pollen and seeds are contained easily because the goal is to harvest leaves; one could mimic commercial tobacco production practices in which flowering in the field is avoided. Potato is also an attractive choice because it is vegetatively propagated, its flowers are often sterile, and its vegetative propagules, the tubers, rarely, if ever, survive the period between seasons in most potato production areas.

Tomato is fertile and fruit production is dependent on the completion of the full reproductive cycle but, like all other fertile species in this family, its heavy pollen is not wind-borne to any significant degree. Thus, pollen spread is dependent on insect pollinators. Current regulations require the use of borders and pollen traps to limit pollen dispersal to within the field trial areas.

The development of regulations and protocols for field testing of transgenic plants has been framed around experiences with these first few tests. Regulations are being formulated either independently or cooperatively in four general theaters (the United States, the United Kingdom, the European Community, and Australia). Through scientific interactions, as well as through the efforts of various private and government channels, cooperation at various times and levels is being achieved. This situation is complex, politically, scientifically, commercially, and from a public awareness perspective, so further clarification of the issues will be needed as commercial production of transgenic plants proceeds. Many of the scientific issues have been addressed in a submission by Calgene (Davis, California) to the U.S. Food and Drug Administration requesting an advisory opinion on the safety of food products of genetically engineered tomato, rapeseed, and cotton containing kanamycin resistance as a selectable marker (Flavell *et al.*, 1992); and in a subsequent submission concerning the safety of genetically engineered tomato (Redenbaugh, *et al.*, 1992).

The USDA periodically publishes reports on the approved field tests of transgenic plants. The report published on January 8, 1991, shows that the number of trials involving solanaceous species is decreasing in the first few years of field testing of transgenic plants. In 1987, all trials approved in the United States were with tobacco or tomato, and in 1988 all but one trial was with one of these species. Testing of potato began in 1989. By 1990, only one-third of the approved trials were with plants in the Solanaceae. This trend is likely to be due both to an increase in the number of nonsolanaceous species that are being transformed and to a concomitant decrease in the number of field tests involving Solanaceae as model systems.

4.3.1.1 Published Results of Field Tests

Few reports have appeared in the scientific literature of results from transgenic plant field tests, either on the performance of the plants or on questions of gene flow beyond the crop species under test. As indicated previously, some plants in the Solanaceae are not very suitable for studies on gene flow into the environment (weeds, noncultivated species, and germ plasm collections) because of the biological and genetic characteristics of the species in this family.

Much of the experimental data on field performance of transgenic plants in the Solanaceae remain unpublished. This situation is not entirely unexpected since most agricultural scientists would generate extensively replicated experiments with sophisticated analysis or multiple years' data before submitting the results of such work. The relatively recent start to such experimentation means that few such experiments have been completed. Moreover, the constraints on experimental design placed by financial resources or the early requirements for physical containment make these kinds of experiments expensive to conduct.

The few reports that have been published show that transgenic traits perform in the field generally as was expected from the careful analysis in the laboratory and greenhouse that preceded field testing. The first such report (Nelson *et al.*, 1988) documented the performance of transgenic tomatoes expressing the capsid protein gene from the common strain of TMV under growth chamber, greenhouse, and field conditions. Two independent transformation events were characterized and evaluated. The expression level of the capsid protein in the leaves of these plants under field conditions was found to be similar to, although slightly lower than, that predicted by the results of greenhouse and growth chamber tests. Virus inoculations were performed twice after planting, about 2 wk apart, at two different inoculum concentrations. The field trial was replicated four times, and the design was a randomized split plot, with inoculation assigned to the main plots and plant genotype to the subplots. Visual symptoms, virus titers, capsid protein accumulation, harvested fruit weight, number of fruit, and maturation dates were documented.

The results indicated that, within the limitations of a single trial at a single site in a single year, the transgenic plants were resistant to the virus. At least 95% of the inoculated transgenic plants lacked virus symptoms and accumulated virus titers that were just above or below the limits of detection of the serological method used to detect the virus. This result contrasts with the results of inoculation of control plants: 99% of the plants showed symptoms and virus titers were 50–200 or more times the limits of detection. Tomato yield and fruit maturation were unaffected by inoculation of transgenic plants with the virus, whereas inoculated control plants had 26–35% lower yield. However, one of the two lines of transgenic tomato produced poor yields, even when not inoculated with the virus. The reason for this yield reduction is not known, but it indicates the need for large scale testing and field selection of transgenic plants.

In 1989, two reports appeared describing results of field trials with transgenic tobacco and potato in Europe (De Greef *et al.*, 1989) and transgenic tomatoes in the United States (Delannay *et al.*, 1989). In the work from Europe, tobacco and potato plants expressing a phosphinothricin acetyltransferase that confers resistance to the herbicide glufosinate were tested. Transgenic plants of both species in the field performed identically to plants of the untransformed controls. Three tobacco genotypes, including two transgenic tobacco lines, were tested under four doses of herbicide in a 3×4 factorial design. Data on leaf length were used to evaluate the performance of the plants. In potato, the trial was replicated minimally because of technical difficulties with the amount of plant material available and the weather. Data on tuber yields were taken and used to evaluate the performance of the transgenic plants in comparison with controls.

The 1989 tomato report (Delannay *et al.*, 1989) discussed plants expressing a gene encoding the insecticidal protein δ-endotoxin from *Bacillus thuringiensis* var. *kurstaki*. The trials reported were conducted in 1987 and 1988 using several different vectors, constructs, and transgenic events at several locations against three different lepidopteran insect pests of tomato (tobacco hornworm, tomato fruit worm, and tomato pinworm). Control of all three insects was reported in terms of control of plant or fruit damage. Comparisons in some trials were made with recommended insecticide controls for the same pests. The transgenic plants were judged to be at least as effective in damage control as was the application of insecticides at the doses used.

A report appeared in 1990 describing the performance of tomatoes with reduced levels of PG as a result of antisense expression (Kramer *et al.*, 1990). Five homozygous lines with PG activities ranging from 8 to 50% of normal were tested in a randomized split-plot design. Plants were harvested weekly over a 7-wk period beginning at 95% ripe. The selection with the lowest PG level had 50% less rotten fruit than controls at all harvests. Comparison of processing characteristics showed that the selection with the lowest PG level had a higher consistency and higher viscosity (both desirable processing characteristics) than control fruit.

One other report in the peer-reviewed scientific literature (March, 1991) on field testing of transgenic Solanaceae presented tests of tobacco expressing an antibiotic resistance gene (encoding CAT) driven by a wound-inducible promoter from a tomato proteinase inhibitor inducing factor gene (*pin*2) (Thornburg *et al.*, 1990). The trial itself was simply a plot of transgenic plants planted in a field in Iowa. Induction of the gene was tested by collecting leaves of plants from the field and infesting these in the laboratory with tobacco hornworm larvae; the results showed that the marker gene was expressed as a result of insect attack. No details of experimental design were given.

4.3.1.2 Future Work with Solanaceae in the Field

Field trials with genetically engineered plants are well past the stage of novelty. Because these trials are expensive to conduct, it is reasonable to conclude that the recurring appearance of a particular trait in a particular crop on the list of approved field trials is some indication that the company seeking approval of the trial has commercial expectations. Under these assumptions, it is reasonable to conclude that commercial products in tomato and potato are under serious consideration by several companies. The traits under development are resistance to lepidopteran larvae and virus infection in potato and tomato and the modulation of ripening or quality characteristics in tomato. From the 1990 list of approved trials, the organizations with approvals for trials in these crops with these traits include Asgrow (a unit of Upjohn), Calgene, and Monsanto.

Field trials are beginning to be proposed from the public sector as well; the

1990 list of approved trials shows that public laboratories moving into testing of transgenic plants are following the course of the private companies. The first such approval was to Iowa State University for the study in tobacco that led to a publication mentioned earlier (Thornburg *et al.*, 1990). Other early recipients of approvals included the USDA for trials with potato and the University of Kentucky for trials with tobacco. Since then, public institution approvals in 1989 and 1990 were granted for trials with cucumber, rice, poplar, and walnut.

Ecologists are beginning to take an interest in the opportunity presented by recombinant DNA techniques, and public interest in predicting consequences of the uses of transgenic plants, for quantitative studies in population biology and genetics (Ellstrand, 1988; Tiedje *et al.*, 1989; Manasse and Kareiva, 1991). For a variety of mostly biological reasons, however, such work with Solanaceae will likely be limited. As already mentioned, the biology of reproduction in the Solanaceae generally limits or virtually precludes (e.g., in potato) the transfer of genes by cross-fertilization in the field. The gene transfer that can occur within the crop can be predicted and controlled using well-established methods such as Association of Official Seed Certifying Agencies protocols that are used routinely to maintain genetic purity during seed production (Anonymous, 1971).

4.3.2 FUTURE RESEARCH DIRECTIONS USING TRANSGENIC SOLANACEAE

For over 40 years, solanaceous plants have been an important experimental tool in the study of plant development. In the new era of transgenic plants, it is likely that the Solanaceae will continue to be an important source of experimental material, in part because of the rich history of previous descriptive, physiological, and genetic work and in part because these species are routinely amenable to gene transfer technologies, placing the production of transgenic plants in easy reach of nonspecialist laboratories. [For a review on the uses of transgenic plants in plant development, see Masterson and Schell (1989).]

Studies on the primary plant hormones, the cytokinins, auxins, gibberellins, and abscissic acid, are examples of developmental work that make use of the Solanaceae. Already, bacterial genes (from *Agrobacterium tumefaciens, A. rhizogenes,* and *Pseudomonas savastanoi*) have been cloned and expressed in transgenic tobacco and tomato. Critical studies are underway on the developmental consequences of such manipulations (see for example, Medford *et al.*, 1989; Smigocki and Owens, 1989; Yusibov *et al.*, 1989). Work to elucidate signaling transduction pathways, including those responsible for hormone messengers, will likely flow from studies of gene induction, for example, the protease inhibitor induction of Ryan and colleagues (Ryan, 1989) and of Willmitzer and colleagues (Koester *et al.*, 1989; Rocha Sosa *et al.*, 1989). Hormone receptor work may also take advantage of the amenability of tobacco and tomato to genetic manipulation.

Tobacco and, to a lesser extent, tomato also have been very important in studies on the reproductive biology of dicotyledonous plants. Goldberg and colleagues have laid an impressive foundation for detailed analysis at the molecular level of flower development in tobacco (see, for example, Goldberg 1988). On this foundation, a wide range of possible studies, for example, to define the developmental program in terms of gene regulation, will be possible. Although many such studies may exploit or require the use of simpler genetic systems, for example, using haploids for genetic analysis, tobacco seems sure to be a model system for much of this work also. Likewise, continuing use of this system for genetic manipulation in work on compatibility during fertilization to define the genetic and biochemical basis of self-incompatibility, some of which has been done in *Nicotiana* species (Ebert and Clark, 1990), will lead to important contributions (Moore and Nasrallah, 1990).

Plants in the Solanaceae are unusual in several aspects of their metabolism, and so are likely to contribute to research in which these metabolic pathways are of interest. Most notably, of course, is the production by many Solanaceae of alkaloids that are of pharmaceutical interest. Although little work has been done using transgenic plants in the study of alkaloid biosynthesis and its regulation, this is an area in which progress may come, now that gene transfer technology and antisense RNA technology are available for probing metabolic pathways.

Finally, it is likely that solanaceous crop species, particularly tomato and potato, will be among the first transgenic plants to result in commercial products.

Note added in proof: This chapter was written based on literature published prior to March, 1991.

References

Abel, P. P., Nelson, R. S., De, B., Hoffman, N., and Rogers, S. G. *et al.* (1986). *Science* **232,** 738–743.

Ainley, W. M., and Key, J. L. (1990). *Plant Mol. Biol.* **14,** 949–968.

Altabella, T., and Chrispeels, M. J. (1990). *Plant Physiol.* **93,** 805–810.

Amasino, R., Taylor, B., White, F., Gordon, M., and Nester, E. (1983). *Plant Physiol.* **72**(Suppl. 1), 147.

An, G., Costa, M. A., Mitra, A., Ha, S. B, and Marton, L. (1988). *Plant Physiol.* **88,** 547–552.

Anderson, E. J., Stark, D. M., Nelson, R. S., *et al.* (1989). *Phytopathology* **79,** 1284–1290.

Andre, D., Colau, D., Schell, J., Van Montagu, M., and Hernalsteens, J. P. (1986). *Mol. Gen. Genet.* **204,** 512–518.

Anonymous (1971). "AOSCA Certification Handbook" (No. 23). Association of Official Seed Certifying Agencies, Raleigh, North Carolina.

Anzai, H., Yoneyama, K., and Yamaguchi, I. (1989). *Mol. Gen. Genet.* **219,** 492–494.

Aoki, S., and Takebe, I. (1969). *Virology* **39,** 439–448.

Baeumlein, H., Meuller, A. J., Schiemann, J., *et al.* (1987). *Biol. Zentralbl.* **106,** 569–575.

Baker, B., Schell, J., Loerz, H., and Fedoroff, N. (1986). *Proc. Natl. Acad. Sci. U.S.A.* **83,** 4844–4848.

Baker, B., Coupland, G., Fedoroff, N., Starlinger, P., and Schell, J. (1987). *EMBO J.* **6,** 1547–1554.
Balazs, E., Bouzoubaa, S., Guilley, H., *et al.* (1985). *Gene* **40,** 343–348.
Barton, K., Binns, A. N., Matzke, A. J. M., and Chilton, M. D. (1983). *Cell* **32,** 1033–1044.
Barton, K. A., Whiteley, H. R., and Yang, N. S. (1987). *Plant Physiol.* **85,** 1103–1109.
Bateman, G. (1985). "Flowering Plants of the World." Prentice Hall, Englewood Cliffs, New Jersey.
Bathgate, B., Goodenough, P. W., and Grierson, D. (1986). *J. Plant Physiol.* **124,** 223–233.
Baumann, G., Raschke, E., Bevan, M., and Scheoffl, F. (1987). *EMBO J.* **6,** 1161–1166.
Beachy, R. N., Chen, Z. L., Horsch, R. B., *et al.* (1985). *EMBO J.* **4,** 3047–3054.
Bellini, C., Chupeau, M. C., Guerche, P., Vastra, G., and Chupeau, Y. (1989). *Plant Sci.* **65,** 63–76.
Belzile, F., Lassner, M. W., Tong, Y., Khush, R., and Yoder, J. I. (1989). *Genetics* **123,** 181–190.
Benfey, P. N., Ren, L., and Chua, N.-H. (1990a). *EMBO J.* **9,** 1685–1696.
Benfey, P. N., Ren, L., and Chua, N.-H. (1990b). *EMBO J.* **9,** 1677–1684.
Bevan, M. W., Flavell, R. B., and Chilton, M. D. (1983). *Nature (London)* **304,** 184–187.
Bird, C. R., Smith, C. J. S., Ray, J. A., *et al.* (1988). *Plant Mol. Biol.* **11,** 651–662.
Boylan, M. T., and Quail, P. H. (1989). *Plant Cell* **1,** 765–774.
Bradley, P. M., and Leith, A. (1979). *Naturwissenschaften* **66,** 111.
Bray, E. A., Naito, S., Pan, N. S., *et al.* (1987). *Planta* **172,** 364–370.
Broglie, K. E., Biddle, P., Cressman, R., and Broglie, R. (1989). *Plant Cell* **1,** 599–608.
Carroll, D. (1972). *Biochem.* **11,** 421–426.
Charest, P. J., Hattori, J., Demoor, J., Iyer, V. N., and Miki, B. L. (1990). *Plant Cell Rep.* **8,** 643–646.
Chen, Z. L., Schuler, M. A., and Beachy, R. N. (1986). *Proc. Natl. Acad. Sci. U.S.A.* **83,** 8560–8564.
Cheon, C. I., Delauney, A. J., and Verma, D. P. S. (1990). *Plant Sci.* **66,** 231–236.
Cheung, A. Y., Bogorad, L., Van Montagu, M., and Schell, J. (1988). *Proc. Natl. Acad. Sci. U.S.A.* **85,** 391–395.
Chilton, M. D., Drummond, M. H., Merlo, D. J., *et al.* (1977). *Cell.* **11,** 263–271.
Chyi, Y. S., Jorgensen, R. A., Goldstein, D., Tanksley, S. D., and Loaiza, F. F. (1986). *Mol. Gen. Genet.* **204,** 64–69.
Coboche, M., and Deshayes, A. (1984). *Can. R. Acad. Sci. Ser. III* **299,** 663–666.
Cocking, E. C. (1960). *Nature (London)* **187,** 962–963.
Cocking, E. C. (1972). *Ann. Rev. Plant Physiol.* **23,** 29–50.
Colot, V., Robert, L. S., Kavanagh, T. A., Bevan, M. W., and Thompson, R. D. (1987). *EMBO J.* **6,** 3559–3564.
Comai, L., Facciotti, D., Hiatt, W. R., *et al.* (1985). *Nature (London)* **317,** 741–744.
Comai, L., Moran, P., and Maslyar, D. (1990). *Plant Mol. Biol.* **15,** 373–381.
Crossway, A., Oakes, J. V., Irvine, J. M., *et al.* (1986). *Mol. Gen. Genet.* **202,** 179–185.
Davey, M. R., Cocking, E. C., Freeman, J., Pearce, N., and Tudor, I. (1980). *Plant Sci. Letters,* **18,** 307–313.
Dean, C., Favreau, M., Bedbrook, J., and Dunsmuir, P. (1989). *Plant Cell.* **1,** 209–216.
De Block, M. (1988). *Theor. Appl. Genet.* **76,** 767–774.
De Block, M., Herrera, E. L., Van Montagu, M., Schell, J., and Zambryski, P. (1984). *EMBO J.* **3,** 1681–1690.
De Block, M., Botterman, J., Vandewiele, M., *et al.* (1987). *EMBO J.* **6,** 2513–2518.
De Clercq, A., Vandewiele, M., De, R. R., *et al.* (1990). *Plant Physiol.* **92,** 899–907.
De Greef, W., Delon, R., De Block, M., Leemans, J., and Botterman, J. (1989). *Bio/Technology* **7,** 61–64.
Deikman, J., and Fischer, R. L. (1988). *EMBO J.* **7,** 3315–3320.

Delannay, X., Lavallee, B. J., Proksch, R. K., *et al.* (1989). *Bio/Technology* **7,** 1265–1269.
Delauney, A. J., Tabaeizadeh, Z., and Verma, D. P. S. (1988). *Proc. Natl. Acad. Sci. U.S.A.* **85,** 4300–4304.
Della Cioppa, G., Bauer, S. C., Taylor, M. L., *et al.* (1987). *Bio/Technology* **5,** 579–584.
Deno, H., Yamagata, H., Emoto, T., *et al.* (1987). *J. Plant Physiol.* **131,** 315–324.
Diakumakos, E. G., Holland, S., and Pecora, P. A. (1970). *Proc. Natl. Acad. Sci. U.S.A.* **65,** 911–918.
Donald, R. G. K., and Cashmore, A. R. (1990). *EMBO J.* **9,** 1717–1726.
Ebert, P. R., and Clark, A. E. (1990). *Plant Mol. Biol.* **14,** 815–824.
Eckes, P., Schmitt, P., Daub, W., and Wengenmayer, F. (1989). *Mol. Gen. Genet.* **217,** 263–268.
Ellis, J. G., Llewellyn, D. J., Dennis, E. S., and Peacock, W. J. (1987). *EMBO J.* **6,** 11–16.
Ellis, J. R., Shirsat, A. H., Hepher, A., *et al.* (1988). *Plant Mol. Biol.* **10,** 203–214.
Ellstrand, N. C. (1988). *In* "Planned Release of Genetically Engineered Organisms" (J. Hodgson and A. M. Sugden, eds.), pp. S30–S31. Elsevier Publications, Cambridge.
Filippone, E., and Lurquin, P. F. (1989). *Plant Cell Rep.* **8,** 370–373.
Fillatti, J. J., Kiser, J., Rose, R., and Comai, L. (1987). *Bio/Technology* **5,** 726–730.
Finnegan, E. J., Taylor, B. H., Craig, S., and Dennis, E. S. (1989). *Plant Cell* **1,** 757–764.
Fischhoff, D. A., Bowdish, K. S., Perlak, F. J., *et al.* (1987). *Bio/Technology* **5,** 807–813.
Flavell, R. B., Dart, E., Fuchs, R. L., and Fraley, R. T. (1992). *Bio/Technology,* **10,** 141–144.
Fraley, R. T., Dellaporta, S. L., and Papahadjopoulos, D. (1982). *Proc. Natl. Acad. Sci. U.S.A.* **79,** 1859–1863.
Fraley, R. T., Rogers, S. G., Horsch, R. B., *et al.* (1983). *Proc. Natl. Acad. Sci. U.S.A.* **80,** 4803–4807.
Frey, M., Tavantzis, S. M., and Saedler, H. (1989). *Mol. Gen. Genet.* **217,** 172–177.
Gabard, J. M., Charest, P. J., Iyer, V. N., and Miki, B. L. (1989). *Plant Physiol.* **91,** 574–580.
Gamborg, O. L. (1981). *Adv. Biochem. Engin.* **19,** 239–263.
Gidoni, D., Bond, N. D., Brosio, P., *et al.* (1988). *Mol. Gen. Genet.* **211,** 507–514.
Giovannoni, J. J., Dellapenna, D., Bennett, A. B., and Fischer, R. L. (1989). *Plant Cell* **1,** 53–64.
Goldberg, R. B. (1988). *Science.* **240,** 1460–1467.
Gordon-Kamm, W., Spencer, T., Mangano, M., *et al.* (1990). *Plant Cell* **2,** 603–618.
Graham, F. L., and Van der Eb, A. J. (1971). *Virology* **52,** 429–436.
Graham, J. S., Hall, G., Pearce, G., and Ryan, C. A. (1986). *Planta* **169,** 399–405.
Greenwood, J. S., and Chrispeels, M. J. (1985). *Plant Physiol.* **79,** 65–71.
Grierson, D., Tucker, G. A., Keen, J., *et al.* (1986). *Nucleic Acids Res.* **14,** 8595–8603.
Guri, A., and Sink, K. C. (1988). *J. Plant Physiol.* **133,** 52–55.
Hain, R., Steinbiss, H. H., and Schell, J. (1984). *Plant Cell Rep.* **3,** 60–64.
Hain, R., Stabel, P., Czernilofsky, A. P., *et al.* (1985). *Mol. Gen. Genet.* **199,** 161–168.
Hamill, J. D., and Rhodes, M. J. C. (1988). *J. Plant Physiol.* **133,** 506–509.
Hamill, J., Parr, A., Robins, R., and Rhodes, M. (1986). *Plant Cell Rep.* **5,** 111–114.
Hamilton, A. J., Lycett, G. W., and Grierson, D. (1990). *Nature (London)* **346,** 284–287.
Hanisch Ten Cate, C. H., Ennik, E., Roest, S., *et al.* (1988). *Theor. Appl. Genet.* **75,** 452–459.
Hanley-Bowdoin, L., Elmer, J. S., and Rogers, S. G. (1989). *Plant Cell* **1,** 1057–1067.
Hanley-Bowdoin, L., Elmer, J. S., and Rogers, S. G. (1990). *Proc. Natl. Acad. Sci. U.S.A.* **87,** 1446–1450.
Haring, M. A., Gao, J., Volbeda, T., *et al.* (1989). *Plant Mol. Biol.* **13,** 189–202.
Harrison, B. D. (1985). *Ann. Rev. Phytopathol.* **23,** 55–82.
Haughn, G. W., Smith, J., Mazur, B., and Somerville, C. (1988). *Mol. Gen. Genet.* **211,** 266–271.
Hehl, R., and Baker, B. (1989). *Mol. Gen. Genet.* **217,** 53–59.
Herrera-Estrella, L., De Block, M., Messens, E., *et al.* (1983a). *EMBO J.* **2,** 987–996.

Herrera-Estrella, L., Depicker, A., Van Montagu, M., and Schell, J. (1983b). *Nature (London)* **303,** 209–213.

Higgins, T. J. V., Newbigin, E. J., Spancer, D., Llewellyn, D. J., and Craig, S. (1988). *Plant Mol. Biol.* **11,** 683–696.

Hilder, V. A., Gatehouse, A. M. R., Sheerman, S. E., Barker, R. F., and Boulter, D. (1987). *Nature (London)* **330,** 160–163.

Hilder, V. A., Barker, R. F., Samour, R. A., *et al.* (1989). *Plant Mol. Biol.* **13,** 701–710.

Hoekema, A., Hirsch, P. R., Hooykaas, P. J. J., and Schilperoort, R. A. (1983). *Nature (London)* **303,** 179–180.

Hoffman, F., and Hess, D. (1972). *Z. Pflanzenphysiol.* **69,** 81–83.

Holdsworth, M. J., Bird, C. R., Ray, J., Schuch, W., and Grierson, D. (1987). *Nucleic Acids Res.* **15,** 731–739.

Holdsworth, M. J., Schuch, W., and Grierson, D. (1988). *Plant Mol. Biol.* **11,** 81–88.

Horsch, R. B., Fraley, R. T., Rogers, S. G., *et al.* (1984). *Science* **223,** 496–498.

Horsch, R. B., Fry, J. E., Hoffmann, N. L., *et al.* (1985). *Science* **227,** 1229–1231.

Houck, C. M., Shintani, D. K., and Knauf, V. C. (1990). *In* "Frontiers in Applied Microbiology" (K. G. Mukerji and V. P. Singh, eds.), pp. 1–17. Rastogi, Meerut, India.

Jaziry, M., Legros, M., Homes, J., and Vanhoelen, M. (1988). *Phytochemistry* **27,** 419–420.

Jefferson, R., Goldbrough, A., and Bevan, M. (1990). *Plant Mol. Biol.* **14,** 995–1006.

Johnson, R., and Ryan, C. A. (1990). *Plant Mol. Biol.* **14,** 527–536.

Johnson, R., Narvaez, J., An, G., and Ryan, C. (1989). *Proc. Natl. Acad. Sci. U.S.A.* **86,** 9871–9875.

Johnson, R., Lee, J. S., and Ryan, C. A. (1990). *Plant Mol. Biol.* **14,** 349–356.

Jones, J. D. G., Dunsmuir, P., and Bedbrook, J. (1985). *EMBO J.* **4,** 2411–2418.

Jones, J. D. G., Dean, C., Gidoni, D., *et al.* (1988). *Mol. Gen. Genet.* **212,** 536–542.

Jones, J. D. G., Carland, F. M., Maliga, P., and Dooner, H. K. (1989). *Science.* **244,** 204–207.

Jordano, J., Almoguera, C., and Thomas, T. L. (1989). *Plant Cell* **1,** 855–866.

Jung, G., and Tepfer, D. (1987). *Plant Sci.* **50,** 145–152.

Kandasamy, M. K., Dwyer, K. G., Paolillo, D. J., *et al.* (1990). *Plant Cell* **2,** 39–50.

Kaulen, H., Schell, J., and Kreuzaler, F. (1986). *EMBO J.* **5,** 1–8.

Kay, R., Chan, A., Daly, M., and McPherson, J. (1987). *Science* **236,** 1299–1302.

Kay, S. A., Nagatani, A., Keith, B., *et al.* (1989). *Plant Cell* **1,** 775–782.

Keil, M., Sanchez Serrano, J. J., and Willmitzer, K. (1989). *EMBO J.* **8,** 1323–1330.

Keith, B., and Chua, N. H. (1986). *EMBO J.* **5,** 2419–2426.

Keller, B., Schmid, J., and Lamb, C. J. (1989). *EMBO J.* **8,** 1309–1314.

Keller, J. M., Shanklin, J., Vierstra, R. D., and Hershey, H. P. (1989). *EMBO J.* **8,** 1005–1012.

Kephalas, T., Mclennan, C., Evans, W., Philipson, J., and Roberta, M. (1989). *J. Pharm. Pharmacol. (Suppl.)* **41,** 68.

Kernan, A., and Thornburg, R. W. (1989). *Plant Physiol.* **91,** 73–78.

Kiernan, J. M., Goldberg, K. B., Young, M. J., Schoelz, J. E., and Shepherd, R. J. (1989). *Plant Sci.* **64,** 67–78.

Klein, T. M., Harper, E. C., Svab, Z., *et al.* (1988). *Proc. Natl. Acad. Sci. U.S.A.* **85,** 8502–8505.

Knopp, E., Strauss, A., and Wehrli, W. (1988). *Plant Cell Rep.* **7,** 590–593.

Koester, T. M., Frommer, W. B., Rocha-Sosa, M., *et al.* (1989). *Mol. Gen. Genet.* **219,** 390–396.

Koester, T. M., Frommer, W. B., Rocha-Sosa, M., and Willmitzer, L. (1990). *Plant Mol. Biol.* **14,** 239–248.

Kramer, M., Sheehy, R. E., and Hiatt, W. R. (1989). *Trends Biotech.* **7,** 191–194.

Kramer, M., Sanders, R. A., Sheehy, R. E., *et al.* (1990). *In* "Horticultural Biotechnology" (A. B. Bennett and S. D. O'Neill, eds.), pp. 347–355. Wiley–Liss, New York.

Krens, F. A., Mans, R. M. W., Van Slogteren, T. M. S., *et al.* (1985). *Plant Mol. Biol.* **5,** 223–234.

Krens, F. A., Molendijk, L., Wullems, G. J., and Schilperoort, R. A. (1982). *Nature (London)* **296,** 72–74.

Kridl, J. C., and Goodman, R. M. (1986). *BioEssays* **4,** 4–8.

Lam, E., and Chua, N. H. (1990). *Science* **248,** 471–474.

Lassner, M. W., Palys, J. M., and Yoder, J. I. (1989). *Mol. Gen. Genet.* **218,** 25–32.

Lee, J. Y., and Davey, M. R. (1988). *Res. Rep. Rural Dev. Adm. (Suweon).*

Lee, K. Y., Townsend, J., Tepperman, J., *et al.* (1988). *EMBO J.* **7,** 1241–1248.

Leisy, D. J., Hnilo, J., Zhao, Y., and Okita, T. W. (1990). *Plant Mol. Biol.* **14,** 41–50.

Li, Y., Hu, Y., Chen, W., and Wang, J. (1989). *Acta Bot. Yunnanica* **11,** 247–253.

Liang, X., Dron, M., Schmid, J., Dixon, R. A., and Lamb, C. J. (1989). *Proc. Natl. Acad. Sci. U.S.A.* **86,** 9284–9288.

Lincoln, J. E., and Fischer, R. L. (1988). *Mol. Gen. Genet.* **212,** 71–75.

Lincoln, J. E., Corder, S., Read, E., and Fischer, R. L. (1987). *Proc. Natl. Acad. Sci. U.S.A.* **84,** 2793–2797.

Liu, W., Parrott, W., Hidebrand, D., Collins, G., and Williams, E. (1990). *Plant Cell Rep.* **9,** 360–364.

Logemann, J., Lipphardt, S., Loerz, H., *et al.* (1989). *Plant Cell* **1,** 151–158.

Lurquin, P. F., and Paszty, C. (1988). *J. Plant Physiol.* **133,** 332–335.

Lyon, B. R., Llewellyn, D. J., Huppatz, J. L., Dennis, E. S., and Peacock, W. J. (1989). *Plant Mol. Biol.* **13,** 533–540.

McBride, K. E., and Summerfelt, K. R. (1990). *Plant Mol. Biol.* **14,** 269–276.

McCabe, D., Swain, W., Martinell, B., and Christou, P. (1988). *Bio/Technology* **6,** 923–926.

McCormick, S., Niedermeyer, J., Fry, J., *et al.* (1986). *Plant Cell Rep.* **5,** 81–84.

Manasse, R., and Kareiva, P. (1991). *In* "Assessing Ecological Risks of Biotechnology" (L. R. Ginzburg, ed.), pp. 215–231. Butterworth-Heinemann, Boston.

Mannino, R. J., Allebach, E. S., and Strohl, W. A. (1979). *FEBS Lett.* **101,** 229–232.

Mano, Y., Ohkawa, H., and Yamada, Y. (1989). *Plant Sci.* **59,** 191–202.

Mansson, P. E., Hsu, D., and Stalker, D. M. (1985). *Mol. Gen. Genet.* **200,** 356–361.

Margossian, L. J., Federman, A. D., Giovannoni, J. J., and Fischer, R. L. (1988). *Proc. Natl. Acad. Sci. U.S.A.* **85,** 8012–8016.

Mariani, C., De Beuckeleer, M., Truettner, J., Leemans, J., and Goldberg, R. B. (1990). *Nature (London)* **347,** 737–741.

Marris, C., Gallois, P., Copley, J., and Kreis, M. (1988). *Plant Mol. Biol.* **10,** 359–366.

Martin, C., Prescott, A., Lister, C., and Mackay, S. (1989). *EMBO J.* **8,** 997–1004.

Martineau, B., McBride, K. E., and Houck, C. M. (1991). *Mol. Gen. Genet.* **228,** 281–286.

Martinez Zapater, J. M., Finkelstein, R., and Somerville, C. R. (1988). *Plant Mol. Biol.* **11,** 601–608.

Marton, L., Wullems, G. J., Molendijk, L., and Schilperoort, R. A. (1979). *Nature (London)* **277,** 129–131.

Masson, P., and Fedoroff, N. V. (1989). *Proc. Natl. Acad. Sci. U.S.A.* **86,** 2219–2223.

Masterson, R. V., and Schell, J. S. (1989). *In* "Nucleic Acids and Molecular Biology" (F. Eckstein and D. Lilley, eds.), Vol. 3, pp. 260–268. Springer, Berlin.

Masterson, R. V., Furtek, D. B., Grevelding, C., and Schell, J. (1989). *Mol. Gen. Genet.* **219,** 461–466.

Maunders, M. J., Holdsworth, M. J., Slater, A., *et al.* (1987). *Plant Cell Environ.* **10,** 177–184.

Medford, J. I., Horgan, R., El, Sawi, Z., and Klee, H. J. (1989). *Plant Cell* **1,** 403–414.

Memelink, J., Wullems, G. J., and Schilperoort, R. A. (1983). *Mol. Gen. Genet.* **190,** 516–522.

Memelink, J., Hoge, J. H. C., and Schilperoort, R. A. (1987). *EMBO J.* **6,** 3579–3584.

Meyer, P. Heidmann, I., Forkmann, G., and Saedler, H. (1987). *Nature (London)* **330,** 677–678.

Millman, R. A., and Lurquin, P. F. (1985). *J. Plant Physiol.* **117,** 431–440.

Moore, H. M., and Nasrallah, J. B. (1990). *Plant Cell* **2,** 29–38.
Morgan, A. J., Cox, P. N., Turner, D. A., *et al.* (1987). *Plant Sci.* **49,** 37–50.
Nagata, T., Okada, K., Takebe, I., and Matsui, C. (1981). *Mol. Gen. Genet.* **184,** 161–165.
Nagy, F., Morelli, G., Fraley, R. T., Rogers, S. G., and Chua, N. H. (1985). *EMBO J.* **4,** 3063–3068.
Nagy, F., Kay, S. A., Boutry, M., Hsu, M. Y., and Chua, N. H. (1986). *EMBO J.* **5,** 1119–1124.
Naito, S., Dube, P. H., and Beachy, R. N. (1988). *Plant Mol. Biol.* **11,** 109–124.
Napoli, C., Lemieux, C., and Jorgensen, R. (1990). *Plant Cell* **2,** 279–290.
Nejidat, A., and Beachy, R. N. (1989). *Virology* **173,** 531–538.
Nejidat, A., and Beachy, R. N. (1990). *Mol. Plant Microbe Interact.* **3,** 247–251.
Nelson, R. S., McCormick, S. M., Delannay, X., *et al.* (1988). *Bio/Technology* **6,** 403–409.
Newbigin, E. J., Delumen, B. O., Chandler, P. M., *et al.* (1990). *Planta* **180,** 461–470.
Odell, J. T., Nagy, R., and Chua, N.-H. (1985). *Nature (London)* **313,** 810–812.
Odell, J. T., Knowlton, S., Lin, W., and Mauvais, C. J. (1988). *Plant Mol. Biol.* **10,** 263–272.
Ohgawara, T., Uchimiya, H., and Harada, H. (1983). *Protoplasma* **116,** 145–148.
Ohyama, K., Gamborg, O. L., and Miller, R. A. (1972). *Can. J. Bot.* **50,** 2077–2080.
Oksman-Caldentey, K., Parkinnen, O., and Hiltunen, R. (1989). *Planta Med.* **55,** 686–687.
Ondrej, M., and Vlasak, J. (1987). *Biol. Plant* **29,** 161–166.
Ono, M., Hashimoto, H., Yoshioka, M., Harada, H., and Uchimiya, H. (1986). *Jpn. J. Genet.* **61,** 491–502.
Ooms, G., Burrell, M. M., Karp, A., Bevan, M., and Hille, J. (1987). *Theor. Appl. Genet.* **73,** 744–750.
Otten, L., De Greve, H., Hernalsteens, J. P., *et al.* (1981). *Mol. Gen. Genet.* **183,** 209–213.
Ow, D. W., Wood, K. V., Deluca, M., *et al.* (1986). *Science* **234,** 856–859.
Paszkowski, J., Shillito, R. D., Saul, M., *et al.* (1984). *EMBO J.* **3,** 2717–2722.
Pear, J. R., Ridge, N., Rasmussen, R., Rose, R. E., and Houck, C. M. (1989). *Plant Mol. Biol.* **13,** 639–651.
Pena-Cortes, H., Sanchez-Serrano, J. J., Mertens, R., Willmitzer, L., and Prat, S. (1989). *Proc. Natl. Acad. Sci. U.S.A.* **86,** 9851–9855.
Pereira, A., and Saedler, H. (1989). *EMBO J.* **8,** 1315–1322.
Poste, G., Papahadjopoulos, D., and Wail, W. J. (1976). *Meth. Cell Biol.* **14,** 33–71.
Potrykus, I., and Hoffmann, F. (1973). *Z. Pflanzenphysiol. Bd.* **69,** 287–289.
Poulsen, C., and Chua, N. H. (1988). *Mol. Gen. Genet.* **214,** 16–23.
Poulsen, C., Fluhr, R., Kauffman, J. M., Boutry, M., and Chua, N. H. (1986). *Mol. Gen. Genet.* **205,** 193–200.
Powell, P. A., Stark, D. M., Sanders, P. R., and Beachy, R. N. (1989). *Proc. Natl. Acad. Sci. U.S.A.* **86,** 6949–6952.
Ray, J., Bird, C., Maunders, M., Grierson, D., and Schuch, W. (1987). *Nucleic Acids Res.* **15,** 10578.
Redenbaugh, K., Hiatt, W., Martineau, B., Kramer, M., Shechy, R., Sanders, R., Houck, C., and Emlay, D. (1992). *In* "Safety Assessment of Genetically Engineered Fruits and Vegetables: A Case Study of the FlavrSavr™ Tomato." CRC Press, Boca Raton, Florida.
Rezaian, M. A., Skene, K. G. M., and Ellis, J. G. (1988). *Plant Mol. Biol.* **11,** 463–472.
Riggs, C. D., and Bates, G. W. (1986). *Proc. Natl. Acad. Sci. U.S.A.* **83,** 5602–5606.
Robert, L. S., Donaldson, P. A., Ladaique, C., *et al.* (1989a). *Plant Mol. Biol.* **13,** 399–410.
Robert, L. S., Thompson, R. D., and Flavell, R. B. (1989b). *Plant Cell* **1,** 569–578.
Rocha Sosa, M., Sonnewald, U., Frommer, W., *et al.* (1989). *EMBO J.* **8,** 23–30.
Rodermel, S. R., Abbott, M. S., and Bogorad, L. (1988). *Cell* **55,** 673–682.
Rogers, S. G., Bisaro, D. M., Horsch, R. B., *et al.* (1986). *Cell* **45,** 593–600.
Rotino, G., and Gleddie, S. (1990). *Plant Cell Rep.* **9,** 26–29.
Ryan, C. A. (1973). *Ann. Rev. Plant Physiol.* **24,** 173–196.

Ryan, C. A. (1989). *BioEssays,* **10,** 20–24.

Sanchez-Serrano, J., Schmidt, R., Schell, J., and Willmitzer, L. (1986). *Mol. Gen. Genet.* **203,** 15–20.

Sanchez-Serrano, J. J., Keil, M., O'Connor, A., Schell, J., and Willmitzer, L. (1987). *EMBO J.* **6,** 303–306.

Sandler, S. J., Stayton, M., Townsend, J. A., *et al.* (1988). *Plant Mol. Biol.* **11,** 301–310.

Sanger, M., Daubert, S., and Goodman, R. M. (1990). *Plant Mol. Biol.* **14,** 433–444.

Schernthaner, J. P., Matzke, M. A., and Matzke, A. J. M. (1988). *EMBO J.* **7,** 1249–1256.

Schoeffl, F., Rieping, M., Baumann, G., Bevan, M., and Angermueller, S. (1989). *Mol. Gen. Genet.* **217,** 246–253.

Schoelz, J., Goldberg, K.-B., and Kiernan, J. M. (1990). *Phytopath.* **18,** 1018.

Seibertz, B., Logemann, J., Willmitzer, L., and Schell, J. (1989). *Plant Cell* **1,** 961–968.

Sengupta Gopalan, C., Reichert, N. A., Barker, R. F., Hall, T. C., and Kemp, J. D. (1985). *Proc. Natl. Acad. Sci. U.S.A.* **82,** 3320–3324.

Shah, D. M., Horsch, R. B., Klee, H. J., *et al.* (1986). *Science* **233,** 478–481.

Shahin, E. A., and Simpson, R. B. (1986). *Hortscience* **21,** 1199–1201.

Shahin, E. A., Sukhapinda, K., Simpson, R. B., and Spivey, R. (1986). *Theor. Appl. Genet.* **72,** 770–777.

Sheehy, R. E., and Lurquin, P. F. (1983). *Plant Physiol.* **72,** 386–390.

Sheehy, R. E., Kramer, M., and Hiatt, W. R. (1988). *Proc. Natl. Acad. Sci. U.S.A.* **85,** 8805–8809.

Sheerman, S., and Bevan, M. W. (1988). *Plant Cell Rep.* **7,** 13–16.

Shirsat, A. H., Wilford, N., and Croy, R. R. D. (1989). *Plant Sci.* **61,** 75–80.

Signs, M., and Flores, H. E. (1989). *Plant Physiol.* **89,** 135.

Simpson, J., Van Montagu, M., and Herrera-Estrella, L. (1986). *Science.* **233,** 34–38.

Slater, A., Maunders, M. J., Edwards, K., Schuch, W., and Grierson, D. (1985). *Plant Mol. Biol.* **5,** 137–147.

Smigocki, A. C., and Owens, L. D. (1988). *Proc. Natl. Acad. Sci. U.S.A.* **85,** 5131–5135.

Smigocki, A. C., and Owens, L. D. (1989). *Plant Physiol.* **91,** 808–811.

Smith, C. J. S., Watson, C. F., Ray, J., *et al.* (1988). *Nature (London)* **334,** 724–726.

Smith, C. J. S., Watson, C. F., Morris, P. C., *et al.* (1990). *Plant Mol. Biol.* **14,** 369–380.

Spano, L., and Costantino, P. (1982). *Z. Pflanzenphysiol.* **106,** 87–92.

Spena, A., and Schell, J. (1987). *Mol. Gen. Genet.* **206,** 436–440.

Spielmann, A., and Simpson, R. B. (1986). *Mol. Gen. Genet.* **205,** 34–41.

Stalker, D. M., McBride, K. E., and Malyj, L. D. (1988). *Science* **242,** 419–423.

Stanford, A., Bevan, M., and Northcote, D. (1989). *Mol. Gen. Genet.* **215,** 200–208.

Stanford, A. C., Northcote, D. H., and Bevan, M. W. (1990). *EMBO J.* **9,** 593–604.

Stark, D. M., and Beachy, R. N. (1989). *Bio/Technology* **7,** 1257–1262.

Steward, F. C., Mapes, M. O., Dent, A. E., and Holsten, R. D. (1964). *Science* **143,** 20–27.

Steward, F. C., Mapes, M. O., and Ammirato, P. V. (1969). *In* "Plant Physiology: A Treatise" (F. C. Steward, ed.), Vol. VB. pp. 329–376. Academic Press, New York.

Stiekema, W. J., Heidekamp, F., Louwerse, J. D., Verhoeven, H. A., and Dijkhuis, P. (1988). *Plant Cell Rep.* **7,** 47–50.

Sunter, G., Gardner, W. E., Rushing, A. E., Rogers, S. G., and Bisaro, D. M. (1987). *Plant Mol. Biol.* **8,** 477–484.

Svab, Z., Hajdukiewicz, P., and Maliga, P. (1990). *Proc. Natl. Acad. Sci. U.S.A.* **87,** 8526–8530.

Takebe, I. (1975). *Ann. Rev. Phytopathol.* **13,** 105.

Tavazza, R., Ordas, R. J., Tavazza, M., Ancora, G., and Benvenuto, E. (1988). *J. Plant Physiol.* **133,** 640–644.

Taylor, B. H., Finnegan, E. J., Dennis, E. S., and Peacock, W. J. (1989). *Plant Mol. Biol.* **13,** 109–118.

Tepfer, D. (1982). *In* "NATO Advanced Science Institute Series: Series A" (P. F. Lurquin and A. Kleinhofs, eds.), pp. 153–264. Plenum, New York.

Thompson, G. A., Hiatt, W. R., Facciotti, D., Stalker, D. M., and Comai, L. (1987). *Weed Sci.* **35** (*Suppl. 1*), 19–23.
Thornburg, R. W., An, G., Cleveland, T. E., Johnson, R., and Ryan, C. A. (1987). *Proc. Natl. Acad. Sci. U.S.A.* **84,** 744–748.
Thornburg, R. W., Kernan, A., and Molin, L. (1990). *Plant Physiol.* **92,** 500–505.
Tiedje, J. M., Colwell, R. K., Grossman, Y. L., *et al.* (1989). *Ecology* **70,** 298–315.
Tomes, D. T., Weissinger, A. K., Ross, M., *et al.* (1990). *Plant Mol. Biol.* **14,** 261–268.
Toneguzzo, G., Hayday, A. C., and Keating, A. (1986). *Mol. Cell. Biol.* **6,** 703–706.
Trinh, T. H., Mante, S., Pua, E. C., and Chua, N. H. (1987). *Bio/Technology* **5,** 1081–1084.
Tumer, N. E., O'Connell, K. M., Nelson, R. S., *et al.* (1987). *EMBO J.* **6,** 1181–1188.
Twell, D., and Ooms, G. (1987). *Plant Mol. Biol.* **9,** 365–376.
Uchimiya, H., Hirochika, H., Hashimoto, H., *et al.* (1986). *Mol. Gen. Genet.* **205,** 1–8.
Ueng, P., Gilili, G., Sapanara, V., *et al.* (1988). *Plant Physiol.* **86,** 1281–1285.
Vaeck, M., Reynaerts, A., Hofte, H., *et al.* (1987a). *Nature (London)* **328,** 33–37.
Vaeck, M., Reynaerts, A., Hofte, H., *et al.* (1987b). *Ann. Soc. Entomol. Bras.* **16,** 427–444.
Van Den Elzen, P. J. M., Townsend, J., Lee, K. Y., and Bedbrook, J. R. (1985). *Plant Mol. Biol.* **5,** 299–302.
Van Der Krol, A. R., Lenting, P. E., Veenstra, J., *et al.* (1988). *Nature (London)* **333,** 866–869.
Van Der Krol, A., Mur, L. A., Beld, M., Mol, J. N. M., and Stuitje, A. R. (1990a). *Plant Cell* **2,** 291–300.
Van Der Krol, A. R., Mur, L. A., De Lange, P., *et al.* (1990b). *Mol. Gen. Genet.* **220,** 204–212.
Van Der Krol, A. R., Mur, L. A., De Lange, P., Mol, J. N. M., and Stuitje, A. R. (1990c). *Plant Mol. Biol.* **14,** 457–466.
Van Harren, M. J. J. and Houck, C. M. (1991). *Plant Mol. Biol.* **17,** 615–630.
Van Tunen, A. J., Mur, L. A., Brouns, G. S., *et al.* (1990). *Plant Cell* **2,** 393–402.
Velten, J., Velten, L., Hain, R., and Schell, J. (1984). *EMBO J.* **3,** 2723–2730.
Vierling, E., Harris, L. M., and Chen, Q. (1989). *Mol. Cell. Biol.* **9,** 461–468.
Visser, R. G. F., Jacobsen, E., Hesseling, Meinders, A., *et al.* (1989). *Plant Mol. Biol.* **12,** 329–338.
Voelker, T., Sturm, A., and Chrispeels, M. J. (1987). *EMBO J.* **6,** 3571–3578.
Waldron, C., Murphy, E. B., Roberts, J. L., *et al.* (1985). *Plant Mol. Biol.* **5,** 103–108.
Wenzler, H., Mignery, G., Fisher, L., and Park, W. (1989a). *Plant Mol. Biol.* **13,** 347–354.
Wenzler, H. C., Mignery, G. A., Fisher, L. M., and Park, W. D. (1989b). *Plant Mol. Biol.* **12,** 41–50.
White, F. F., Ghidossi, G., Gordon, M. P., and Nester, E. W. (1982). *Proc. Natl. Acad. Sci. U.S.A.* **79,** 3193–3197.
Wohllenben, W., Arnold, W., Broer, I., *et al.* (1988). *Gene* **70,** 25–38.
Wong, T. K., and Neumann, E. (1982). *Biochem. Biophys. Res. Commun.* **107,** 584–587.
Wright, R., McClure, B., and Guilfoyle, T. (1989). *J. Plant Growth Regul.* **8,** 283–292.
Yang, F., and Simpson, R. B. (1981). *Proc. Natl. Acad. Sci. U.S.A.* **78,** 4151–4155.
Yang, M. S., Espinoza, N. O., Nagpala, P. G., *et al.* (1989). *Plant Sci.* **64,** 99–112.
Yang, N. S., and Russell, D. (1990). *Proc. Natl. Acad. Sci. U.S.A.* **87,** 4144–4148.
Yoder, J. I., Palys, J., Alpert, K., and Lassner, M. (1988). *Mol. Gen. Genet.* **213,** 291–296.
Yusibov, V. M., Pogosyan, G. P., Andrianov, V. M., and Piruzyan, E. S. (1989). *Mol. Genet. Mikrobiol. Virusol.* **7,** 11–13.
Zaitlin, M. (1959). *Science* **184,** 1002–1003.
Zambryski, P., Joos, H., Genetello, C., *et al.* (1983). *EMBO J.* **2,** 2143–2150.
Zhang, Y. L. (1988). *Acta Bot. Sin.* **30,** 368–372.

5

Transgenic Plants from Legumes

Ning-Sun Yang

Department of Mammalian Genetics
Agracetus, Inc.
Middleton, Wisconsin, 53562

Grain legumes provide a leading source of plant proteins for human food consumption. Soybean products provide a key nutritional food supply for Asian countries; various beans and peas are major food supplies for African and Latin American countries. In the United States, 5% of the total proteins in food consumption is derived from grain legumes (United States Department of Agriculture, 1989) and up to 65% of the oil used by the food processing industry comes from soybean oil (Iowa Agriculture & Home Economics Experimental Station, 1990). Alfalfa and clover, forage legumes, are used extensively as animal feeds in European and North American countries. Therefore it is important that genetic engineers be capable of improving various legume crops by effectively applying the developing technology of plant gene transfer. This area of research has become active in the past 5 years, and progress has been made in both industry and academic research laboratories. Breakthroughs in soybean transformation were achieved by employing the particle bombardment technology (McCabe *et al.*, 1988; Christou *et al.*, 1990); a similar method has been shown to be applicable to beans (D. Russell, personal communication, 1992). Soybeans, peas, alfalfa, and clover also have been transformed using the *Agrobacterium* infection method by Hinchee *et al.* (1988), Puonti-Kaerlas *et al.* (1990), Shahin *et al.* (1986), and White and Greenwood (1987), respectively. These results have laid the foundation for future genetic engineering of legume plants. Systematic application and optimization of newly developed transformation techniques can begin to generate agronomically useful transgenic legume plants, especially soybeans. Commercial introduction of soybeans improved by plant genetic engineering is no longer an uncertain future possibility, as was thought only a few years ago, but may soon become a reality (Christou *et al.*, 1990).

Assessing and identifying useful genes that can be transferred into soybean or other grain legume crops has become urgent. Desirable agronomic traits include herbicide resistance, nematode resistance, high methionine seed proteins, high quality soybean oil, decreased sensitivity to "combined nitrogen sources," fungal resistance, highly proliferative root systems, and early maturity. Although most of these agronomic traits have been defined genetically, only a very limited number of the genes is now identified or cloned to serve the need of legume plant transformation. The scarcity of identified useful genes for the transformation of crop plants is even more critical for legumes than for other crops. The fact that transformation and molecular cloning technologies are far ahead of the basic sciences of plant biochemistry and physiology has become a genuine concern for plant genetic engineers. Obviously, the future challenge is not so much the legume transformation technology as the biochemical and physiological characterizations of the various agronomically useful gene traits already defined in legume crops.

In addition to crop plant improvement, plant transformation technology may serve other needs in biotechnology. Transgenic plant crops may have advantages over bacterial or mammalian cell reactors for the production of valuable therapeutic drugs, antibodies, and vaccines (Hiatt *et al.*, 1989; Vandekerckhove *et al.*, 1989; Krebbers and Vandekerckhove, 1990; Sijmons *et al.*, 1990). Alfalfa, which is rich in leaf metabolic proteins and very efficient for mechanical farming and harvesting, seems to be an excellent target crop for this aspect of plant biotechnology. The fast-growing protein-rich alfalfa plant also is considered by some researchers to be a highly desirable transgenic host crop for production of industrial enzymes (R. Burgess and E. Bingham, personal communication, 1991).

In this chapter, I will discuss breakthroughs and the uncertain approaches in legume plant transformation, bottlenecks and advantages of different gene transfer methods, examples of foreign gene expression and inheritance in transgenic soybeans, agronomically useful genes for legumes, and future prospects for improvement of legume crops via genetic engineering technology.

5.1 Methods for Transformation of Legume Plants

5.1.1 TRANSFORMED TISSUES COMPARED WITH TRANSGENIC PLANTS

The literature contains many claims that soybeans or other legume crops were "transformed" successfully by various methods. However, only some of these reports provided complete evidence or follow-up studies to confirm that stably transformed legume plants had been obtained. An interesting feature of legume plant transformation technology today is that it is fairly easy to obtain transformed plant tissues in various forms, but it is not easy to obtain stable, inheritable, transgenic plants. The term "stable transformation" also seems to have been used so loosely that different investigators are employing different criteria. Therefore, we need to make a clear distinction between success in obtaining transformed plant tissues and successful generation of *bona fide* stably transformed plants. The failure to make such a distinction has resulted in many confusing claims in the literature, to the detriment of the field.

5.1.2 TARGET TISSUES FOR DNA DELIVERY

Protoplast cultures of gain legume crops can be established easily using a variety of source tissues; callus regeneration often can be obtained from them (Baldes *et al.*, 1987; Christou *et al.*, 1987; Eapen *et al.*, 1987; Lin *et al.*, 1987; Puonti-Kaerlas *et al.*, 1989). Attempts have been made to use protoplasts as target cells for gene transfer via either *Agrobacterium* infection (Baldes *et al.*, 1987; Eapen *et al.*, 1987) or electroporation (Christou *et al.*, 1987; Lin *et al.*, 1987), and transformed callus tissues have been obtained. However, effective regeneration of transgenic plants from protoplast-derived callus cultures has proven to be difficult or irreproducible (Christou *et al.*, 1987; Lin *et al.*, 1987; Wei and Xu, 1988). In the rare experiments that were successful, plant regeneration was found to be highly genotype dependent (Wei and Xu, 1988). Alternatively, various tissue explants, including stem sections, epicotyls, hypocotyls, leaf disks, or callus tissues, have been tested for plant regeneration or transformation using the standard methods of tissue culture manipulation and *Agrobacterium* infection (Facciotti *et al.*, 1985; Owens and Cress, 1985; Wright *et al.*, 1986; Byrne *et al.*, 1987; Wright *et al.*, 1987a,b). For soybeans and beans, virtually all such attempts have failed to produce transgenic plants. Hence, the difficulty is having the transformation and regeneration events occur in the same cell types of the target tissues. Therefore, the conventional "tissue culture pathway" for plant transformation and regeneration, which works efficiently for explant or callus cultures derived from plants of the family Solanaceae, has not been shown to be useful for the transformation of most grain legume crops.

Several investigators (Christianson *et al.*, 1983; Ranch *et al.*, 1985; Lazzeri *et al.*, 1987; Hartweck *et al.*, 1988; Christou and Yang, 1989) have shown that immature zygotic embryos of soybean seeds can serve as useful tissue sources for establishing somatic embryogenic callus cultures. Biochemically and histologically, somatic embryos in culture have been found to exhibit characteristics similar to those observed for zygotic embryos (Hartweck *et al.*, 1988; Christou and Yang, 1989). Apparently normal and fertile plants have been regenerated from such somatic embryogenic cultures (Ranch *et al.*, 1985; Barwale *et al.*, 1986; Barwale and Widholm, 1987; Lazzeri *et al.*, 1987; Parrott *et al.*, 1989b).

Thus, these cultures were considered to be a good source tissue for gene transfer and regeneration of transgenic soybean plants. Parrott *et al.* (1989a) reported that primary transformants of soybean plants were obtained by infecting these somatic embryos in culture with *Agrobacterium*. However, in their study, none of the R_1 progeny of the three putative primary transformants (R_0) were found to carry or to express the reporter gene(s). Since these results may be due to the presence of *Agrobacterium* sequestered in the plant tissues of putatively transformed plants (see subsequent text), it has yet to be demonstrated that this approach can generate stable transgenic soybean plants.

Another approach to transformation of soybeans was taken by Hinchee *et al.* (1988). They showed that, for the 'Peking' cultivar, explants of mature cotyledon nodes can be used as target tissue for *Agrobacterium* infection, and that transgenic plants can be regenerated from this tissue under shoot-induction culture conditions. Progeny plants derived from this study are fertile and carry and express the foreign gene(s).

A new approach that has been tested extensively is to use the shoot apical meristem of mature seed as target tissue for direct gene transfer using particle bombardment (McCabe *et al.*, 1988; Christou *et al.*, 1989,1990). This developing technology was designed to circumvent the conventional tissue culture and plant regeneration problems. It employs a fast multiple-shooting, organ culture system for the regeneration of plants. The method has proven reproducible for soybean transformation and applicable to all soybean cultivars tested so far, and has been improved to serve as a routine method for commercial production of transgenic soybeans (Christou *et al.*, 1990). Therefore, conceptually, this transformation system has indicated a new direction for plant genetic engineering. Since similar shoot meristem culture systems for plant regeneration also have been established for peas (Ahmed *et al.*, 1987; Puonti-Kaerlas *et al.*, 1990) and beans (D. Russell, personal communication), there is good reason to believe that the particle bombardment transformation method developed for soybeans (McCabe *et al.*, 1988; Christou *et al.*, 1990) may be applied to these crops as well.

Compared with grain legumes, alfalfa and clover are much more effective in plant regeneration from callus cultures derived from various tissue explants, thus making them amenable to conventional gene transfer and plant regeneration methods (Shahin *et al.*, 1986; White and Greenwood, 1987).

5.1.3 *AGROBACTERIUM* INFECTION

Disarmed bipartite Ti plasmids constructed with cauliflower mosaic virus (CaMV)-35S or nopaline synthase (*nos*) promoter and various reporter genes (*nptII*, *cat*, or *gus*) have been used often in expression vectors for gene transfer via *Agrobacterium* infection of legume plant tissues. Several studies have shown that not all *Agrobacterium* strains can infect soybean tissues readily (Facciotti *et al.*, 1985; Owens and Cress, 1985; Byrne *et al.*, 1987). Infectivity was found highest for strains C58, A208, and A281 on most soybean genotypes tested (Owens and Smigocki, 1988; Delzer *et al.*, 1990); some phenolic compounds (e.g., acetosyringone) were found to promote the infectivity of less virulent strains for soybean cultivars (Owens and Smigocki, 1988). The range of cultivar response to virulent *Agrobacterium* strains was found to vary considerably also; cultivars 'Maple,' 'Presto,' 'Peking,' and 'Delmer' were the most responsive (Owens and Cress, 1985; Byrne *et al.*, 1987; Hinchee *et al.*, 1988). Although less information on *Agrobacterium*-mediated gene transfer is available for peas and beans, limited reports (Jackson *et al.*, 1989; Puonti-Kaerlas *et al.*, 1989) suggest high specificity of infectivity between *Agrobacterium* strains and plant cultivars. This strain/cultivar specificity restricts the use of agronomically important grain legume cultivars in transformation experiments.

As mentioned earlier, systematic and extensive efforts by Hinchee *et al.* (1988) have generated transgenic soybean plants of the cultivar 'Peking' successfully. In this case, mature cotyledon node explants were used for infection with *Agrobacterium*. In B_5BA shoot-induction medium, numerous adventitious shoots were produced not from the existing axillary buds, but adjacent to one another in a *de novo* fashion from the nodal tissue of the cotyledon explants. In this culture system, kanamycin selection of *nptII* gene-transformed cotyledon tissues and the derived shoots was found to be "leaky" but helpful (Hinchee *et al.*, 1988). Transgenic soybean plants cotransformed with a mutated petunia 5-enolpyruvylshikimate-3-phosphate (EPSP) synthase gene were found to express EPSP synthase activity. The R_1 and R_2 progeny of these plants have been field tested for efficiency in glyphosate tolerance. Commercialization of herbicide-resistant, transgenic soybeans is expected by the mid-late 1990s (J. Schillinger, personal communication, 1992). Because the number of mature transgenic plants obtained in this study is small and has been derived only from the 'Peking' cultivar, it is not clear whether this soybean transformation system can be routinely useful either for experimental purposes or for commercial production of transgenic soybean seed stocks.

Transformation of soybean plants also has been reported by Chee *et al.* (1989). Their method involves the use of *Agrobacterium* infection directed at the plumule, cotyledon node, and adjacent cotyledon tissues of germinating seeds. To obtain transformed soybean plants, *Agrobacterium* cells in liquid suspension were injected, with the aid of a syringe fitted with a thin-gauge needle, into target tissues of imbibed soybean seeds. Inoculated seeds were germinated in soil under normal conditions and reporter gene activities assayed in grown-out R_0 and R_1 plants. The frequency of obtaining putatively transformed soybean lines was determined as 0.07%. In this study, whereas enzyme activities and DNA sequences of reporter gene *nptII* were detectable in putative transformants, the investigators have not been able to establish unambiguously the integrative mode of stable transformation in test plants, for example they have not reported Mendelian segregation patterns in subsequent generations. Similar to the somatic embryo infection experiments reported by Parrott *et al.* (1989a), results of this experiment still raise the possibility that the detected signals of reporter genes were contributed by the rearranged Ti plasmid DNA present in *Agrobacterium* cells that were sequestered in test tissues of putative transformants. Detection of reporter gene signals by polymerase chain reactions (PCR) or by Southern blot analysis, without employing extensive control tests, may not be sufficient to distinguish such contamination from *bona fide* stable transformation. Therefore, although a similar seed infection method has been well established by Feldman *et al.* (1987,1989) for obtaining transformed *Arabidopsis thaliana* plants, the efficacy of the soybean seed transformation system described by Chee *et al.* (1989) has yet to be demonstrated fully.

Transformation of pea, *Pisum sativum* L., by *Agrobacterium* infection has been assessed by several groups. Jackson *et al.* (1989) tested *Agrobacterium* infectivity in stems of whole plants *in vivo* and in immature leaf explants *in vitro,* finding specificity in genotype/strain interaction. Mariotti *et al.* (1989) tested various organ explants for transformation of *Pisum, Phaseolus,* and *Cicer.* Eapen *et al.* (1987) examined the transformation rate of moth bean (*Vigna*) protoplast cultures after cocultivation with *Agrobacterium.* Transformed calluses or tissues were obtained in these studies. Puonti-Kaerlas *et al.* (1990) have recovered transgenic pea plants using explants from axenic shoot cultures and seedling epicotyls cocultivated with *Agrobacterium.* Transformation of calluses and regenerated plants was confirmed by DNA analysis.

Alfalfa (*Medicago sativa* L.) is a highly valued forage legume that is cultivated extensively in temperate and cool subtropical regions. An effective transformation system for cultivated alfalfa was established by Shahin *et al.* (1986). Stem segments of mature plants were infected with strain LBA4404 harboring disarmed Ti plasmid. Transformed tissues were selected as kanamycin-resistant calluses; then embryoids and transgenic plants were regenerated from callus tissues. Between 16 and 28 independently transformed alfalfa plants were generated from every 100 inoculated stem slices. The system was developed for alfalfa 'CUF101,' a leading commercial cultivar grown in California, and was shown later by other investigators to be readily applicable to other alfalfa cultivars (Deak *et al.*, 1986; Halk *et al.*, 1989; Matheson *et al.*, 1990). This system should provide a simple and efficient method for the genetic transformation of most alfalfa cultivars. Since highly productive cultivars of alfalfa were regenerable also from callus cultures derived from petiole (Wan *et al.*, 1988), immature ovary (Ray and Bingham, 1989), and cotyledon tissues (Chen *et al.*, 1987), for specific experimental purposes, these tissue explants presumably could be used also as alternative target tissues for gene transfer and regeneration of alfalfa plants.

White clover (*Trifolium repens* L.) and red clover (*Trifolium prafense* L.)

are two other useful forage legumes. At present, only a portion of tested genotypes gives satisfactory plant regeneration from callus cultures (White and Greenwood, 1987; MacLean and Nowak, 1989; Myers *et al.*, 1989). However, because clovers are outcrossing species, a foreign gene could be introduced into a single recipient genotype and the gene subsequently could be moved into an appropriate genetic background by conventional breeding practice. Kanamycin-resistant callus cultures were obtained by culturing *Agrobacterium*-inoculated stolon internode segments, and were regenerated into transformed plants (White and Greenwood, 1987). Since certain genotypes of white (Myers *et al.*, 1989) and red clover (Yamada, 1989) also can regenerate plants from protoplast cultures, direct gene transfer methods presumably could be applied for transformation of clovers also, although this has yet to be demonstrated.

5.1.4 DIRECT GENE TRANSFER METHODS

Electroporation has been used to transform protoplast cultures of soybean (Christou *et al.*, 1987; Lin *et al.*, 1987). Transformed cells grown as callus tissues or organized structures have been useful for some experimental purposes, including verification of promoter activity in specific legume plant cells, usefulness and efficiency of antibiotic resistant marker genes in transformed tissues, and levels of foreign gene expression in legume cells (Christou *et al.*, 1987; Lin *et al.*, 1987). However, none of these studies have been successful in producing transgenic plants for grain legumes.

Leminger-Mertens and Jacobsen (1989) have regenerated fertile plants from pea protoplasts via somatic embryogenesis. High level of auxins and increased osmolarity of the medium were found necessary for embryo induction in two genotypes. More recently, they reported that polyethylene glycol (PEG)-mediated gene transfer was applied successfully to this culture system, and transgenic fertile pea plants were regenerated from transformed protoplasts (Jacobsen, 1990).

Microinjection once was considered to be a useful method for transforming alfalfa protoplasts (Reich *et al.*, 1986), but the high level of skill and labor required for this technique has made this method impractical for transformation work in general, including transforming forage legumes.

5.1.5 PARTICLE BOMBARDMENT

Particle bombardment technology, a means for direct gene transfer, has enabled plant biologists to transform not only soybean plants (McCabe *et al.*, 1988; Christou *et al.*, 1990), but also other important crop plants, including corn

(Fromm *et al.*, 1990; Gordon-Kamm *et al.*, 1990) beans (D. Russel, personal communication, 1992), peanuts (G. Brar, personal communication, 1992), rice (Christou *et al.*, 1991), and cotton (Finer and McMullen, 1990; D. McCabe and B. Martinell, personal communication, 1992). The method has been shown also to be applied usefully for *in vivo* or *in vitro* gene transfer to mammalian systems (Yang *et al.*, 1990; Williams *et al.*, 1991), demonstrating the wide application of this technology for gene delivery. Two research groups, Sanford and co-workers (Klein *et al.*, 1987; Sanford, 1988) and McCabe and co-workers (McCabe *et al.*, 1988; Christou *et al.*, 1990), have developed this technology independently and have designed two different devices for particle acceleration and bombardment of target plant tissues. The electric discharge mediated bombardment method developed by McCabe's group is now identified as Accell™ technology. The gun powder or gas-mediated bombardment method established by Sanford's group has been named the Biolistics™ method. Since transgenic soybean plants were first obtained by McCabe *et al.* (1988) using the electric discharge apparatus, and because transformed soybean plants have been produced continuously by this group (Christou *et al.*, 1989,1990), this method will be discussed as a model in this review.

The basic design of a high voltage electric discharge device that accelerates DNA-coated gold particles into living cells has been reported by Christou *et al.* (1990), as shown in Fig. 5.1. The procedure for using such a device involves several basic steps. Plasmid DNA in solution first is precipitated onto gold beads, which are chemically inert in most, if not all, biological systems. The DNA-coated particles then are resuspended in ethanol, and evenly deposited onto a thin mylar film. The mylar sheet is placed adjacent to two closely spaced electrodes. An electric arc, generated by a high-voltage discharge, provides the motive force. The force accelerates the DNA-coated gold particles to a high velocity, enabling efficient penetration of target organs, tissues, or cells. With this particle bombardment device, one can adjust the velocity and resulting distribution of gold particles in various target tissues finely by varying the discharge voltage (3000–25,000 V), bead density, and size of the employed gold beads. For most studies, 1–3-μm gold beads at a density of 0.1 mg beads/cm^2 were used; the velocity of the beads was adjusted by using different discharge voltages (Christou *et al.*, 1990). Plasmid DNA can be loaded onto beads at a rate of 1 ng to 10 μg DNA per mg beads, corresponding to 10–100,000 copies of a 5-kb DNA molecule per gold particle (1–3 μm). The particle bombardment procedure involves minimal manipulation of target organs, tissues, or cells and is versatile, efficient, and flexible.

Using Accell technology, McCabe *et al.* (1988) demonstrated that bombardment of shoot apical meristem tissues (isolated either from immature embryos or from matured seeds) resulted in the generation of clonal or chimeric soybean plants. Molecular analysis of such transformed clonal plants and their progeny showed that plasmid DNA was integrated into the soybean nuclear genome and

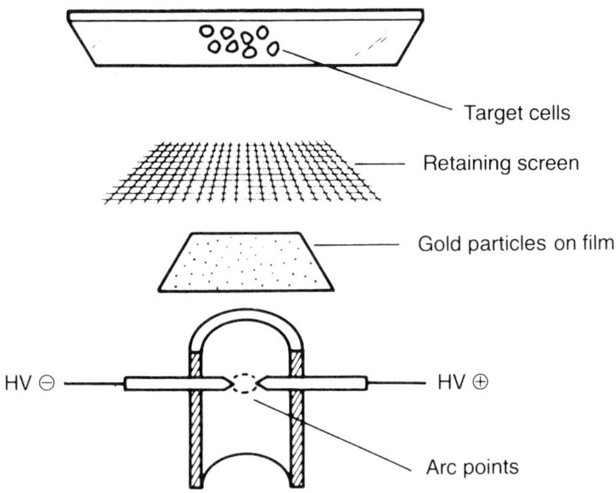

Figure 5.1 The motive force is generated in a spark discharge chamber containing two electrodes. A 10-μl water droplet is placed between the electrodes and a high-voltage capacitor is discharged through the water droplet, which vaporizes instantly, creating a shock wave. We have found that a polyvinyl chloride pipe with an internal diameter of 13 mm is adequate as a spark discharge chamber. The electrodes are located opposite each other, and are protected at the tips with an arc-resistant alloy. The gap between the two electrodes can be adjusted by appropriately threading them into or out of the spark chamber. A spacer ring that, in a fixed apparatus for transformations of a single crop species, can be a vertical extension of the spark discharge chamber is placed above the spark chamber. However, a removable spacer ring allows the distance from the spark discharge to the carrier sheet to be varied, so the force of the shock wave can be adjusted. The motive force also can be adjusted by varying the voltage of the discharge. The carrier sheet on which the DNA-coated gold particles are precipitated is placed on top of the spacer; the function of this sheet is to transfer the force of the shock wave from the spark discharge into acceleration of the carrier particles. Located above the carrier sheet is a 100-mesh stainless steel screen that retains the sheet so it does not proceed to the target tissue. The target tissue can be placed on a water–agar plate so, when the plate is inverted over the retaining screen, the tissue is in the direct path of the gold particles. The whole assembly is under a partial vacuum to minimize aerodynamic drag. (Reprinted with permission of the publisher from Christou *et al.*, 1990.)

was transmitted to progeny plants according to Mendelian inheritance. Continued studies from this group further proved that the Accell method is applicable to various soybean cultivars in general, and can be employed as an effective means for commercial production of transgenic soybean plants (Christou *et al.*, 1990).

A key feature of the particle bombardment method in transforming soybean is that shoot apical meristems were used effectively as the target site for DNA delivery. As shown in Fig. 5.2 (color plate), it was suggested that, for certain plant species, specific cell compartments termed as L1, L2, and L3 tissues in a young shoot apical meristem provide defined cell lineages that develop into various tissues and organs of a mature plant, including the germline or reproductive tissues (Poethig and Sussex, 1985; Poethig, 1987,1989; Drews and Goldberg, 1989). Whereas the exact physical arrangement and the derivative cell lineages of L1, L2, and L3 tissues of a shoot apical meristem are not completely clear, and the interpretations of such tissue organization may vary among different investigators (Poethig, 1987,1989; Drews and Goldberg, 1989), it is quite clear, however, that the top several cell layers of a shoot apical meristem tissue contain all the necessary progenitor cells for the development of the aerial portion of a mature plant. Bombardment of the shoot apical meristems of 1-day seedlings showed transformation occurring in meristematic cells competent to produce germline tissues. When bombarded meristems were transferred subsequently to shoot-induction medium and grown in an organ culture mode, multiple shoots were regenerated quickly in 3–4 wk. Transformation and regeneration in the same cell type(s) occurred at a high enough frequency that clonal shoots were able to be generated without applying selection pressure (McCabe *et al.*, 1988; Christou *et al.*, 1989,1990). The final transformation frequency obtained by this method is 1–3 clonal transgenic plant per 10^3 meristem explants (P. Christou, B. Martinell, and D. McCabe, personal communication, 1992). Kanamycin selection of transgenic shoots in this highly proliferative organ culture system was found to be ineffective (McCabe *et al.*, 1988), as one would expect for large tissue explants containing complex tissue organizations. However, by

Figure 5.2 (A)Schematic representation of shoot apical meristems of higher plants. The vegetative shoot apical meristem is divided into two regions, the tunica and the corpus. The tunica contains the L_1, L_2, and L_3 germ layers. The corpus is divided into three zones: central mother cells, peripheral meristem, and pith-rib meristem. A portion of the peripheral meristem is derived from cells in the tunica. Cells in the central mother zone divide infrequently. (Reprinted with permission of the publisher from Drews and Goldberg, 1989.) (B) Histochemical staining of β-glucuronidase activity and gold bead penetration in particle-bombarded soybean apical meristem. Shoot apical meristems of 1-day-old soybean seedlings were bombarded using gold particles coated with β-GUS DNA. Meristems were cultivated for 8 days; thin-tissue sections were stained for GUS activity. Note that small gold beads had penetrated several cell layers into the apical dome structure, covering the L_1, L_2, and L_3 layers as shown in A. Expression of GUS activity can be seen in leaf primordia and in a bud-like structure, presumably representing a *de novo* organized shoot meristem dome. (Courtesy of P. Christou and D. McCabe.)

Figure 5.3 Chimerism of transgenic soybean plants. Regenerating shoots were harvested 3–6 weeks after particle bombardment; GUS activity was stained histochemically. Degrees of chimerism ranged from plantlets exhibiting activity at organ, tissue, and "subtissue" levels. Completely "blue" plants were recovered also; these plants were shown later to be clonal. (Reprinted with permission of the publisher from Christou *et al.*, 1990.)

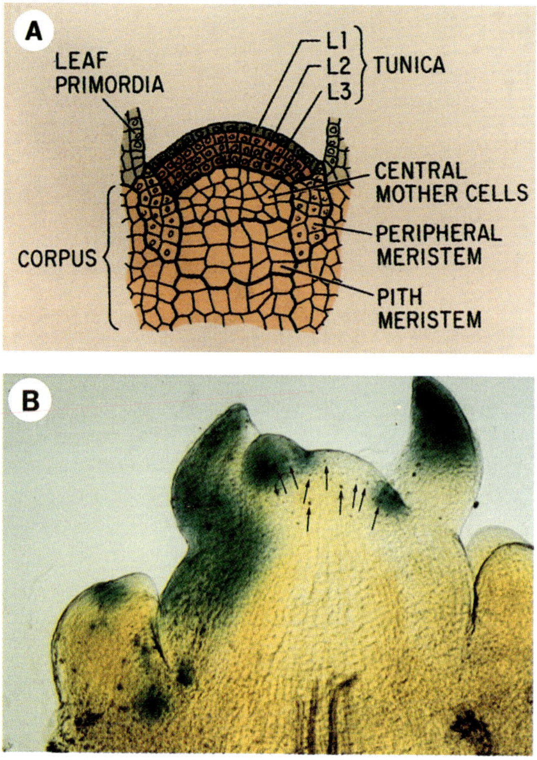

Figure 5.2

Figure 5.3

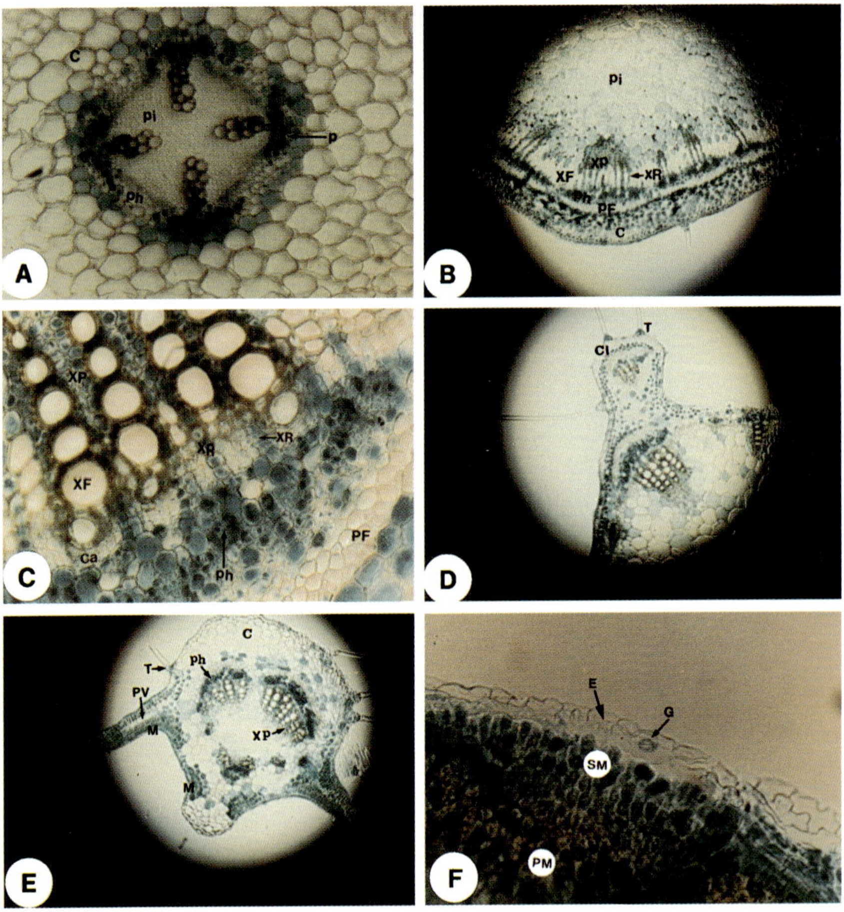

Figure 5.6 Expression of 35S-GUS activity in tissue cross sections of transgenic soybean plants: (A) root, (B) stem, (C) vascular tissue of stem, (D) leaf petiole, (E) leaf midrib, and (F) leaf mesophyll tissues. C, cortex; Ca, vascular cambium; Cl, collenchyma; E, epidermic; G, guard cells; M, mesophyll cells; P, pericycle cells and derivatives; PF, phloem fiber; Ph, phloem; Pi, pith; PM, palisade mesophyll cells; PV, paraveinal cells; SM, spongy mesophyll cells; T, trichome; XF, xylem fiber; XP, xylem parenchyma; XR, xylem ray cells. (Reprinted with permission of the publisher from Yang and Christou, 1990.)

using histochemical staining of β-glucuronidase (GUS) activity, these investigators have developed an effective screening procedure for the identification of clonal, chimeric, and nontransformed soybean shoots in developing organ cultures.

Obviously, an element of art is involved in the particle bombardment technology, since similar results have not been reported successfully by other groups who employed the Biolistic process for soybean transformation, whereas continued progress has been reported by Christou *et al.* (1990). This difference could be due to variation in protocols involving the culture conditioning of source plants, parameters of particle bombardment in DNA delivery, and recovery of bombardment-wounded tissues.

The particle bombardment technique developed for soybean allows the recovery of transformed plants from tissues and lines that cannot be transformed using conventional methods, such as *Agrobacterium* or electroporation. In addition, the technique bypasses the need for time-consuming and labor-intensive tissue culture manipulations involving protoplast or callus cultures. The soybean transformation system developed by McCabe, Christou, and co-workers may serve as a useful model for other legume crops. D. Russel *et al.* (personal communcation, 1992) and G. Brar *et al.* (personal communication, 1992) showed that a similar approach is indeed applicable to beans and peanuts.

5.1.6 VARIANT PHENOTYPES OF TRANSGENIC PLANTS

Genetically engineered plants commonly exhibit variant phenotypes. Some of the anomalous phenotypes result from epigenetic effects generated by tissue culture manipulations. Other variant phenotypes, called somaclonal variations, are caused by stable genetic changes that occur in cultured tissues and are not related to the gene transfer procedure. In soybean, extensive somaclonal variation in plants regenerated from embryogenic cultures was reported by Barwale and Widholm (1987). Finally, some genetic changes are associated directly with the integration of foreign DNA into the host plant nuclear genome. This integration may cause an insertional mutation in a functionally important genetic locus, effecting phenotype changes. More details on the various aspects of phenotype and genotype changes in variant plants have been discussed previously by Yang (1988). In soybean transformation experiments using particle bombardment (McCabe *et al.*, 1988; Christou *et al.*, 1990), because minimal and brief organ culture manipulations were employed to generate transformed plants, somaclonal variation and epigenetic effects on phenotype changes in transgenic plants were expected to occur at a much reduced rate. This reduced somaclonal variation was observed by Christou *et al.* (1990; P. Christou, personal communication, 1992).

5.2 Evaluation and Proof of Stable Transformation

5.2.1 TRANSIENT, LONG-TERM, AND INTEGRATIVE TRANSFORMATION

When functional DNA is delivered into plant cells using direct gene transfer methods, the introduced genes often are expressed at a readily detectable level that usually lasts for several days. However, depending on the plant species, tissue type, culture conditions, and reporter gene(s) used, the level and duration of this transient gene expression may vary considerably. For some reporter genes (e.g., *gus*), expression of transient activity can continue for a relatively long period of time (3–4 wk; N. S. Yang, unpublished data, 1990). Such relatively long-term but nonetheless transient activity can be attributed to one or a combination of several possibilities: (1) foreign DNA is degraded slowly in target cells, (2) mRNA transcribed from input gene(s) is relatively stable, and (3) marker enzyme translated from mRNA of a reporter gene is highly stable in target cells. Long-term GUS activity often can be observed in tissue clusters in quiescent or slow-growing tissue cultures, or in tissue sectors or streaks in differentiating organs in cultures. Although the extent of such long-term expression may be useful as an indicator for effective DNA delivery and expression, it often may not be indicative of a stable integrative transformation. When antibiotic resistance genes are used for selection of stable transformants, similar long-term expression also could result in the appearance of antibiotic-resistant calluses or embryos during the early selection period, but later fail to confer *bona fide* stable transformations.

In addition to the confusion and complexities resulting from this unstable but relatively long-term reporter gene expression, there are other potential problems in interpreting preliminary results on stable transformation. Feldmann *et al.* (1989) showed that *Agrobacterium* cells used to infect *Arabidopsis* seeds can remain in plant tissues throughout plant development. Proliferated or sequestered *Agrobacterium* cells residing in the reproductive tissues can infect germline or reproductive cells, resulting in integrative transformation in some of the derived progeny in subsequent generations. Therefore, in transformation experiments, sustained presence in regenerated plants of sequestered *Agrobacterium,* used as a vector for initial transformation, is a concern unless it can be proven that the reporter gene(s) follows a Mendelian inheritance pattern in sexually derived progeny plants (R_1 and R_2). For plants generated or regenerated by a direct gene transfer method, there is also the concern that endogenous endophyte bacteria (e.g., *Pseudomonas maltophilia, Acinetobacter calcoaceticus*) could be co-transformed by the same method and could survive in the plant tissues throughout plant development and even greenhouse breeding for several generations (Konstantinov *et al.*, 1990). These investigators showed that such bacteria can be isolated from putatively transformed plant tissues, and can account for the observed Southern signals and marker enzyme activity.

In addition to these potential problems in transformation procedures, assay background, endogenous marker enzyme-like activities (Hu *et al.*, 1990), or the casual contamination of test plants or DNA samples with plasmid DNA in molecular analysis (e.g., PCR) also can generate false positives during the confirmation of stably transformed plants or tissues. Hence, it is important that stable transformation is defined stringently. To prove stable transformation, we need to demonstrate (1) a tight correlation between molecular data on integration of foreign gene(s) (e.g., Southern blot) and phenotypic expression of integrated genes (e.g., enzyme assays of reporter genes), (2) transmission and expression of integrated foreign genes in sexual offspring, and (3) use of appropriate controls in the various assays to rule out false positives and contamination problems often encountered in experimentation. These criteria have been reviewed in detail by Potrykus (1990). In the demonstration of a new transformation method, lack of a complete fulfillment of these three conditions could lead to spurious claims of stable transformation.

5.2.2 COMPLICATIONS USING GUS GENE AS SCREENING MARKER

Although the GUS marker gene has the advantage of detecting transgenic gene expression at the cellular level, it does have some significant problems. Endogenous GUS-like enzyme activities have been shown in certain plant species (Hu *et al.*, 1990). Many types of fungus and bacteria, which often do not overgrow plant tissues in culture or under greenhouse conditions, will stain dark blue in GUS reaction buffer. In addition, depending on the promoter gene employed (including the CaMV 35S promoter), GUS activity in transgenic plants can be highly specific in cell or tissue type expression (Yang and Christou, 1990). Further, the pattern of stable expression for certain promoters may not correlate to the expression patterns observed for various tissues in transient assays. In general, GUS activity staining can be used effectively only if appropriate negative and positive controls are employed in specific and stringent assay conditions (Yang and Russell, 1990).

5.2.3 UTILITY OF ANTIBIOTIC-RESISTANCE GENES AS SELECTABLE MARKERS

In cell or friable callus cultures of legume plants, kanamycin and geneticin (G418) are useful antibiotics for selection of tissues transformed with the neomycin phosphotransferase (*nptII*) gene (Christou *et al.*, 1987). However, they are much less effective in selection of transformed cell clusters or tissue sectors present within large heterogeneous tissue clumps or organ segments such as embryos and shoots. Therefore, although use of kanamycin in soybean cotyledon node explant cultures can enrich the chance of promoting growth for transgenic tissues (Hinchee *et al.*, 1988), it is too "leaky" or too harsh to be practical for soybean meristem organ cultures (McCabe *et al.*, 1988; Christou *et al.*, 1990). Hygromycin has been reported to differ drastically from kanamycin in its efficacy as a selectable marker for some transformation systems. In peas, shoot organogenesis could be induced reproducibly on hygromycin-resistant calluses, but not on the calluses selected for kanamycin resistance (Puonti-Kaerlas *et al.*, 1990). Similar results were reported by Christou *et al.* (1991) in embryogenic callus cultures of rice. As an alternative to antibiotic selection, the method of stringent and systematic screening for clonal expression of GUS activity, established by McCabe *et al.* (1988), may be applicable to grain legume transformation systems in general.

5.2.4 CHIMERIC AND CLONAL TRANSGENIC PLANTS

When apical shoot meristem is used as the target tissue for particle bombardment (McCabe *et al.*, 1988; Christou *et al.*, 1990), chimeric shoots are obtained at a much higher rate than are clonal shoots, because locally transformed cells and their progeny cells in a meristem are recruited during the organogenesis step in multiple shooting. Later, only tissues derived from these cell lineages are transformed stably, generating chimeric plants. The chimeric pattern can be demonstrated by a tight correlation between expression of reporter gene activity and the presence of the reporter gene sequence in transformed as opposed to untransformed tissue sectors within a single chimeric plant (P. Christou, personal communication, 1991). Chimerism was observed to be detectable not only at the organ level (e.g., one shoot branch as opposed to another), but also at the suborgan and tissue level (Fig. 5.3, color plate). At the organ level, transformed shoot branches can produce stably transformed seeds as progeny, but the untransformed branches cannot. Therefore, for certain experimental purposes, researchers can take advantage of certain chimeric plants to obtain isogenic lines that differ only in the defined transgene(s).

5.3 Inheritance and Expression of Transgenes in Transformed Soybeans

5.3.1 PATTERN OF GENE INTEGRATION

The lineages of integrated foreign DNA in transgenic soybean plants have been studied by Christou *et al.* (1989). These investigators demonstrated that (1) foreign DNA introduced into soybean plants via particle bombardment can be inherited in a Mendelian manner, (2) there was a 100% concordance between presence of the foreign gene sequences and expression of reporter gene(s) (*gus* or *nptII*), (3) tandem marker genes cotransformed into soybeans can be inherited as closely linked genes in subsequent generations, (4) although some plants contain a single copy of the foreign gene(s), others contain multiple copies (Fig. 5.4), and (5) the high copy number of foreign genes present in some transgenic plants often is inherited as a single locus, even when cotransformed with two separate plasmids (Fig. 5.5). Moreover, results of their molecular and genetic analyses suggest that whole plants can be derived from single transformed cells by a *de novo* organogenic pathway. This is a surprising observation when compared with the widely held view that organogenesis, as opposed to somatic embryogenesis, follows only a multicellular regeneration pathway. Results of Christou *et al.* (1989,1990) on the clonal nature of the R_0 plants suggest that a single transformed cell in the apical meristem tissue underwent extensive cell proliferation, resulting in a population of uniformly transformed cells from which organogenesis was initiated. The frequency for transformation and clonal organogenesis is high enough to practically produce 1–3 clonal transgenic plants per 1000 meristem explants (P. Christou and D. McCabe, personal communication, 1991).

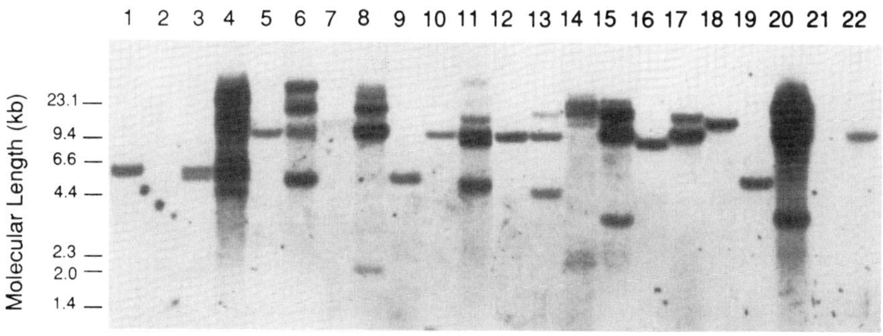

Figure 5.4 Southern analysis of transgenic soybean plants. Lanes 1 and 22 are one gene copy per genome reconstructions for two different GUS plasmids, prepared by mixing plasmid DNA and nontransformed soybean DNA at a rate corresponding to one gene copy per genome, based on a genome size of 6×10^9 nucleotide pairs. These lanes show the position of the internal GUS fragment for each plasmid. The copy standard in lane 1 corresponds to the input GUS plasmid in the transgenic plants shown in lanes 3 and 4; the standard in lane 22 corresponds to the input GUS plasmid for lanes 5–20. Lanes 2 and 21 contain samples from nontransformed soybean. Lanes 3–20 contain samples from independently transformed soybean plants. (Reprinted with permission of the publisher from Christou *et al.*, 1990.)

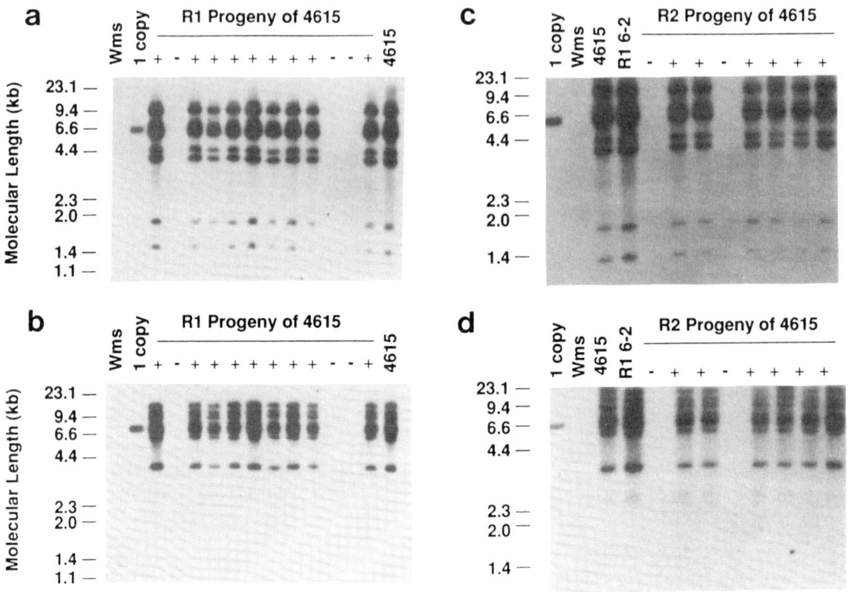

Figure 5.5 Southern analysis of R_1 and R_2 progeny of plant 4615. Southern blots were made as described in Fig. 5.4, except digestion was with *Bam*HI; 1-copy reconstructions included both GUS and NPTII plasmids. (a, c) Samples from self-pollinated R_1 and R_2 plants, respectively. The membranes were hybridized with a ^{32}P-labeled RNA probe corresponding to the minus strand of the GUS coding region. (b, d) The GUS probe was removed and the membranes were rehybridized with a ^{32}P-labeled RNA probe corresponding to the minus strand of the NPTII coding region. Lanes labeled Wms contains DNA from a nontransformed soybean (cv. Williams) plant. Lanes 4615 (a, b) and 6-2 (c, d) contain DNA samples from the parental R_0 or R_1 plants, respectively. Remaining lanes contain DNA samples from self-pollinated R_1 (a, b) or R_2 (c, d) progeny and indicate the presence (+) or absence (−) of GUS and NPTII enzyme activity in leaf samples from the same plant. The positions and lengths, in kilobase pairs, of the molecular size markers are indicated. (Reprinted with permission of the publisher from Christou *et al.*, 1989.)

5.3.2 TISSUE-SPECIFIC EXPRESSION OF SOYBEAN PROMOTERS IN HETEROLOGOUS HOSTS

Regulation of gene expression during soybean embryogenesis and seedling development has been investigated systematically in the past few years by two main research groups, Goldberg and co-workers (Goldberg, 1988; Drews and Goldberg, 1989; Goldberg *et al.*, 1989) and Beachy and co-workers (Beachy *et al.*,

1985, 1986; Chen *et al.*, 1986,1989; Bray *et al.*, 1987). Results obtained on endogenous and transgenic expression of soybean seed-storage proteins or their promoter genes, at both the cellular and the molecular levels, have provided important baseline information and guidelines for future work requiring seed tissue-specific expression of transgenes in legume plants. Faithful reproduction of endogenous expression patterns generally has been found for transgenes expressed in seeds of heterologous plant systems. Identified soybean promoters useful for targeting agronomically useful transgenes into soybean seed tissues include those of β-conglycinin, α-subunit, β-subunit, and lectin genes (see Benfey and Chua, 1989, for review). Tapetum-specific promoter is useful in targeting gene expression to specific cell types in anther tissues (Goldberg, 1988; Mariani *et al.*, 1989). Other tissue-specific promoter sequences isolated from grain legumes include those of pea *rbcS* and *Cab* genes, soybean nodulin *N23* and leghemoglobin genes, pea legumin gene, bean lectin gene, and French bean β-phaseolin gene (Benfey and Chua, 1989).

5.3.3 CELL TYPE EXPRESSION OF CaMV-35S PROMOTER IN TRANSGENIC SOYBEAN PLANTS

Benfey and Chua (1989) first demonstrated that the nominally constitutive CaMV 35S promoter is specifically expressed in tobacco and petunia tissues. Yang and Christou (1990) have examined activity of this promoter at the cellular level in transformed soybean plants. Figure 5.6 (color plate) shows that cell-type specificity and developmental regulation of the 35S promoter can be illustrated clearly in soybean plants, and that some new features of this promoter were observed. Three categories of 35S promoter activity were observed for transgenic soybeans: (1) the highest expression was seen consistently in the parenchyma cells of xylem and phloem tissues of nonroot organs and in pericycle cells and their derivatives in roots, (2) little or no expression was seen in epidermal cells with the exception of guard cells and trichomes, and (3) medium and variable levels of expression were seen in stem, cortex, leaf, and petal mesophyll tissues.

Since specificity of cell-type expression for the CaMV 35S promoter has been shown to vary from one plant species to another (Benfey and Chua, 1989), and since 35S promoter is currently the most commonly used promoter for plant genetic engineering, the results obtained by Yang and Christou (1990) may serve as useful guidelines for finely tuned transgene expression of desirable agronomic traits in vegetative tissues of soybeans and, perhaps, of other grain legume plants.

5.4 Extension and Optimization of New Techniques

Different gene transfer methods and techniques should be able to be combined to provide new means or optimized conditions for plant transformation. Dekeyser *et al.* (1990) have shown that electroporation confers gene delivery and transient expression in intact and organized rice tissues. It may be useful to determine if this method is applicable to the apical meristem system of grain legumes, alone or in combination with particle bombardment, perhaps effecting more efficient transformation than either method alone. Gilbertson *et al.* (1990) have shown that particle bombardment greatly enhances infectivity of golden mosaic virus in bean plants. Bidney *et al.* (1992) showed that a similar enhancing of *Agrobacterium* infection of tobacco and sunflower tissue explants can be achieved by employing particle bombardment for intratissue delivery of bacterial cells. Many studies have shown that PEG-mediated gene transfer is as effective as electroporation for protoplast cultures. Because intact cells and tissues are not necessarily a barrier for electroporation (Dekeyser *et al.*, 1990), this could mean that the PEG method may be applicable for gene transfer into intact and organized tissues also. In legume transformation, can particle bombardment of meristem tissue be combined with the PEG method to improve efficiency of transformation?

5.5 Agronomically Important Traits for Legume Crops

Many traits of commercial legume crops could be improved by genetic engineering. Glyphosate or other herbicide-tolerance traits are expected to improve legume plants in general. EPSP gene-transformed soybean plants, derived from particle bombardment, have been field tested (Petrie, 1990). Efficacy for tolerance to applied herbicide was established for these transgenic soybeans in field trials (J. Schillinger, personal communication, 1992). High-methionine or other sulfur-rich proteins would compensate for the nutritional deficiencies of grain legume seeds, in general, and alfalfa leaf tissues. Target seed protein genes that have been considered for legume crops include genes encoding the 2S seed proteins of Brazil nut (Altenbach *et al.* 1989) and zein proteins of corn (Williamson *et al.*, 1988). Casein, the milk protein that is one of the most important animal proteins in nutritional value, also may be engineered to improve the protein quality of legumes. Obviously, future studies need to show that high level expression and packaging of casein as protein bodies in embryo or seed tissues would not interfere with the normal physiology or development of legume seeds.

More than 65% of the food oil used in cooking, shortening, and other processing steps for the United States food industry is derived from soybean oil (Iowa Agriculture & Home Economics Experimental Station, 1990). Elevation of oleic acid at the expense of linoleic and linolenic acids is expected to render soybean oil more resistant to oxidation, thus improving its utility in the food industry. Also, reducing the concentrations of palmitic and stearic acids (saturated fatty acids) in soybean oil is desirable for nutritional reasons. Specific desaturase enzymes are needed for the metabolic conversion of saturated fatty acids to unsaturated fatty acids in oil seeds. *Arabidopsis* mutants deficient in desaturation of palmitic acid in leaf lipids were obtained (Kunst *et al.*, 1989; Browse *et al.*, 1990). Knutzen *et al.* (1991) and others (Knauf, 1992) obtained cDNA clones encoding stearoyl-acyl carrier protein (ACP) desaturase, lauroyl-ACP thioesterase and two other enzymes. Transgenic plants of rapeseed expressing sense or antisense genes for these enzymes were found to redirect the profile of fatty acids produced, resulting in high stearate or high laurate rapeseed (Knauf, 1992). Increased expression of acyl-carrier protein, a key component in the fatty acid biosynthetic pathway, may be a useful strategy for increasing total lipid biosynthesis in transgenic oil seeds. However, transgenic ACP is not expected to change the overall oil quality in engineered oil seeds.

Bacillus thuringiensis toxin genes are not expected to benefit most grain legume crops, because lepidopteran and coleopteran insects in general do not pose a major problem for these crops. A few years ago, new strains of bean common mosaic virus caused significant damage to bean crops in the United States, but the problems were regional and did not persist long enough to raise serious concerns. Bean golden mosaic viruses (BGMV) are significant pathogens for various bean crops, often causing severe bean damage in developing countries. BGMV viral coat protein gene(s) (Howarth *et al.*, 1985), when engineered into transgenic bean plants, may be useful in protecting bean crops against BGMV or related viral diseases. It is not clear whether alfalfa mosaic virus (AMV) causes significant damage to alfalfa crops because virus-free alfalfa cultivars are not available as control sets for field tests (E. Bingham, personal communication, 1991). Transgenic alfalfa plants expressing AMV coat proteins have been generated by Halk *et al.* (1989), but efficacy of these plants in field tests has not been reported.

Cyst nematode is a major cause of yield loss of soybean crops, especially in the southern United States (Brewer, 1981). Although soybean cultivars have been bred for resistance against some major races of cyst nematodes, the minor races quickly increase in population and cause significant crop damage. Genetic analyses show that resistance genes for different races are located at different chromosomal loci, and multiple genes probably are involved in conferring resistance against a single race of nematode (Palmer and Kilen, 1987; Anand and Rao-

Arelli, 1989). Since little information is available for the biochemical and molecular basis of these race-specific nematode-resistant soybean genes, there is no clear strategy for attacking this problem by the transgenic plant approach.

Compared with nematodes, fungal pathogens cause much less damage in grain legume crops. *Phytophthora* and *Rhizoctonia* cause root and stem rot. Since one or two pairs of genes are involved in pathogen/host gene–gene interactions, investigators are using transposon-tagging, chromosome-walking, and subtraction cDNA cloning techniques to identify the resistant gene(s) in host plants as well as the avirulent genes in fungal pathogens (Bennetzen *et al.*, 1988; Ellis *et al.*, 1988; Michelmore, 1990; Orbach *et al.*, 1990). These lines of study eventually may identify useful fungus-resistance genes for transformation of plants including legumes. Some fungi contain chitins as cell wall component of their mycellium. Studies by Broglie *et al.* (1991) showed that tobacco plants transformed with a chitinase gene exhibit enhanced resistance to the fungal pathogen Rhizoctonia-Solani. This strategy for introducing fungal resistance to transgenic legume plants remains to be applied.

In addition to various types of disease resistance, certain morphological and physiological traits also may prove valuable for transgenic legumes. Single or double stem character is expected to provide a more upright and narrow canopy, improving mechanical farming of dry beans. The early maturity trait, which increases the efficiency of nutrient translocation from leaves to the maturing seeds, is desirable for late maturity soybean cultivars. Grain legume root systems are sensitive to "combined nitrogen source," that is, soybean or bean plants grown with applied nitrogen fertilizers, when normally nodulated with *Rhizobium,* do not fix more nitrogen than plants grown with fertilizers alone. Insensitivity to "combined nitrogen sources" is, therefore, a highly desirable trait for grain legumes, since it would improve utilization of nitrogen fertilizers. Specific genotypes of common beans (F. Bliss, personal communication, 1991) and soybeans (Schuller *et al.*, 1988) have been identified to exhibit significantly reduced sensitivity to high levels of "combined nitrogen source." Augmented nitrate reductase activity in root systems also has been considered to have potential for improving nitrogen assimilation in legume plants. In addition to the various genetic traits mentioned here, many other economically desirable traits for soybean or other grain legumes have been described in a special report (Iowa Agriculture & Home Economics Experimental Station, 1990).

Most of these morphological or physiological traits are probably under multiple-gene control. However, we do not necessarily have to avoid this potential problem. Yeast artificial chromosomes (YAC) have been demonstrated to carry large (300–790 kb) fragments of genomic DNA (Green and Olson, 1990). Also, the particle bombardment system can deliver large genomic DNA into plant and animal cells (N.-S. Yang, unpublished data, 1991). Particle bombardment is thus likely to be useful in the delivery of YAC vectors constructed with multiple long fragments of genomic DNA. With the new techniques increasingly available for both construction and delivery of large genomic DNA, one may predict that, in the future, plant transformation work involving multiple genes or large genomic DNA will not be a major technical barrier for the genetic engineering of plants.

5.6 Concluding Remarks and Prospects

Plant genetic engineers have made important progress during the past few years. A key milestone was the transformation of soybeans. The particle bombardment technology for soybean transformation has become practical for routine production of soybeans intended for commercial use (Christou *et al.*, 1990). Field tests are being conducted to establish the efficacy of herbicide tolerance in soybean plants. Similar bombardment procedures are being applied to transform beans, peanuts and other grain legumes, with promising results (D. Russell and G. Brar, personal communication, 1992), indicating that the particle bombardment technology may be applicable generally for grain legume transformation. Meanwhile, routine and efficient *Agrobacterium* transformation procedures have been established for the two key forage legumes, alfalfa and clover. Overall, led by innovative technology, extensive progress has been and is continually being made in legume transformation. Based on these successes, we can project optimistically that, in the next few years, there may be a real demand for systematic and scaled up production of various transgenic legume plants, aiming for commercial production of agronomically improved legume crops. Glyphosate tolerance and high-methionine seed proteins are two primary target traits to be introduced into commercial seed stocks of soybeans.

Because of the increased level of sophistication in plant transformation technology, as observed in the developing techniques of particle bombardment, many components of the transformation technology have become proprietary information or specific arrangements belonging to industry specialty laboratories. On one hand, this situation is a strong indication that powerful technology is being employed to build a new "transformation" industry. On the other hand, the high level of complexity and team work often restricts the availability of such techniques, and the technology as a whole, to general research laboratories. Although this long has been a general practice for many other fields of industry involving high technology, it is, however, a new experience for the field of plant biology. In recognition of such a new research environment, it seems to be very important that research cooperations between industrial and academic laboratories be coordinated appropriately. Extensive and organized collaboration in the form of consortium, contract, and project-oriented research would benefit not only the participating research laboratories, but also the future development of the plant genetic engineering field as a whole.

With this perspective, we can begin to feel the excitement and see the challenges lying ahead of us. Because only a few limited genes are now identified to have potential agronomic use for the improvement of legume crops, the key challenge will be the identification of more of these genes from grain legume crops. With the recent advances in PCR cloning, pulse-field gel electroporesis, and restriction fragment length polymorphism mapping technologies, systematic and large-scale screening may become practical in the search for specific genes involved in nematode resistance, fungus resistance, and other desirable cell physiological traits for legume plants.

Acknowledgments

I thank D. Russell, P. Christou, D. McCabe, G. Brar, F. Bliss, R. Burgess, and E. Bingham for providing unpublished results as personal communications. I also thank P. Christou, W. Swain, F. Bliss, D. McCabe, D. Russell, G. Brar, P. Umbeck, M.-M. Qin, K. Barton, P. Ludden, and E. Bingham for many useful discussions. I am especially grateful to C. De Luna for careful editing of the manuscript. I also thank R. Putman for an excellent job on typing and final preparation of the manuscript.

References

Ahmed, R., Dutta Gupta, S., and Ghosh, P. (1987). *Plant Breed.* **98,** 306–311.
Altenbach, S., Pearson, K., Meeker, G., Staraci, L., and Sun, S. (1989). *Plant Mol. Biol.* **13,** 513–522.
Anand, S., and Rao-Arelli, A. (1989). *Crop Sci.* **29,** 1181–1184.
Baldes, R., Moos, M., and Geider, K. (1987). *Plant Mol. Biol.* **9,** 135–145.
Barton, K., Whiteley, H., and Yang, N.-S. (1987). *Plant Physiol.* **85,** 1103–1109.
Barwale, U., Kerns, H., and Widholm, J. (1986). *Planta* **167,** 473–481.
Barwale, U., and Widholm, J. (1987). *Plant Cell Rep.* **6,** 365–368.
Beachy, R., Chen, Z.-L., Horsch, R., *et al.* (1985). *EMBO J.* **4,** 3047–3053.
Beachy, R., Chen, Z.-L., Bray, B., *et al.* (1986). *In* "Molecular Biology of Seed Storage Proteins and Lectins" (L. Shannon and M. Chrispeels, eds.), pp. 193–201. American Society of Plant Physiology, Rockville, Maryland.
Benfey, P., and Chua, N.-H. (1989). *Science* **244,** 174–181.
Bennetzen, J., Qin, M., Ingels, S., and Ellingboe, A. (1988). *Nature (London)* **332,** 369–370.
Bray, E., Naito, S., Pan, N., *et al.* (1987) *Planta* **172,** 364–370.
Brewer, F. (1981). *In* "Special Assessment of the Soybean Cyst Nematode *Heterodera glycines* Problem: Joint Planning and Evaluation Staff Paper," pp. 9–11. United States Department of Agriculture, Washington, D.C.
Browse, J., Lemieux, B., Guerra, D., and Sommerville, C. (1990). *In* "UCLA Symposium on Molecular and Cellular Biology: New Series" (C. Lamb, and R. Beachy, eds.), Vol. 129, pp. 301–310. Wiley-Liss, New York.
Byrne, M., McDonnell, R., Wright, M., and Carnes, M. (1987). *Plant Cell Tissue Organ Cult.* **8,** 3–15.
Chee, P., Fober, K., and Slightom, J. (1989). *Plant Physiol.* **91,** 1212–1218.
Chen, T., Marowitch, J., and Thompson, B. (1987). *Plant Cell Tissue Organ Cult.* **8,** 73–81.
Chen, Z.-L., Schuler, M., and Beachy, R. (1986). *Proc. Natl. Acad. Sci. U.S.A.* **83,** 8560–8564.
Chen, Z.-L., Naito, S., Nakamura, I., and Beachy, R. (1989). *Dev. Genet.* **10,** 112–122.

Christianson, M., Warnick, D., and Carlson, P. (1983). *Science* **222,** 632–634.

Christou, P., and Yang, N.-S. (1989). *Ann. Bot.* **64,** 225–234.

Christou, P., Murphy, J., and Swain, W. (1987). *Proc. Natl. Acad. Sci. U.S.A.* **84,** 3962–3966.

Christou, P., Swain, W., Yang, N.-S., and McCabe, D. (1989). *Proc. Natl. Acad. Sci. U.S.A.* **86,** 7500–7504.

Christou, P., McCabe, D., Martinell, B., and Swain, W. (1990). *Trends Biotechnol.* **8,** 145–151.

Deak, M., Kiss, G., Koncz, C., and Dudits, D. (1986). *Plant Cell Rep.* **5,** 97–100.

Dekeyser, R., Claes, B., DeRycke, R., *et al.* (1990). *Plant Cell* **2,** 591–602.

Delzer, B., Somers, D., and Orf, J. (1990). *Crop Sci.* **30,** 320–322.

Drews, G., and Goldberg, R. (1989). *Trends Genet.* **5,** 256–261.

Eapen, S., Kohler, F., Gerdemann, M., and Shieder, O. (1987). *Theor. Appl. Genet.* **75,** 207–210.

Ellis, J., Lawrence, G., Peacock, W., and Pryor, A. (1988). *Ann. Rev. Phytopathol.* **26,** 245–263.

Facciotti, D., O'Neal, J., Lee, S., and Shrewmaker, C. (1985). *Bio/Technology* **3,** 241–246.

Feldmann, K., and Marks, D. (1987). *Mol. Gen. Genet.* **208,** 1–9.

Feldmann, K., Marks, M., Christianson, M., and Quatrano, R. (1989). *Science* **243,** 1351–1354.

Finer, J., and McMullen, M. (1990). *Plant Cell Rep.* **8,** 586–589.

Fromm, M., Morrish, F., Armstrong, C., *et al.* (1990). *Bio/Technology* **8,** 833–839.

Gilbertson, R., Faria, J., Hanson, S., *et al.* (1990). Manuscript submitted.

Goldberg, R. (1988). *Science* **240,** 1460–1467.

Goldberg, R., Barker, S., and Peraz-Grau, L. (1989). *Cell* **56,** 149–160.

Gordon-Kamm, W., Spencer, T., Mangano, M., *et al.* (1990) *Plant Cell* **2,** 603–618.

Green, E., and Olson, M. (1990). *Science* **250,** 94–98.

Halk, E., Merlo, D., Liao, L., *et al.* (1989). *In* "Molecular Biology of Plant–Pathogen Interactions" (B. Staskawicz, P. Ahlquist, and O. Yoder, eds.), pp. 283–296. Liss, New York.

Hartweck, L., Lazzeri, P., Cui, D., Collins, G., and Williams, E. (1988). *In Vitro Cell. Dev. Biol.* **24,** 821–828.

Hiatt, A., Cafferkey, R., and Bowdish, K. (1989). *Nature (London)* **342,** 76–78.

Hinchee, M., Connor-Ward, D., Newell, C., *et al.* (1988). *Bio/Technology* **6,** 915–922.

Howarth, A., Caton, J., Bossert, M., and Goodman, R. (1985). *Proc. Natl. Acad. Sci. U.S.A.* **82,** 3572–3576.

Hu, C.-Y., Chee, P., Chesney, R., *et al.* (1990). *Plant Cell Rep.* **9,** 1–5.

Iowa Agriculture & Home Economics Experimental Station (1990). *In* "Economic Implications of Modified Soybean Traits" (C. Greiner, ed.), pp. 29–56. Iowa Soybean Promotion Board & American Soybean Association, Ames.

Jackson, J., Hobbs, S., and Mahon, J. (1989). *J. Cell Biochem. Suppl.* **13,** 261.

Jacobsen, H.-J. (1990). Paper presented at the Workshop on Advanced *Phaseolus* Research Network, Cali, Columbia, September 11–14, 1990.

Klein, T., Wolf E., Wu, R., and Sanford, J. (1987). *Nature (London)* **327,** 70–73.

Komatsuda, K., and Ohyama, K. (1988). *Theor. Appl. Genet.* **75,** 695–700.

Konstantinov, K., Mladenovic, S., Denic, M., and Plecas, M. (1990). *In* "NATO Advanced Study Institute: Plant Molecular Biology (1990)" (R. Herrmann and B. Larkins, eds.), p. 99. Federation of European Biochemical Scientists, Elmau, Germany.

Krebbers, E., and Vandekerckhove, J. (1990). *Trends Biotechnol.* **8,** 1–3.

Kunst, L., Browse, J., and Somerville, C. (1989). *Plant Physiol.* **90,** 943–947.

Lazzeri, P., Hilderbrand, D., and Collins, G. (1987). *Plant Cell Tissue Organ Cult.* **10,** 197–208.

Leminger-Mertens, R., and Jacobson, H.-J. (1989). *Plant Cell Rep.* **8,** 379–382.

Lin, W., Odell, J., and Schreiner, R. (1987). *Plant Physiol.* **84,** 856–861.

McCabe, D., Swain, W., Martinell, B., and Christou, P. (1988). *Bio/Technology* **6,** 923–926.

MacLean, N., and Nowak, J. (1989). *Plant Cell Rep.* **8,** 395–398.

Mariani, C., Leemans, J., DeGreef, W., and Beuckeleer, M. (1989). W. O. Patent No. 8,910,396, November 2, 1989.

Mariotti, D., Fontana, G., and Santini, L. (1989). *J. Cell Biochem. Suppl.* **13,** 273.
Matheson, S., Nowak, J., and MacLean, N. (1990). *Euphytica* **45,** 105–112.
Michelmore, R. (1990). *In* "NATO Advanced Study Institute: Plant Molecular Biology (1990)," p. 126. Federation of European Biochemical Scientists, Elmau, Germany.
Myers, J., Crosser, J., Taylor, N., and Collins, G. (1989). *Plant Cell Tissue Organ Cult.* **19,** 113–127.
Orbach, M., Farrall, L., Valent, B., and Chumley, F. (1990). *In* "NATO Advanced Study Institute: Plant Molecular Biology (1990)," p. 42. Federation of European Biochemical Scientists, Elmau, Germany.
Owens, L., and Cress, D. (1985). *Plant Physiol.* **77,** 87–94.
Owens, L., and Smigocki, A. (1988). *Plant Physiol.* **88,** 570–573.
Palmer, R., and Kilen, T. (1987). *In* "Soybeans: Improvement, Production, and Uses" 2d Ed. (J. Wilcox, ed.), p. 150. American Society of Agronomy, Madison, Wisconsin.
Parrott, W., Hoffman, L., Hildebrand, D., Williams, E., and Collins, G. (1989a). *Plant Cell Rep.* **7,** 615–617.
Parrott, W., Williams, E., Hildebrand, D., and Collins, G. (1989b). *Plant Cell Tissue Organ Cult.* **16,** 15–21.
Petrie, M. (1990). *Fed. Reg.* **55,** 32279.
Poethig, S. (1987). *Am. J. Bot.* **74,** 581–594.
Poethig, S. (1989). *Trends Genet.* **5,** 273–277.
Poethig, S., and Sussex, I. (1985). *Planta* **165,** 170–184.
Potrykus, I. (1990). *Bio/Technology* **8,** 535–542.
Puonti-Kaerlas, J., Stabel, P., and Eriksson, T. (1989). *Plant Cell Rep.* **8,** 321–324.
Puonti-Kaerlas, J., Eriksson, T., and Engstrom, P. (1990). *Theor. Appl. Genet.* **80,** 246–252.
Ranch, J., Oglesby, L., and Zeilinski, A. (1985). *In Vitro Cell Dev. Biol.* **21,** 653–657.
Ray, I., and Bingham, E. (1989). *Crop Sci.* **29,** 1545–1548.
Reich, T., Iyer, V., and Miki, B. (1986). *Bio/Technology* **4,** 1001–1004.
Sanford, J. (1988). *Trends Biotechnol.* **6,** 299–302.
Schuller, K., Minchin, F., and Gresshoff, P. (1988). *In* "Nitrogen Fixation: Hundred Years After" (H. Bothe, F. de Bruijn, and W. Newton, eds.), pp. 664. Gustav Fischer, New York.
Shahin, E., Spielmann, A., Sukhapinda, K., Simpson, R., and Yashar, M. (1986). *Crop Sci.* **26,** 1235–1239.
Sijmons, P., Dekker, M., Schrammeijer, B., *et al.* (1990). *Bio/Technology* **8,** 217–221.
United States Department of Agriculture (1989). *In* "Agricultural Statistics, 1989," p. 496. U. S. Government Printing Office, Washington, D.C.
Vandekerckhove, J., Van Damme, J., Van Lijse-bettens, M., *et al.* (1989). *Bio/Technology* **7,** 929–932.
Wan, Y., Sorensen, E., and Liang, G. (1988). *Euphytica* **39,** 3–9.
Wei, Z.-M., and Xu, Z.-H. (1988). *Plant Cell Rep.* **7,** 348–351.
White, D., and Greenwood, D. (1987). *Plant Mol. Biol.* **8,** 461–469.
Williamson, J., Galili, G., Larkins, B., and Gelvin, S. (1988). *Plant Physiol.* **88,** 1002–1007.
Wright, M., Koehler, S., Hinchee, M., and Carnes, M. (1986). *Plant Cell Rep.* **5,** 150–154.
Wright, M., Ward, D., Hinchee, M., Carnes, M., and Kaufman, R. (1987a). *Plant Cell Rep.* **6,** 83–89.
Wright, M., Williams, M., Pierson, P., and Carnes, M. (1987b). *Plant Cell Tissue Organ Cult.* **8,** 83–90.
Yamada, T. (1989). *Euphytica* **44,** 181–186.
Yang, N.-S. (1988). *Trends Biotechnol.* **6,** 521–522.
Yang, N.-S., and Christou, P. (1990). *Dev. Genet.* **11,** In press.
Yang, N.-S., and Russell, D. (1990). *Proc. Natl. Acad. Sci. U.S.A.* **87,** 4144–4148.
Yang, N.-S., Burkholder, J., Roberts, B., Martinell, B., and McCabe, D. (1990). *Proc. Natl. Acad. Sci. U.S.A.* In press.

6

Transgenic Oilseeds and Vegetables

Brian L. A. Miki

Plant Research Centre
Agriculture Canada
Ottawa, Ontario, Canada K1A 0C6

Bin Huang

Allelix Crop Technologies
Mississauga, Ontario
Canada L4V 1P1

Pierre J. Charest

Petawawa National Forestry Institute
Forestry Canada
Chalk River, Ontario, Canada K0J 1J0

A technical discussion of transgenic vegetable and oilseed crops is a challenging assignment for a single chapter because of the tremendous diversity of species represented. Each species differs in the ease of achieving stable transformation and the experimental approaches for regenerating transgenic plants in culture. We will therefore focus on the genus *Brassica,* which includes a variety of important crops such as the vegetable cole crops, mustards, and oilseed rape. For comparison, we will discuss the progress made in vegetables such as carrot, cucumber, and melon, as well as oilseeds such as sunflower and flax. Major crops such as potato, tomato, corn, and soybean have been the subject of intense research and are discussed in other chapters.

6.1 Agronomic Objectives

Among the *Brassica* species, the tissue culture and transformation technologies for *Brassica napus* rapeseed varieties has progressed significantly (Tables 6.1–

6.4). Transgenic plants have been generated in several laboratories by *Agrobacterium*-mediated transformation and several small-scale field trials have been conducted in Canada over the last few years. Two cultivars of canola or rapeseed with low levels of erucic acid and glucosinolates currently occupy more than 90% of the 7 million acres planted in western Canada. Introduction of agronomically important genes into the cultivars *B. napus* cv. 'Westar' and *Brassica campestris* cv. 'Tobin' or comparable varieties could have a significant economic impact. As an example, losses due to weeds account for 12–14% of canola production after optimal cultured practices are implemented (Chandler *et al.*, 1984). Since the crop is valued at over $1 billion annually, the development of high-yielding herbicide-resistant cultivars would have a significant impact on production after optimal cultural practices are implemented (Chandler *et al.*, 1984). Since the crop is valued at over $1 billion annually, the development of addressing resistance to a variety of chemical herbicides. In the near future, we would expect to find resistance to fungal diseases and insect damage as high priorities for transgenic canola research. At this time, however, transgenic canola lines have not been incorporated into breeding nurseries.

Flax is an important source of nonedible oils and fiber. The total annual production may not be on the same scale as canola in Canada; however, it is equivalent in value to several of the vegetable crops grown in the United States and Canada. Losses due to weeds are particularly severe in flax and may be as high as 16% of potential production (Chandler *et al.*, 1984). Transgenic field trials of herbicide-resistant lines have begun in Canada.

In North America, the vegetable *Brassica* species such as cabbage, cauliflower, and broccoli are important, but the total annual production value is far below that of canola and the major vegetables, such as tomato or potato. Transgenic vegetable *Brassica* lines have not yet advanced to small-scale field trials, whereas trials of transgenic tomato or potato are relatively extensive and transgenic cucumber, cantaloupe, and squash trials have been initiated. Although the tissue culture capabilities needed to generate transgenic vegetable *Brassica* are relatively advanced compared with those for other vegetable species (Tables 6.1–6.3), research on genetic transformation strategies or gene expressions in transgenic plants has been sparse (Table 6.4). At this time, the overall objectives for breeding of transgenic vegetable *Brassica* appear to be similar to those for other crops, for example, resistance to viral diseases through mechanisms similar to viral cross-protection.

An examination of transgenic field trials approved in Canada (provided by the Seeds Division, Agriculture Canada) and the United States (provided by the Freedom of Information Office, Department of Agriculture) revealed that most trials dealt with resistance to various herbicides, viral diseases, or insect pests. Although strategies for enhancing yield and production are extremely important and viable short-term objectives, crop quality will be the long-term objective.

For canola, the amount and composition of seed-storage lipid and protein are the primary issues. An understanding of the metabolic and biosynthetic pathways that terminate in seed-storage products is essential for the development of genetic approaches. Experimental manipulations of these pathways must incorporate the principles of developmental genetics and gene regulation to modulate the desired temporal, spatial, and tissue-specific patterns of expression in developing seeds. Genes that are absent or unidentified in canola may be taken from other organisms. Several examples have been discussed. The identification and transfer of *Nasturtium* genes responsible for attachment of erucic acid (22:1) to the second position of glycerol 3-phosphate as well as to the first and third positions would increase the yield of erucic acid for industrial uses (Knauf, 1987). Altering the levels of plastid glycolytic enzymes using *Castor* bean genes may divert carbon and reducing power to specific biosynthetic pathways, thereby influencing the quantity of specific products (D. Dennis, personal communication, 1989). The possibility of modifying the amino acid profile of protein meal with the Brazil nut 2S storage protein has been considered (Guerche *et al.*, 1990). Transgenic plants are, thus, more than recipients of genetic change but are also experimental systems for understanding the basic genetic and physiological principles that dictate and control crop composition and quality.

6.2 Culture Systems for Transformation

Efficient plant cell and tissue culture systems are essential for the regeneration of transgenic plants from transformed cells. Regeneration may proceed preferentially through organogenesis or embryogenesis depending on the species, cell and tissue origins, and culture conditions. Both differentiation events have been characterized in several *Brassica* species and embryogenesis in carrot has been well studied. For genetic transformation, a variety of recipient cell types already have proven useful, yet others remain untested. Tables 6.1–6.3 review tissue culture systems that may prove to be important for research in transgenic oilseeds and vegetables relevant to this chapter.

6.2.1 PROTOPLAST CULTURE

Protoplasts are versatile recipients for genetic transformation and transient expression analyses. They retain all the genetic information contained in the organelles and nuclei of normal cells and have the potential to express totipotency by eventually developing into plants. Protoplasts are strictly single cells, whereas cell-suspension cultures originally developed for this purpose usually comprise cell clusters. Due to the lack of cell walls, protoplasts can take up DNA molecules in various forms. Several methods have been developed to facilitate

DNA uptake. These include the use of polyethylene glycol (PEG) and other chemicals (Paszkowski *et al.*, 1986), microinjection (Neuhaus and Spangenberg, 1990), particle bombardment (Sanford, 1990), and laser microbeam perforation (Weber *et al.*, 1988, 1989a, 1989b, 1990). DNA uptake, mediated by chemical treatments or electroporation, has been limited to protoplasts, whereas other methods of DNA delivery, such as microinjection, microprojection, and laser microbeam perforation, can be used in protoplasts and intact cells. In addition to direct DNA uptake, protoplasts are amenable to *Agrobacterium*-mediated transformation.

Regeneration of plants from isolated protoplasts has been the focus of a large number of laboratories since the 1960s. As a consequence, it is now possible to obtain plants, embryoids, or calluses from many plant species. As shown in Table 6.1, plants have been regenerated from several *Brassica* species as well as from carrot, cucumber, lettuce, and sunflower. In other vegetables and oilseed crops in which plant regeneration from protoplasts is yet to be achieved, cell division and callus formation are often feasible, thereby allowing the transfer of genes for studies on transient expression and gene expression in calluses (Table 6.1).

Studies on regeneration from protoplasts have been mostly empirical. Table 6.1 lists the recommended conditions; however, these conditions may not be optimal for all genotypes of the same species. Indeed, optimal conditions may vary among laboratories and scientists. Factors affecting efficiency of regeneration are many and their interactions are complicated. Generally, protoplasts can be released from most, if not all, plant tissues by manipulating the source, concentration, and combination of the commercially available cellulases and pectinases. Despite this feat, successful protoplast culture, characterized by cell division, callus formation, and plant regeneration, is dependent on the use of "responsive" genotypes and donor tissues, which may vary from whole plant organs to undifferentiated cultured cells (Table 6.1). Protoplasts derived from cell-suspension cultures generally are viable and divide readily. Hence, they can be efficient systems for analysis for transient gene expression. However, regeneration more often is obtained in protoplasts isolated from differentiated tissues such as leaf and hypocotyl tissues. *In vitro* shoots, hypocotyls, and cotyledons are favored by some laboratories since sterile tissues can be obtained. In many species, particularly *Brassica* and cucumber, hypocotyls and protoplasts derived from these tissues possess high regeneration ability.

Culture conditions play important roles in regeneration of many tissues. Media formulations devised by Murashige and Skoog (1962;MS), Gamborg *et al.* (1968; B5), Kao and Michayluk (1975), and others have been used extensively as basal media for protoplast cultures of vegetables and oilseed crops. Modifications to and exchange of various components among these media (e.g., MS salt plus B5 vitamins) have produced media superior to the original. In addition to basal medium, osmoticum and supplements of growth regulators to

TABLE 6.1
Protoplast Isolation and Plant Regeneration of Cultivated Vegetable and Oilseed Crops

Family, genus, species	Common name	Culture medium and growth regulators (mg/liter)[a]	Cells and tissues[b] Source	Recovered	References
Asteraceae					
Lactuca sativa	Lettuce	mMS + 2,4-D(0.1) + BAP(0.3)	1	pl	Engler and Grogan (1983)
		KP	1,ct,r	pl	Berry *et al.* (1982)
		mNT	1 + EP	pl	Chupeau *et al.* (1989)
Brassicacea (Cruciferae)					
Brassica oleracea	Cabbage	mDPD + 2,4-D(0.5) + KIN(1)	1	pl	Fu *et al.* (1985)
		KMP8/KM8	r	pl	Xu *et al.* (1982)
		TM2 + 2,4-D(1) + BAP(0.5) + NAA(0.1)	h	pl	Lillo and Shahin (1986)
	Cauliflower	KMP8 + 2,4-D(1) + NAA(0.1) + BAP(0.5)	h	pl	Glimelius (1984)
		KMP8	ct	pl	Vatsya and Bhaskaran (1982)
	Broccoli		1	pl	Robertson and Earle (1986)
	Kale	B5 + 2,4-D(1) + BAP(0.5) + NAA(0.5)	1	c	Gatenby and Cocking (1977)
chinensis	Chinese cabbage	mNNM + BAP(0.1) + NAA(0.1) + 2,4-D(0.25)	1	c,r	Guo and Scheider (1983)
alboglabra	Chinese kale	MS + 2,4-D(0.2) + BAP(0.2)	s	pl	Pua (1987)
napus	Rapeseed	MI + 2,4-D(0.5) + NAA(0.5) + BAP(0.5)	1	pl	Li and Kohlenbach (1982)
		KMP8 + NAA(0.1) + BAP(0.4) + 2,4-D(0.1)	h	pl	Barsby *et al.* (1986)
		NLN	sc	pl	Klimaszewska and Keller (1987)
juncea	Mustard	V47 + 2,4-D(1) + NAA(0.1) + BAP(0.4)	h	pl	Kirti and Chopra (1989)
		MI	1	pl	Eapen *et al.* (1989)
Chenopodiaceae					
Spinacia oleracea	Spinach	MS + NAA(3) + BAP(5)	cs	c	Nakagawa *et al.* (1985)
Beta vulgaris	Beet	KM + NAA(1) + IAA(2)	cs,1	c	Bhat *et al.* (1986)

(*continues*)

TABLE 6.1
(Continued)

Family, genus, species	Common name	Culture medium and growth regulators (mg/liter)[a]	Cells and tissues[b]		References
			Source	Recovered	
Cucurbitaceae					
Cucumis sativus	Cucumber	mDPD + 2,4-D(0.5) + KIN(1)	ct	pl	Jia *et al.* (1986)
		MS + 2,4-D(0.2) + BAP(0.5)	ct	pl	Punja *et al.* (1990)
		MS + 2,4-D(1) + BAP(1)	l	pl	Punja *et al.* (1990)
Umbelliferae (Apiaceae)					
Daucus carota	Carrot	mMS or B_5 + 2,4-D(0.5) + NAA(0.2) + ZEA(0.1)	ecs	e,pl	Dudits (1984)
Foeniculum vulgare	Fennel	LS + 2,4-D(2.2) + KIN(2.2)	ecs	e,pl	Miura and Tabata (1986)

[a]Culture media formulations: B_5 (Gamborg *et al.*, 1968), DPP (Durand *et al.*, 1973), 8E (Zapata *et al.*, 1981), KM, KMP8, KM8 (Kao and Michayluk, 1975), KP (Kao, 1977), LS (Linsmaier and Skoog, 1965), MI (Li and Kohlenbach, 1982), MS (Murashige and Skoog, 1962), NLN (Lichter, 1981), NNM (Nitsch and Nitsch, 1969), NT (Nagata and Takeba, 1971), SH (Schenk and Hildebrandt, 1972), TM2 (Shahin, 1985), UM (Uchimiya and Murashige, 1974), V47 (Binding, 1974); modified (m). Growth regulators: BAP (6-benzylaminopurine), 2,4-D (2,4-dichlorophenoxyacetic acid), GA3 (gibberellic acid), IAA (3-indolacetic acid), IBA (3-indolebutyric acid), KIN (kinetin), NAA (naphthaleneacetic acid), ZEA (zeatin).

[b]c, Callus; cs, cell suspension; cspp, cell suspension protoplasts; ct, cotyledon; ctp, cotyledonary petioles; e, embryoid; ecs, embryogenic cell suspension; ep, epicotyl; h, hypocotyl; hpp, hypocotyl protoplasts; l, leaf; lpp, leaf protoplasts; me, microspore embryoids; p, petiole; pp, protoplasts; pl, plants; r, roots; s, stems; sc, stem cortex; sh, shoots; tcl, thin cell layer; tm, tumor tissue; tr, teratomas.

the medium are also crucial to successful protoplast culture. Growth regulators are so important that, by manipulating their concentration and combination alone, one basal medium can often be made suitable for a variety of different plant species. Culture density is another important factor. When insufficient numbers of viable or dividing protoplasts are available, feeder cultures of the same (D. Simmonds and W. Keller, personal communication, 1989; *Brassica* feeder for *Brassica* protoplasts) or different (Yarrow *et al.*, 1986; tobacco feeder for *Brassica* protoplasts) origin can be used to improve plating efficiency.

Culture media listed in Table 6.1 are those that have been used to induce cell division and callus formation of protoplast cultures of various species. For plant regeneration, calluses are transferred onto MS or B5 media solidified with agarose or agar. Cytokinins at 1–5 mg/liter usually are supplemented at this stage to induce green shoots. Shoots then are cultured on media containing 0.1–1 mg/liter auxin for root development. Although it has been possible to induce embryoids and plants in one culture medium (Li and Kohlenbach, 1982), the three-stage protocol (callus, shoot, roots) is the most common practice. In fact, with some species many more steps of manipulation are required to achieve optimal regeneration from protoplast culture.

6.2.2 CELL, TISSUE, AND ORGAN CULTURE

Intact plant cells, tissues, and organs of various origin also have been used as recipient cells for transformation studies. Regeneration of plants from cells is generally less technically demanding than regeneration from protoplasts, since fewer steps are involved. The number and range of species that can be regenerated in this manner are, therefore, much greater. The cell wall presents a barrier for direct DNA uptake; however, it can be overcome by physical means such as microinjection or microprojection and laser microbeam perforation or by biological means such as *Agrobacterium*-mediated DNA delivery. Indeed, most transgenic plants undergoing field trials have been produced in this manner.

Plants have been regenerated from cultured cells and tissues for many vegetables and oilseed crops (Table 6.2). Genotype is one of the most important factors affecting efficiency of regeneration. For species that do not respond very well to tissue culture, donor plants derived from tissue culture experiments have been shown to increase the efficiency of regeneration (Robertson *et al.*, 1988). Plant regeneration can be achieved from cells and tissues as diverse as seedling organs, leaves, stems, roots, and immature and mature seeds (Table 6.2). Again, *in vitro* shoots and seedling organs are some of the favorite starting materials since they are maintained aseptically and usually possess high regeneration ability.

Cell-suspension cultures that are embryogenic or organogenic are potential hosts for genetic transformation, but surprisingly few successes have been re-

TABLE 6.2
Tissue Culture and Plant Regeneration of Cultivated Vegetables and Oilseed Crops

Family, genus, species	Common name	Culture medium and growth regulators (mg/liter)[a]	Cells and tissues[b] Source	Recovered	References
Asteracea					
Lactuca sativa	Lettuce	SH + KIN(0.05) + ZEA(0.05)	ct	pl	Michelmore *et al.* (1987)
Brassicacea (Cruciferae)					
Brassica alboglabra	Kale	MS + BAP(1) + KIN(1)	h	pl	Wong and Loh (1987)
		MS + BAP(1)	ct,s	pl	Wong and Loh (1987)
hirta		MS + ZEA(1) + NAA(0.1)	ct	pl	Jain *et al.* (1989)
juncea	Mustard	MS + KIN(1) + NAA(0.1)	h,ct	pl	Murata and Orton (1987)
		MS + BAP(1)	ct	pl	Murata and Orton (1987)
napus	Rapeseed	MS + BAP(1)	s	pl	Kartha *et al.* (1974)
		MS + BAP(1)	s	pl[c]	Fry *et al.* (1987)
		MS + BAP(10)	tcl	pl	Klimaszewska and Keller (1987)
		MS + BAP(10)	tcl	pl[c]	Charest *et al.* (1988)
		MS + KIN(1) + NAA(0.1)	ct,h	pl	Murata and Orton (1987)
		MS + BAP(1)	ct	pl[c]	Moloney *et al.* (1989)
		mMS + KIN(0.1) + IAA(0.2)	r	pl	Sharma and Thorpe (1989)
oleracea					
var. *acephala*	Ornamental kale	MS + ZEA(2) + NAA(0.1)	l	pl[c]	Hosoki *et al.* (1989)
var. *femmifera*	Brussels sprouts	mMS + KIN(0.1) + 2,4-D(0.1)	h	pl	Williams *et al.* (1990)
var. *italica*	Broccoli	mMS + BAP(0.1) + IAA(0.1)	h	pl	Lazzeri and Dunwell (1986a,b)
		mMS + BAP(5) + IAA(5)	l	pl	Lazzeri and Dunwell (1986a,b)
		mMS + KIN(0.075) + IAA(0.025)	r	pl	Lazzeri and Dunwell (1986a,b)
		MS	l	pl[c]	Srivastava *et al.* (1988)
Cucurbitaceae					
Cucumis sativus	Cucumber	MS	h	pl[c]	Trulson *et al.* (1986)
			h	pl[c]	Chee (1990)

Umbelliferae (Apiaceae)					
Apium graveolens	Celery	B_5 + KIN(0.05)	p	pl[c]	Catlin *et al.* (1988)
Daucus carota	Carrot	B_5 + 2,4-D(1.5)	h	pl[c]	Thomas *et al.* (1989)
		UM	ecs	pl[c]	Scott and Draper (1987)
	Flax	MS + BAP(1) + NAA(0.2)	h	pl[c]	McHughen *et al.* (1989)
	Sunflower	MS + BAP(1)	h,ct	pl	Greco *et al.* (1984)
		MS + BAP(1) + 2,4-D(1)	h,ct	c	Greco *et al.* (1984)
		mMS + BAP(1) + NAA(1) + GA3(0.1)	h	pl	Paterson and Everett (1985)

[a]Abbreviations for culture media are given in Table 6.1.
[b]Abbreviations for cells and tissues are given in Table 6.1.
[c]Transgenic plants.

ported. Carrot cell-suspension cultures possess high regeneration capacity and have been well studied as a model for somatic embryogenesis. The isolation and microinjection of single proembryogenic cells has been achieved (Nomura and Komamine, 1986). Protoplasts derived from cell-suspension cultures have been used for electroporation and direct DNA uptake (see Table 6.7). The regeneration of transgenic plants through the process of somatic embryogenesis is an area that requires more research to assess fully its potential for the vegetables and oilseeds.

In contrast to protoplast culture media, which often contain many amino acids and vitamins, media formulations for plant tissue and organ culture based on that devised by Murashige and Skoog (1962) usually suffice. Cytokinins are commonly at 0.1–10 mg/liter to induce shoots with or without a callus phase. If the callus phase is desired to maximize the number of shoots per explant, tissues can be cultured on a medium containing auxin at 0.1–1 mg/liter in addition to the cytokinins. Well-established shoots can be excised and cultured on either basal medium or medium containing 0.1–0.5 mg/liter 1-naphthaleneacetic acid (NAA) for root development.

For some plant species (Thomas *et al.*, 1976), shoots have been shown to originate from single cells *in vitro*. The fact that many transformed plants are not chimeric provides further evidence for their single cell origins. Appropriate timing of shoot induction and DNA delivery should allow the recovery of homogeneous transformants from multicellular structures. For *Agrobacterium*-mediated transformation, the wounding response is presumed to be an important signal for induction of virulence; however, degeneration of wounded cells and tissues results in the release of compounds that may be deleterious to viable cells. Addition of 1–5 mg/liter silver nitrate to the culture medium was found to be beneficial to *Brassica* tissues in transformation experiments (De Block *et al.*, 1989), possibly by inhibition of ethylene action. There is still much room for improvement in many of the tissue culture systems currently available.

6.2.3 MICROSPORE AND ANTHER CULTURE

For analysis of transgenic plants and incorporation into breeding programs, homozygous lines must be produced rapidly and in large numbers. When diploid cells and tissues of somatic origin are used as recipients for gene transfer, transformants usually are heterozygous for the integrated foreign gene. After self-pollination and segregation, at least one generation is required for the recovery of homozygous transformants. Plants with multiple gene insertions may require many generations to reach homozygosity.

One approach for rapidly achieving homozygosity is the use of haploid cells or tissues as transformation hosts. In higher plants, haploids can be obtained (1) from male (anther or microspore culture) or female (ovule culture) gametophytes or their immediate precursors (Dunwell, 1986), (2) by hybridization between appropriate parents (Kao and Kasha, 1969), (3) by selecting naturally existing haploid seedlings (Stringham, 1977), and (4) by the use of *hap* genes (Hagberg and Hagberg, 1980). Of all the methods available, anther and microspore culture is the most versatile and efficient. Under appropriate culture conditions, isolated microspores and microspores in cultured anthers divide repeatedly to form multicellular structures that develop into embryoids or calluses (microspore embryogenesis). Plants regenerated from these embryoids or calluses are usually haploid and become homozygous diploids on colchicine treatment. This approach has been tested in over 200 plant species and has yielded haploid and homozygous diploid plants in many (Dunwell, 1986). In some plant species, anther and microspore culture is used routinely in breeding programs (Hu and Huang, 1987).

Brassica are among the most responsive species for anther and microspore culture (Table 6.3). In the well-studied variety, *B. napus* cv. 'Topas,' over 10% of the cultured microspores will develop into embryoids and, eventually, plants (Fan *et al.*, 1988; Pechan and Keller, 1988). This high efficiency of haploid production requires the use of donor plants grown at low temperatures (10–15°C), isolation and wash of microspores in B5 medium, and microspore culture in NLN medium at 32°C for at least 4 days. Both wash and culture medium

TABLE 6.3
Plant Regeneration from Haploid Cell and Tissue Culture[a]

Genus, species	Common name	Cells and tissues[b] Source	Recovered	References
Brassica campestris	Chinese cabbage	m	pl	Sato *et al.* (1989)
	Rapeseed	a	pl	Keller *et al.* (1975)
femmifera	Brussels sprouts	a	pl	Biddington and Robinson (1990)
juncea	Mustard	a	pl	Sharma and Bhojwani (1985)
napus	Rapeseed	m	pl	Lichter (1981, 1982)
nigra	Mustard	m	pl	Lichter (1989)
oleracea	Broccoli	a	pl	Arnison *et al.* (1990)
	Cauliflower	a	pl	Phippen and Ockendon (1990)
Raphanus sativa	Radish	m	pl	Lichter (1989)
Daucus carota	Carrot	a	pl	Anderson (1984)

[a]For culture medium and conditions, see Section 6.2.3.
[b]a, Anthers; m, isolated microspores; pl, plants.

contain 13% sucrose. Deviation from these conditions reduces the yield of microspore-derived embryoids. Although this protocol can be applied to other *Brassica* species with little modification, the frequencies of microspore embryogenesis are much lower. This is another example of the strong genotypic influence on tissue culture efficiency. Nevertheless, sufficient numbers of microspore-derived embryoids or haploid plants may be obtained in many *Brassica* genotypes for transformation experiments.

Anther culture of carrots also yielded microspore-derived embryoids (Anderson, 1984). Production of haploid plants from many other plant species, including most vegetable and oilseed crops, however, is far from efficient. Many empirical studies need to be carried out to determine optimal conditions for donor plant growth, developmental stages of microspores suitable for microspore embryogenesis, and appropriate pretreatments (e.g., starvation, temperature stress) and culture conditions (medium composition, culture density, temperature).

Although transformation of haploids is a relatively new concept, homozygous diploid transformants already have been produced in some plant species, including rapeseed (*Agrobacterium*-mediated transformation of microspore embryoids; Swanson and Erickson, 1989) and rice (DNA uptake by protoplasts; Datta *et al.*, 1990). Other haploid targets for transformation include various tissues and organs of haploid plant and haploid cell-suspension cultures.

In addition to the use of haploid cells as transformation hosts, the concept and techniques of haploidy and homozygosity can be used to obtain homozygous diploids from the heterozygous transgenic plants via microspore culture and subsequent chromosome doubling of the microspore-derived embryoids and plants.

6.3 Transgenic Plants

Evidence presented in this chapter and others shows that progress toward the efficient recovery of transgenic plants varies considerably among species. As discussed earlier, this variation is governed to a large extent by the status of the tissue culture technology for plant regeneration once genetic transformation has been achieved. When examining the literature, it is also important to consider the experimental objectives of the research. For instance, carrot protoplasts have been used frequently for transient expression studies in which plant regeneration was not a high priority (see Table 6.7). Sunflower has been used for preliminary examination of gene expression at the callus level (see Table 6.6). For canola, flax, and several vegetable species, transgenic plants that have not been altered in agronomic properties are essential for plant breeding. Tables 6.4–6.8 summarize selected literature that reflects the current status of the field.

6.3.1 *BRASSICA* SPECIES

The transgenic canola lines undergoing field evaluation in Canada all have been produced by *Agrobacterium*-mediated transformation of tissue explants. Although *Brassica napus* cv. 'Westar' has been particularly successful in these studies (Table 6.4), other cultivars such as 'Profit' (Miki *et al.*, 1990) and 'Topas' (Swanson and Erickson, 1989) also have been examined. Genotype differences in the efficiency of transgenic canola production exist, but these problems appear to be waning with the emergence of refined methodologies (Charest *et al.*, 1989; De Block *et al.*, 1989; Moloney *et al.*, 1989).

For transgenic *B. napus*, a variety of explant sources including hypocotyls, stems, cotyledonary petioles, and microspore-derived embryos have been used for cocultivation with *Agrobacterium tumefaciens* (Table 6.4). Specific culture conditions have been determined to enhance inoculation and genetic transformation (Charest *et al.*, 1988; De Block *et al.*, 1989). The efficiency of transgenic plant recovery per explant varied from 0.5 to 55%. The use of *in vitro* grown material (Moloney *et al.*, 1989) provides several experimental advantages and seems to be very efficient (Table 6.4). The use of microspore embryos adds another dimension to these experiments because haploid plantlets may be converted to homozygous diploids in a single generation (Swanson and Erickson, 1989).

Chimeric neomycin phosphotransferase genes and selection for kanamycin resistance have been very successful; however, the levels of kanamycin used are generally lower than those used in other species (15–100 μg/ml) and selection is less efficient (De Block *et al.*, 1989). Other selectable marker genes, including chimeric phosphinotricin acetyltransferase (PAT) genes which convey resistance to the herbicide phosphinotricin (PPT), also have been shown to be useful in the recovery of transgenic canola (De Block *et al.*, 1989).

Transgenic canola has been produced with vectors derived from three Ti plasmid classes. Among the *Brassica* species, *B. napus* is unique because the octopine Ti plasmids are less virulent than either the nopaline or agropine Ti plasmids (Charest *et al.*, 1989; Holbrock and Miki, 1985). Vectors derived from pTiT37, pTiC58, and pTi Bo542 have been particularly effective in canola transformation (Charest *et al.*, 1988; Fry *et al.*, 1987; Pua *et al.*, 1987; Radke *et al.*, 1988; De Block *et al.*, 1989; Moloney *et al.*, 1989). The virulence of octopine Ti plasmids in *B. napus* can be enhanced by treatment with acetosyringone or cocultivation of *B. napus* explants with tobacco leaf disks (Charest *et al.*, 1989).

Transgenic plants produced by *Agrobacterium rhizogenes*-mediated transformation have been generated in several laboratories (Ooms *et al.*, 1985; Guerche *et al.*, 1987a; Vartanian and Berkaloff, 1989); however, the agronomic value of these plants is questionable because of the severe changes in mor-

TABLE 6.4
Transformation of *Brassica* Species

Species	Cultivar/variety[a]	Method	Cells and tissues[b]		Selection[c]	Frequencies[d] (%)	References
			Source	Recovered			
B. napus	cv 'Westar,' 'R8494'	*A. tumefaciens*	h	pl	km,ppt	30–41	De Block *et al.* (1989)
	cv 'Westar'	*A. tumefaciens*	h	pl	km	0.4–2.5	Radke *et al.* (1988)
	cv 'Westar'	*A. tumefaciens*	ctp	pl	km	55	Moloney *et al.* (1989)
	cv 'Westar'	*A. tumefaciens*	s	pl	km	7	Fry *et al.* (1987)
	cv 'Westar'	*A. tumefaciens*	s	pl	mtx	11–14	Pua *et al.* (1987)
	cv 'Profit'	*A. tumefaciens*	s	pl	–	–	Miki *et al.* (1990)
	cv 'Westar'	*A. tumefaciens*	tcl	pl	km	2	Charest *et al.* (1988)
	cv 'Westar'	*A. tumefaciens*	tcl	pl	km	2	Misra (1990), Misra and Gedamu (1989)
	cv 'Topas'	*A. tumefaciens*	me	pl	km	0.5	Swanson and Erickson (1989)
	cv 'Brutor'	*A. rhizogenes*	s	pl	aut	0–90	Guerche *et al.* (1987a)
	cv 'Jet neuf,' 'Willi,' 'Rafal,' 'Brutor'	*A. rhizogenes*	s	pl	–	10–70	Ooms *et al.* (1985)
	cv 'Darmor'	*A. rhizogenes*	s	pl	–	–	Vartanian and Berkaloff (1989)
	–	Microinjection	me	pl	–	27–51	Neuhaus *et al.* (1987)
	cv 'Brutor'	Electroporation	pp	pl	pm	2 total	Guerche *et al.* (1987b)
B. campestris	var *oleifera*	*A. tumefaciens*	hpp	c	km	–	Ohlsson and Eriksson (1988)
	var *rapa*	Direct DNA uptake	lpp	c	km, G418	–	Paszkowski *et al.* (1986)
	var *rapa*	*A. rhizogenes*	r	r	–	–	Tanaka *et al.* (1985)
B. juncea	cv 'Varuna'	*A. tumefaciens*	s	sh	–	–	Mathews *et al.* (1985, 1986)
B. nigra	–	*A. tumefaciens*	lpp	c	hm	–	Sacristan *et al.* (1989)
	–	Direct DNA uptake	lpp	c	hm	–	Sacristan *et al.* (1989)

B. oleracea	var *botrytis*	*A. tumefaciens*	l	pl	–	–	Srivastava *et al.* (1988)
	var *botrytis*	*A. tumefaciens*	h	pl	km, ppt	8–30	De Block *et al.* (1989)
	var *botrytis*	*A. rhizogenes*	h	pl	–	–	David and Tempé (1988)
	var *acephala*	*A. rhizogenes*	l	pl	–	–	Hosoki *et al.* (1989)

[a]*B. napus* cultivars are all var *oleifera*.

[b]Abbreviations for cells and tissues are given in Table 6.1.

[c]Selective agents were kanamycin (km), phosphinotricin (ppt), methotrexate (mtx), paramomycin (pm), hygromycin (hm), or hormone autonomous growth (aut).

[d]Recovery of transgenic plants was estimated as a percentage of the number of explants inoculated.

phology and fertility. Typically, the leaves are wrinkled, the internodes are very short, the roots are plagiotropic, and flower development is severely reduced. These features are transmitted to viable progeny. The severity of the symptoms is species dependent; symptoms are much less severe in transgenic *B. oleracea* var. *botrytis* (David and Tempé, 1988).

Transgenic *B. oleracea* lines also have been generated efficiently with disarmed Ti plasmid vectors by inoculation of hypocotyl explants (De Block *et al.*, 1989) or leaf explants (Srivastava *et al.*, 1988). Other *Brassica* species including *B. campestris, B. juncea,* and *B. nigra* have been transformed readily by *Agrobacterium* but the efficient recovery of transgenic plants has not yet been reported (Table 6.4).

Direct gene transfer methods have been used to generate transgenic *B. napus* (Table 6.4). Electroporation has been employed to transform *B. napus* protoplasts (Guerche *et al.*, 1987b). Two plants were regenerated from 4×10^8 protoplasts. PEG-induced uptake of DNA was demonstrated with *B. campestris* protoplasts (Paszkowski *et al.*, 1986). Calluses resistant to kanamycin were recovered; however, transgenic plants were not. Liposome delivery of DNA also was attempted (Hussain *et al.*, 1985) but no proof of transformation was shown. Microinjection of *B. napus* microspore-derived embryoids at the 4- to 8-cell stage and the recovery of transgenic plants has been demonstrated (Neuhaus *et al.*, 1987). Regeneration through secondary embryogenesis was necessary because of the chimeric nature of the transformants. The transformation frequencies ranged from 27 to 51% and large numbers of plants were regenerated. Transmission of the introduced gene to the progeny was not demonstrated; therefore the practical value of this approach cannot be assessed.

6.3.2 FLAX

Small-scale field trials of transgenic flax have been initiated in Canada. The experimental approaches are similar to those for transgenic canola (Table 6.5). *Agrobacterium*-mediated transformation of hypocotyl explants has been successful with vectors derived from pTiT37 or pTiB6S3. Selection for resistance to kanamycin and glyphosate was achieved with chimeric genes encoding NPTII (Basiran *et al.*, 1987) and petunia 5-enolpyruvylshikimate-3-phosphate (EPSP) synthase (Jordan and McHughen, 1988a). The transformation efficiency varied from 2 to 10%. In separate experiments, resistance to the herbicide chlorsulfuron was achieved also (McHughen, 1989).

Research with transgenic flax revealed the potential for false positives in the recovery of transgenic lines and the need for targeting transformation to cells with the potential for plant regeneration (Jordan and McHughen, 1988b; McHughen *et al.*, 1989). As in canola, *A. rhizogenes*-mediated transformation produced highly abnormal transgenic plants (Table 6.5).

TABLE 6.5
Linum usitatissimum (Flax) Genetic Transformation

Method	Cells and tissues[a] Source	Recovered	Purpose	References
A. tumefaciens	ep	tm	T-DNA analysis	Hepburn *et al.* (1983)
	h	sh	Transformation	Basiran *et al.* (1987)
	h	sh	Transformation	McHughen and Jordan (1989)
	h	pl	Herbicide resistance	McHughen (1989)
	h	pl	Herbicide resistance	Jordan and McHughen (1988a)
A. rhizogenes	c	pl	Transformation	Zhan *et al.* (1988)

[a]Abbreviations for cells and tissues are given in Table 6.1.

6.3.3 SUNFLOWER

In many early studies, sunflower tumor calluses (Table 6.6) generated with *Agrobacterium tumefaciens* was used as a system for studying gene expression. Transformed calluses acted as plant tissue to assess the β-galactosidase gene as a marker (Helmer *et al.*, 1984), the neomycin phosphotransferase gene as a selectable marker (Nutter *et al.*, 1987), T-DNA promoters (Moyne *et al.*, 1989), and

TABLE 6.6
Helianthus annuus (Sunflower) Genetic Transformation

Method	Cells and tissues[a] Source	Recovered	Purpose	References
A. tumefaciens	s	tm	Zein expression	Goldsbrough *et al.* (1986)
	s	tm	Zein expression	Matzke *et al.* (1984)
	s	tm	Growth hormone expression	Barta *et al.* (1986)
	s	tm	*nptII* expression	Nutter *et al.* (1987)
	s	tm	T-DNA integration	Ursic (1985)
	h	tm	Phaseolin expression	Murai *et al.* (1983)
	h	tm	β-galactosidase expression	Helmer *et al.* (1984)
	h	tm	Heat shock expression	Schoffl and Baumann (1985)
	h	pl	Transformation	Everett *et al.* (1987)
Direct DNA uptake	hpp	c	Transformation	Moyne *et al.* (1989)

[a]Abbreviations for cells and tissues are given in Table 6.1.

T-DNA insertions (Ursic, 1985). Apparently, the regeneration of transgenic plants was not essential for the objectives of these studies. Despite the long history of plant transformation research with sunflower, reports of transgenic plant recovery are few. *Agrobacterium*-mediated transformation of hypocotyl explants was shown to yield stable transgenic plants (Everett *et al.*, 1987).

6.3.4 CARROT

As discussed earlier, carrot cell and tissue culture has been well studied as a system for somatic embryogenesis and plant development; however, progress toward the efficient generation of transgenic plants has been slow (Table 6.7).

TABLE 6.7
Daucus carota (Carrot) Genetic Transformation

Method	Cells and tissues[a]		Purpose	References
	Source	Recovered		
A. tumefaciens	h	tm	Hormone metabolism	Ishikawa *et al.* (1988a,b)
	r	tm	Bacterial attachment	Ryder *et al.* (1985)
	h	cs	Transformation	Thomas *et al.* (1989)
	c	c	Transformation	Wurtele and Burke (1989)
	cs	c	Transformation	Sen *et al.* (1986)
	cs	c	Transformation	Zhi *et al.* (1987)
	cs	pl	Transformation	Scott and Draper (1987)
A. rhizogenes	cs	cs	Bacterial attachment	Sykes and Matthysse (1988)
	cs	cs	Bacterial attachment	Mathysse (1987)
	cs	cs	Bacterial attachment	Matthysse and Gurlitz (1982)
	cs	c	Transformation	Sen *et al.* (1986)
	r	r	Herbicide metabolism	Mugnier (1988)
	r	r	T-DNA analysis	Byrne *et al.* (1983)
	r	r	T-DNA analysis	David *et al.* (1988)
	r	r	T-DNA analysis	Combard and Boucher (1988)
	r	r	T-DNA analysis	Petit *et al.* (1986)
	r	r	pRi analysis	Das *et al.* (1987)
	r	r	pRi analysis	Capone *et al.* (1989)
	r	r	pRi analysis	Estramareix *et al.* (1986)
	r	r	pRi analysis	Ryder *et al.* (1985)
	r	r	Transformation	Zhi *et al.* (1987)
	r	pl	Transformation	Tepfer (1984)
Electroporation	cspp	pp	Transient expression	Fromm *et al.* (1985)
	cspp	pp	Transient expression	Ballas *et al.* (1987)
	cspp	pp	Transient expression	Hauptmann *et al.* (1987b)
	cspp	pp	Transient expression	Boston *et al.* (1987)
	cspp	tr	Transformation	Langridge *et al.* (1985)

[a]Abbreviations for cells and tissues are given in Table 6.1.

Several papers have appeared in which carrot has been used to study *Agrobacterium* attachment, T-DNA structure and function, Ri and Ti plasmid gene functions, hormone metabolism, herbicide metabolism, and transient gene expression (Table 6.7). The development of microinjection technology for proembryogenic carrot cells has been demonstrated also (Nomura and Komamine, 1986). At this time, however, only preliminary reports of transgenic carrot plants have appeared. *Agrobacterium tumefaciens*-mediated transformation of proembryogenic cell-suspension cultures yielded transgenic plants (Scott and Draper, 1987). *Agrobacterium rhizogenes*-mediated transformation of root disks yielded transgenic plants with the morphological abnormalities associated with the Ri plasmid T-DNA genes (Tepfer, 1984).

6.3.5 OTHER SPECIES

Reports of transgenic plants among diverse vegetable species are appearing rapidly (Table 6.8). Generally, these are technical reports; evaluations of the agronomic properties of the plants have not been completed. A number of factors have contributed to these successes, including the broad host range of *Agrobacterium tumefaciens,* the availability of several vector systems, and the relatively modest tissue culture requirements for regenerable explants that may be subjected to cocultivation with *A. tumefaciens* strains. Transgenic plants have been demonstrated with lettuce, celery, cucumber, muskmelon, and strawberry (Table 6.8). For cucumber, *A. rhizogenes*-mediated transformation yielded morphologically normal plants (Trulson *et al.*, 1986). Cytological analysis of transgenic celery revealed a high frequency of chromosomal aberrations (Catlin *et al.*, 1988). Clearly, detailed genetic and physiological analysis under laboratory and field conditions is required to assess these transgenic plants fully.

6.4 Conclusions

Many transgenic oilseeds and vegetables have been generated. This large number reflects the creativity of scientists in harnessing the genetic mechanisms of gene transfer, integration, and expression revealed in other species and combining them with plant regeneration strategies suited to the species and genotype of interest. The underlying principles involved in this complex process are barely understood; therefore, a great deal of research still is required to achieve the efficiencies and precision that will be required to meet the long-term objectives of breeding with transgenic plants. The recovery of transgenic plants is only the first step in this process. For many of the species described, the consequences to the plant of genetic transformation and regeneration in culture have not been examined in the field or laboratory. Frequently, plants with obvious morphological abnormalities or infertility are discarded, yet these may harbor the

TABLE 6.8
Transformation of Selected Vegetables

Species	Common name	Method	Cells and tissues[a]		Selection[b]	Frequencies[c] (%)	References
			Source	Recovered			
Armoracia lupathifolia	Horseradish	*A. rhizogenes*	l	pl	aut	NA	Noda *et al.* (1987)
Apium graveolens	Celery	*A. tumefaciens*	p	pl	km	20 total[d]	Catlin *et al.* (1988)
Cucumis sativus	Cucumber	*A. rhizogenes*	h	pl	km	22 total[e]	Trulson *et al.* (1986)
Cucumis melo	Muskmelon	*A. tumefaciens*	c	pl	km	3–7	Fang and Grumet (1990)
Fragaria × *anassa*	Strawberry	*A. tumefaciens*	l	pl	km	0.6	Nehra *et al.* (1990)
		A. tumefaciens	l,p	pl	km	0.1–19.9	James *et al.* (1990)
Lactuca sativa	Lettuce	*A. tumefaciens*	c	pl	km	6.5	Michelmore *et al.* (1987)
		A. tumefaciens	c	pl	km	4.3–16.8	Enomoto *et al.* (1990)
		Electroporation	lpp	pl	km	0.1	Chupeau *et al.* (1989)
Rubus species	Raspberry	*A. tumefaciens*	l,s	pl	km	4	Graham *et al.* (1990)

[a]Abbreviations for cells and tissues are given in Table 6.1.
[b]Selection for resistance to kanamycin (km) was encoded in chimeric *nptII* genes; hormone autonomous growth (aut) also was used for selection.
[c]Recovery of transgenic plants was estimated as a percentage of the numbers of the explants or protoplasts inoculated; NA, data not available.
[d]Total of 20 transgenic plants were recovered.
[e]Total of 22 transgenic plants were recovered from 691 root sections.

information needed to understand the factors that govern genetic transformation and the preferred chromosomal sites of insertion. Among the oilseeds and vegetables, it will be interesting to follow the data that emerge from analysis of *Brassica napus* canola cultivars for which testing has progressed the farthest.

References

Anderson, S. B. (1984). *In* "Proceedings of the First Nordic Symposium on Cell and Tissue Culture," p. 36. Frostavallen, Sweden.

Arnison, P. G., Donaldson, P., Jackson, A., Semple, C., and Keller, W. (1990). *Plant Cell Tissue Organ Cult.* **20,** 217–222.

Ballas, N., Zakai, N., and Loyter, A. (1987). *Exp. Cell Res.* **170,** 228–234.

Barsby, T. L., Yarrow, S. A. and Shepard, J. F. (1986). *Plant Cell Rep.* **5,** 101–103.

Barta, A., Sommergruber, K., Thompson, D., Hartmuth, K., Matzke, M. A., and Matzke, A. J. M. (1986). *Plant Mol. Biol.* **6,** 347–357.

Basiran, N., Armitage, P., Scott, R. J., and Draper, J. (1987). *Pant Cell Rep.* **6,** 396–399.

Berry, S. F., Lu, D. Y., Pental, D., and Cocking, E. C. (1982). *Z. Pflanzenphysiol.* **108,** 31–38.

Bhat, S. R., Ford-Lloyd, B. V., and Callow, J. A. (1986). *J. Plant Physiol.* **124,** 419–424.

Biddington, N. L., and Robinson, H. T. (1990). *Plant Cell Tissue Organ Cult.* **22,** 48–54.

Binding, H. (1974). *Z. Pflanzenphysiol.* **74,** 327–356.

Boston, R. S., Becwar, M. R., Ryan, R. O., Goldsbrough, P. B., Larkins, B. A., and Hodges, T. K. (1987). *Plant Physiol.* **83,** 742–746.

Byrne, M. C., Kaplan, J., David, C., Tempé, J., and Chilton, M. D. (1983). *J. Mol. Appl. Genet.* **2,** 201–209.

Capone, I., Spano, L., Cardelli, M., Bellincampi, D., Petit, A., and Constantino, P. (1989). *Plant Mol. Biol.* **13,** 43–52.

Catlin, D., Ochoa, O., McCormick, S., and Quiros, C. F. (1988). *Plant Cell Rep.* **7,** 100–103.

Chandler, J. M., Hamill, A. S., and Thomas, A. G. (1984). "Crop Losses Due to Weeds in Canada and the United States." Weed Science Society of America, Champagne, Illinois.

Charest, P. J., Holbrook, L. A., Gabard, J., Iyer, V. N., and Miki, B. L. (1988). *Theor. Appl. Genet.* **75,** 438–445.

Charest, P. J., Iyer, V. N., and Miki, B. L. (1989). *Plant Cell Rep.* **8,** 303–306.

Chee, P. P. (1990). *In* "Progress in Plant Cellular and Molecular Biology" (H. J. J. Nijkamp, L. H. N. van der Plas, and J. van Aartrijk, eds), pp. 85–90. Kluwer, The Netherlands.

Chupeau, M. C., Bellini, C., Guerche, P., Maisonneuve, B., Vastra, G., and Chupeau, Y. (1989). *Bio/Technology* **7,** 503–508.

Combard, A., and Boucher, M. F. (1988). *Plant Mol. Biol.* **10,** 499–509.

Das, S., Bhaumik, C., and Sen, S. K. (1987). *Ind. J. Exp. Biol.* **25,** 371–374.

Datta, S. K., Peterhans, A., Datta, K., and Potrykus, I. (1990). *Bio/Technology* **8,** 736–740.

David, C., and Tempé, J. (1988). *Plant Cell Rep.* **7,** 88–91.

David, C., Petit, A., and Tempé, J. (1988). *Plant Cell Rep.* **7,** 92–95.

De Block, M., deBrouwer, D., and Tenning, P. (1989). *Plant Physiol.* **91,** 694–710.

Dudits, D. (1984). *In* "Cell Culture and Somatic Genetics" (I. K. Vasil, ed.), Vol. 1, pp. 391–397. Academic Press, New York.

Dunwell, J. M. (1986). *In* "Plant Tissue Culture and Its Agricultural Applications" (L. A. Withers and P. G. Alderson, eds.), pp. 375–404. Butterworths, Boston.

Durand, J., Potrykus, I., and Donn, G. (1973). *Z. Pflanzenphysiol.* **64,** 26–34.

Eapen, S., Abraham, V., Gerdemann, M., and Scheider, O. (1989). *Ann. Bot.* **63,** 369–372.

Engler, D. E., and Grogan, R. G. (1984). *Plant Sci. Lett.* **28,** 223–229.
Enomoto, S., Itoh, H., Ohshima, M., and Ohashi, Y. (1990). *Plant Cell Rep.* **9,** 6–9.
Estramareix, C., Ratet, P., Boulanger, F., and Richard F. (1986). *Plasmid* **15,** 245–247.
Everett, N. P., Robinson, K. E. P., and Mascarenhas, D. (1987). *Bio/Technology* **5,** 1201–1204.
Fan, Z., Armstrong, K., and Keller, W. A. (1988). *Protoplasma* **147,** 191–199.
Fang, G., and Grumet, R. (1990). *Plant Cell Rep.* **9,** 160–164.
Fromm, N., Taylor, L. P., and Walbot, V. (1985). *Proc. Natl. Acad. Sci. U.S.A.* **82,** 5824–5828.
Fry, J., Barnason, A., and Horsch, R. B. (1987). *Plant Cell Rep.* **6,** 321–325.
Fu, Y. Y., Jia, S.-R., and Lin, Y. (1985). *Theor. Appl. Genet.* **71,** 495–499.
Gamborg, O. L., Miller, R. A., and Ojima, K. C. (1968). *Exp. Cell Res.* **50,** 151–158.
Gatenby, A. A., and Cocking, E. C. (1977). *Plant Sci. Lett.* **8,** 275–280.
Glimelius, K. (1984). *Physiol. Plant.* **61,** 38–44.
Goldsbrough, P. B., Gelvin, S. B., and Larkins, B. A. (1986). *Mol. Gen. Genet.* **202,** 374–381.
Graham, J., McNicol, R. J., and Kumar, A. (1990). *Plant Cell Tissue Organ Cult.* **20,** 35–39.
Greco, B., Tanzarella, O. A., Carrozzo, G., and Blanco, A. (1984). *Plant Sci. Lett.* **36,** 73–77.
Guerche, P., Jouanin, L., Tepfer, D., and Pelletier, G. (1987a). *Mol. Gen. Genet.* **206,** 382–386.
Guerche, P., Charbonnier, M., Jouanin, L., Tourneur, C., Paszkowski, J., and Pelletier, G. (1987b). *Plant Sci.* **52,** 111–116.
Guerche, P., De Almeida, E. R. P., Schwarztein, M. A., Gander, E., Krebber, E., and Pelletier, G. (1990). *Mol. Gen. Genet.* **221,** 306–314.
Guo, B. J., and Scheider, O. (1983). *Z. Pflanzenphysiol.* **110,** 375–377.
Hagberg, A., and Hagberg, G. (1980). *Hereditas* **93,** 341–343.
Hauptmann, R. M., della-Cioppa, G., Smith, A. G., Kishore, G. M., and Widholm, J. M. (1987a). *Mol. Gen. Genet.* **206,** 357–363.
Hauptmann, R. M., Ozias-Akins, P., Vasil, V., Tabaeizadeh, Z., Rogers, S. G., Horsch, R. B., Vasil, I. K., and Fraley, R. T. (1987b). *Plant Cell Rep.* **6,** 265–270.
Helmer, G., Casadaban, M., Bevan, M., Kayes, L., and Chilton, M. D. (1984). *Bio/Technology* **2,** 520–527.
Hepburn, A. G., Clarke, L. E., Blundy, K. S., and White, J. (1983). *J. Mol. Appl. Genet.* **2,** 211–224.
Holbrook, L. A., and Miki, B. L. (1985). *Plant Cell Rep.* **4,** 329–332.
Hosoki, T., Shiraishi, K., Kigo, T., and Ando, M. (1989). *Scientia Hort.* **40,** 259–266.
Hu, H., and Huang, B. (1987). *Int. Rev. Cytol.* **107,** 293–313.
Hussain, M. M., Melcher, U., and Essenberg, R. C. (1985). *Plant Cell Rep.* **4,** 58–62.
Ishikawa, K., Kamada, H., and Harada, H. (1988a). *Plant Cell Physiol.* **29,** 461–466.
Ishikawa, K., Kamada, H., and Harada, H. (1988b). *Plant Cell Physiol.* **29,** 1171–1175.
Jain, R. K., Brune, U., and Fredt, W. (1989). *Euphytica* **43,** 153–163.
James, D. T., Passey, A. J., and Barbara, D. J. (1990). *Plant Sci.* **69,** 79–94.
Jia, S. R., Fu, Y. Y., and Lin, Y. (1986). *J. Plant Physiol.* **124,** 393–398.
Jordan, M. C., and McHughen, A. (1988a). *Plant Cell Rep.* **7,** 281–284.
Jordan, M. C., and McHughen, A. (1988b). *Plant Cell Rep.* **7,** 285–287.
Jounanin, L., Guerche, P., Pamboukdjian, N., Tourneur, C., Casse-Delbart, F., and Tourneur, J. (1987). *Mol. Gen. Genet.* **206,** 387–392.
Kao, K. N. (1977). *Mol. Gen. Genet.* **150,** 225–230.
Kao, K. N., and Kasha, K. J. (1969). *Barley Genetics* **2,** 82–87.
Kao, K. N., and Michayluk, M. R. (1975). *Planta* **126,** 105–110.
Kartha, K. K., Gamborg, O. L., and Constabel, F. (1974). *Physiol. Planta* **31,** 217–220.
Keller, W. A., Rajhathy, T., and Lacapra, J. (1975). *Can. J. Genet. Cytol.* **17,** 655–666.
Kirti, P. B., and Chopra, V. L. (1989). *Plant Breeding* **102,** 73–78.
Klimaszewska, K., and Keller, W. A. (1985). *Plant Cell Tissue Organ Cult.* **4,** 183–197.

Klimaszewska, K., and Keller, W. A. (1987). *Plant Cell Tissue Organ Cult.* **8,** 225–233.
Knauf, V. C. (1987). *Tibtech* **Feb,** 40–47.
Krens, F. A., Zijlstra, C., Molen, W. V. D., Jamar, D., and Huizing, H. J. (1988). *Euphytica* **5,** 185–194.
Langridge, W. H. R., Li, B. J., and Szalay, A. A. (1985). *Plant Cell Rep.* **4,** 355–359.
Lazzeri, P. A., and Dunwell, J. M. (1986a). *Ann. Bot.* **58,** 689–697.
Lazzeri, P. A., and Dunwell, J. M. (1986b). *Ann. Bot.* **58,** 696–710.
Li, L., and Kohlenbach, H. W. (1982). *Plant Cell Rep.* **1,** 209–211.
Lichter, R. (1981). *Z. Pflanzenphysiol.* **103,** 229–237.
Lichter, R. (1982). *Z. Phlanzenphysiol.* **105,** 427–434.
Lichter, R. (1989). *Plant Breeding* **103,** 119–123.
Lillo, C. and Shahin, E. A. (1986). *HortScience* **21,** 315–317.
Lindsay, K., and Jones, M. G. K. (1989). *Plant Cell Rep.* **8,** 71–74.
Lindsay, K., and Jones, M. G. K. (1990). *Physiol. Plant* **79,** 168–172.
Linsmaier, E. M., and Skoog, F. (1965). *Physiol. Plant* **18,** 100–127.
Mathews, V. H., Bhatia, C. R., Mitra, R., Krishna, T. G., and Rao, R. S. (1985). *Plant Science* **39,** 49–54.
Mathews, V. H., Rao, P. S., and Bhatia, C. R. (1986). *J. Genet.* **65,** 37–44.
Matthysse, A. G. (1987). *J. Bacteriol.* **169,** 313–323.
Matthysse, A. G., and Gurlitz, R. H. G. (1982). *Physiol. Plant Pathol.* **21,** 381–387.
Matzke, M. A., Susani, M., Binns, A. N., Lewis, E. D., Rubenstein, I., and Matzke, A. J. M. (1984). *EMBO J.* **3,** 1525–1531.
McHughen, A. (1989). *Plant Cell Rep.* **8,** 445–449.
McHughen, A., and Jordan, M. C. (1989). *Plant Cell Rep.* **7,** 611–614.
McHughen, A., Jordan, M., and Feist, G. (1989). *J. Plant Physiol.* **135,** 245–248.
Michelmore, R., Marsh, E., Seely, S., and Landry, B. (1987). *Plant Cell Rep.* **6,** 439–442.
Miki, B. L., Labbé, H., Hattori, J., Ouellet, T., Gabard, J., Sunohara, G., Charest, P. J., and Iyer, V. N. (1990). *Theor. Appl. Genet.* **80,** 449–458.
Misra, S. (1990). *J. Exp. Bot.* **41,** 269–275.
Misra, S., and Gedamu, L. (1989). *Theor. Appl. Genet.* **78,** 161–168.
Miura, Y., and Tabata, M. (1986). *Plant Cell Rep.* **5,** 310–313.
Moloney, M. M., Walker, J. M., and Sharma, K. K. (1989). *Plant Cell Rep.* **8,** 238–242.
Moyne, A. L., Tagu, D., Thor, V., Bergouniaux, C., Freyssinet, G., and Gadal, P. (1989). *Plant Cell Rep.* **8,** 97–100.
Mugnier, J. (1988). *J. Exp. Bot.* **39,** 1045–1056.
Murai, N., Sutton, P. W., Murray, M. G., Slightom, J. L., Merlo, D. J., Reichert, N. A., Sengupta-Gopalan, C., Stock, C. A., Barker, R. F., Kemp, J. D., and Hall, T. C. (1983). *Science* **222,** 477–481.
Murashige, T., and Skoog, F. (1962). *Physiol. Plant* **15,** 473–497.
Murata, M., and Orton, T. J. (1987). *Plant Cell Tissue Organ Cult.* **11,** 111–123.
Nagata, T., and Takeba, I. (1971). *Planta* **99,** 12–20.
Nakagawa, N., Tanaka, H., Oba, T., Ogura, N., and Izuka, M. (1985). *Plant Cell Rep.* **4,** 148–150.
Nehra, N. S., Chibbar, R. N., Kartha, K. K., Datla, R. S. S., Crosby, W. L., and Stushnoff, C. (1990). *Plant Cell Rep.* **9,** 10–13.
Neuhaus, G., and Spangenberg, G. (1990). *Physiol. Planta* **78,** 213–217.
Neuhaus, G., Spangenberg, G., Mittlesten Scheid, O., and Schweiger, H. G. (1987). *Theor. Appl. Genet.* **75,** 30–36.
Nitsch, J. P., and Nitsch, C. (1969). *Science* **163,** 85–87.
Noda, T., Tanaka, N., Mano, Y., Nabeshima, S., Ohkawa, H., and Matsui, C. (1987). *Plant Cell Rep.* **6,** 283–286.

Nomura, K., and Komamine, A. (1986). *Plant Sci.* **44,** 53–58.
Nutter, R., Everett, N., Pierce, D., Panganiban, L., Okubara, P., Lachmansingh, R., Mascarenhas, D., Welch, H., Mettler, I., Pomeroy, L., Johnson, J., and Howard, J. (1987). *Plant Physiol.* **84,** 1185–1192.
Ohlsson, M., and Ericksson, T. (1988). *Hereditas* **108,** 173–177.
Ooms, G., Bains, A., Burrell, M., Karp, A., Twell, D., and Wilcox, E. (1985). *Theor. Appl. Genet.* **71,** 325–329.
Paszkowski, T., Pisan, B., Shillito, R. D., Hohn, T., Hohn, B., and Potrykus, I. (1986). *Plant Mol. Biol.* **6,** 303–312.
Paterson, K. E., and Everett, N. P. (1985). *Plant Sci.* **42,** 125–132.
Pechan, P. M., and Keller, W. A. (1988). *Physiol. Plant* **74,** 377–384.
Petit, A., Berkaloff, A., and Tempé, J. (1986). *Mol. Gen. Genet.* **202,** 388–393.
Phippen, C., and Ockendon, D. J. (1990). *Theor. Appl. Genet.* **79,** 33–38.
Pua, E. C. (1987). *Plant Sci.* **50,** 153–160.
Pua, E. C., Mehra-Palta, A., Nagy, F., and Chua, N.-H. (1987). *Bio/Technology* **5,** 815–817.
Punja, Z. K., Tang, F. A., and Sarmento, G. G. (1990). *Plant Cell Rep.* **9,** 61–64.
Radke, S. E., Andrews, B. M., Moloney, M. M., Crouch, M. L., Kridl, J. C., and Knauf, V. C. (1988). *Theor. Appl. Genet.* **75,** 685–694.
Robertson, D., and Earle, E. D. (1986). *Plant Cell Rep.* **5,** 61–64.
Robertson, D., Earle, E. D., and Mutshler, M. A. (1988). *Plant Cell Tissue Organ Cult.* **14,** 15–24.
Ryder, M. H., Tate, M. E., and Kerr, A. (1985). *Plant Physiol.* **77,** 215–221.
Sacristan, M. D., Gerdemann-Knorck, M., and Scheider, O. (1989). *Theor. Appl. Genet.* **78,** 194–200.
Sanford, J. C. (1990). *Physiol. Plant* **79,** 206–209.
Sato, T., Nishio, T., and Hirai, M. (1989). *Plant Cell Rep.* **8,** 486–488.
Schoffl, F., and Baumann, G. (1985). *EMBO J.* **4,** 1119–1124.
Schenk, R. U., and Hildebrandt, A. C. (1972). *Can. J. Bot.* **50,** 199–204.
Scott, R. J., and Draper, J. (1987). *Plant Mol. Biol.* **8,** 265–274.
Sen, P., Chatterjee, G., Kumar, P. M., and Sen, S. K. (1986). *Ind. J. Exp. Biol.* **24,** 153–155.
Shahin, E. A. (1985). *Theor. Appl. Genet.* **69,** 235–240.
Sharma, K. K., and Bhojwani, S. S. (1989). *Biol. Plant* **31,** 276–279.
Sharma, K. K., and Thorpe, T. A. (1989) *Plant Cell Tissue Organ Cult.* **18,** 129–141.
Srivastava, V., Reddy, A. S., and Guha-Mukherjee, S. (1988). *Plant Cell Rep.* **7,** 504–507.
Stringham, G. R. (1977). *Z. Pflanzenphysiol.* **92,** 459–462.
Swanson, E. B., and Erickson, L. R. (1989). *Theor. Appl. Genet.* **78,** 831–835.
Sykes, L., and Matthysse, A. G. (1988). *Phytopathol.* **78,** 1322–1326.
Tanaka, N., Hayakawa, M., Mano, Y., Ohkawa, H., and Matsui, C. (1985). *Plant Cell Rep.* **4,** 74–77.
Tepfer, D. (1984). *Cell* **37,** 959–967.
Thomas, E., Hoffman, F., Potrykus, I., and Wenzel, G. (1976). *Mol. Gen. Genet.* **145,** 245–247.
Thomas, J. C., Guiltinan, M. J., Bustos, M. J., Thomas, T., and Nessler, C. (1989). *Plant Cell Rep.* **8,** 354–357.
Trulson, A. J., Simpson, R. B., and Shahin, E. A. (1986). *Theor. Appl. Genet.* **73,** 11–15.
Uchimiya, H., and Murashige, T. (1974). *Plant Physiol.* **56,** 936–944.
Ursic, D. (1985). *Biochem. Biophys. Res. Commun.* **131,** 152–159.
Vartanian, N., and Berkaloff, A. (1989). *Plant Cell Environ.* **12,** 197–204.
Vatsya, B., and Bhaskaran, S. (1982). *Protoplasma* **113,** 161–163.
Weber, G., Monajembashi, S., Greulich, K. O., and Wolfrum, J. (1988). *Plant Cell Tissue Organ Cult.* **12,** 219–222.
Weber, G., Monajembashi, S., Wolfrum, J., and Greulich, K. O. (1989a). *Ber. Bunsenges. Phys. Chem.* **93,** 252–254.

Weber, G., Monajembashi, S., Greulich, K. O., and Wolfrum, J. (1989b). *Eur. J. Cell Biol.* **49,** 73–79.

Weber, G., Monajembashi, S., Greulich, K. O., and Wolfrum, J. (1990). *Physiol. Plant* **79,** 190–193.

WhiteMan Runs Him, K., Scheffer, R. J., and Strobel, G. A. (1988). *Can. J. Bot.* **66,** 642–644.

Williams, J., Pink, D. A. C., and Beddington, N. L. (1990). *Plant Cell Tissue Organ Cult.* **21,** 61–66.

Wong, K. W., and Loh, C. S. (1987). *Plant Cell Tissue Organ Cult.* **10,** 143–148.

Wurtele, E. S., and Burke, K. (1989). *Plant Sci.* **61,** 253–262.

Xu, Z. H., Davey, M. R., and Cocking, E. C. (1982). *Plant Science Lett.* **24,** 117–121.

Yarrow, S. A., Wu, S. C., Barsby, T. L., Kemble, R. J., and Shepard, J. F. (1986). *Plant Cell Rep.* **5,** 415–418.

Zapata, F. J., Sink, K. C., and Cocking, E. C. (1981). *Plant Sci. Lett.* **23,** 41–46.

Zhan, X., Jones, D. A., and Kerr, A. (1988). *Plant Mol. Biol.* **11,** 551–559.

Zhi, L., Mengkang, F., and Cheng, C. (1987). *Scientia Sinica (Ser. B)* **30,** 1145–1151.

7

Transgenic Woody Plants

Abhaya M. Dandekar and Gale H. McGranahan

Department of Pomology
University of California
Davis, California 95616

David J. James

Horticulture Research International
East Malling, Maidstone,
Kent, ME19 6BJ United Kingdom

7.1 Woody Plants: Sources of Germ Plasm

The term woody plants is used here to describe a range of crop species from perennial shrubs and vines to trees. These plants are an integral part of modern society because they provide the fruit, nuts, flowers, wood, and paper products that we have grown to depend on. The study of horticulture and forestry has had a significant impact on the domestication of a wide and seemingly endless variety of woody plants, creating important industries and a demand for their products. The 21st century brings with it new challenges, since the increasing demand for horticultural and forestry products must be met using production practices that will sustain yield and quality but not damage the environment or the food chains vital to human health. The technology to produce transgenic plants will have an important and powerful impact not only on some of the immediate problems, such as disease and pest resistance, but also on future problems, if this technology is integrated with conventional technologies of plant breeding, plant propagation, production physiology, integrated pest management, and postharvest technology.

The genetic modification of most woody crops by traditional breeding methods is both space and time consuming. In traditional breeding, entire sets of genes as chromosomes are introduced through the sexual process and hybridized among sexually compatible species (Fig. 7.1). This process has been and will be important for the identification and use of polygenic variation, or major genes, from related woody species. However, genetic transformation certainly will en-

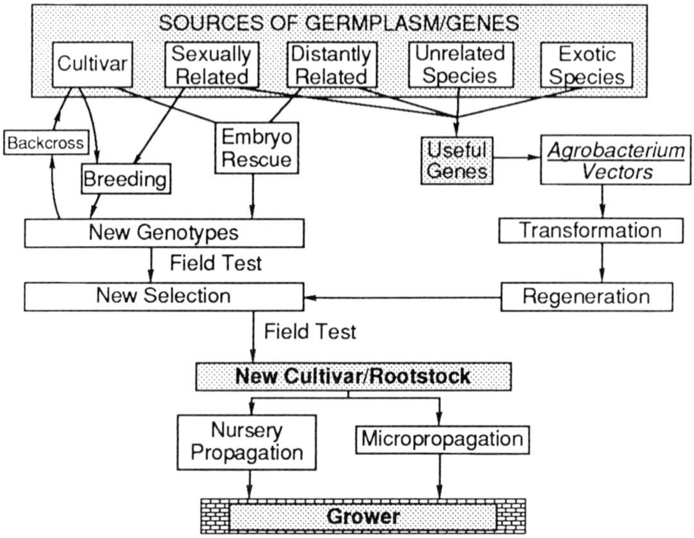

Figure 7.1 Sources of germ plasm for genetic modification of woody plants.

hance ongoing breeding efforts when small specific changes need to be made in the genome through the addition of one or a few genes. This goal is achieved currently by repeated backcrossing of the F_1 generation to the recipient species, accompanied by progeny selection for the trait of interest.

The procedure of repeated backcrossing takes several decades in woody tree crops and results in the "dilution" of the undesirable donor genes and the retention of the desirable gene. This technique, also referred to as introgression, is therefore, of necessity, limited by the generation time of the crop in question. Another limitation of this process is the phenomenon called linkage drag, in which selection for a certain allele results in the simultaneous selection for others linked to it. Sometimes these linked alleles may be detrimental to the market value of the plant.

With generation times of years rather than months, woody species are good examples of plant species for which an introgression approach to transfer a single trait becomes impractical, especially if juvenile periods are extremely long. For this reason, production of new woody cultivars has resulted mostly from the selection of chance seedlings, hybrids, and clonal variants that showed some desirable characteristic. Such processes are somewhat random; when a more predictable change in a cultivar is desired, genetic transformation becomes indispensable. A gene of interest can be inserted, by one of several techniques, into the genome of a single cell. The single cell can then be cultured to ultimately develop into a transgenic plant, all cells of which carry the gene. This strategy is not limited by generation time but rather by regeneration time (the time it takes to produce a transgenic plant from the single cell). For most plants, regeneration time is less than generation time, representing a significant saving in time, if one considers the number of generations required for introgression of a single trait. Moreover, the phenomenon of linkage drag does not occur in transgenic plants. Transgenic woody cultivars have potential for commercial success and grower acceptance because the overall genome of the cultivar has not been disturbed, as it would be in a traditional sexual crossing program.

As shown in Fig. 7.1, transformation, at least in theory, allows the introduction of any gene from any source, greatly enhancing the germ plasm that can be used for a particular woody plant species. The major limitation, apart from gene availability, is the development of transformation or gene transfer technologies for the production of transgenic woody plants. In this chapter, the following areas will be discussed: (1) the nature of the techniques that have been used successfully to transfer foreign genes into woody plant tissues, (2) the techniques that have resulted in the isolation of transgenic woody plants, and (3) limitations and scope of achieving transformation in woody plants and some of its impact on the environment.

The subject of plant transformation and its potential applications has been reviewed intensively (Gheysen *et al.*, 1985; Fraley *et al.*, 1986; Lichtenstein and Fuller, 1987; Zambryski *et al.*, 1989; Gasser and Fraley, 1989; Schuerman and Dandekar, 1990; Dandekar, 1992). Most of the successful gene transfer experiments have resulted from *Agrobacterium*-mediated gene transfer technologies that were developed previously and used in annual crops such as tobacco and tomato.

7.2 Transformation of Woody Plants

7.2.1 *AGROBACTERIUM*-MEDIATED TRANSFORMATION: CENTRAL QUESTIONS

The ability to transform tissue and to regenerate transgenic woody plants successfully using the natural gene transfer system of *Agrobacterium* is dependent on three factors: (1) the virulence of *Agrobacterium* cells that permits the transformation of a majority of cells in the target tissue, (2) the efficiency of selection that would allow growth of all the transformed cells, and (3) the frequency of regeneration that occurs among the transformed cell population. Shown in Fig. 7.2 is a representation of the problem and potential solutions. The circles represent the subset of the total population of cells in an explant in any given *Agrobacterium*-mediated transformation experiment. There are populations of cells that are transformed and there are populations of cells that are not transformed but

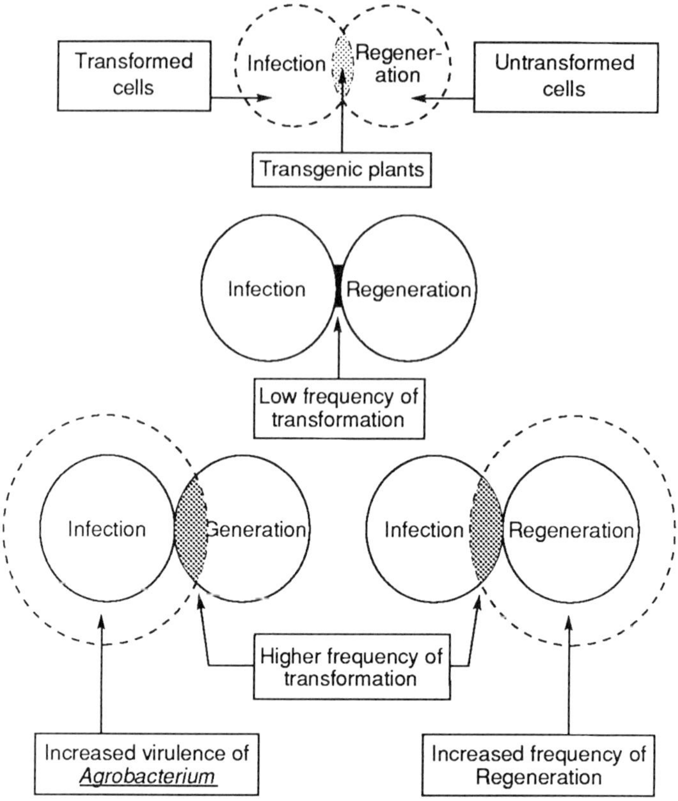

Figure 7.2 Key factors influencing transformation and, ultimately, regeneration of transgenic plants.

regenerate (Fig. 7.2). Transgenic plants only can be regenerated from the population of cells that are transformed and also regenerate, represented by shaded area of overlap in Fig. 7.2. In most woody plants, this area is small because of low transformation frequencies and low rates of regeneration. Therefore, higher transformation frequencies can be expected if the frequency of infection is improved through the use of more virulent *Agrobacterium* or, alternatively, if regeneration rates increased (Fig. 7.2). To accomplish this goal, the following questions need to be answered through further research.

What is the nature of *Agrobacterium*–woody plant interaction? (1) Are there quantitative differences among *Agrobacterium* strains in their ability to transform individual cells in the target explant? (2) Can *Agrobacterium* transfer DNA efficiently into the cells that have the capacity to regenerate? What is the nature of regeneration events and their relationship to woody plant transformation? (1)

Can we select transformed cells within infected tissue and regenerate transgenic plants? (2) Do the observable regeneration events in the target tissue explant have a single cell origin?

7.2.2 NATURE OF *AGROBACTERIUM*–WOODY PLANT INTERACTION

A wide variety of perennial crops have been shown to be susceptible to both *Agrobacterium tumefaciens* and *A. rhizogenes* (De Cleene and De Ley, 1976, 1981). Strains obtained from natural infections in the field can be divided into three biotypes (group 1, 2, and 3) based on different metabolic and physiological traits (Keane *et al.*, 1970; Kerr and Panagopoulos, 1977). These bacteria harbor the tumor-inducing Ti plasmid or root-inducing Ri plasmid, respectively, causing the formation of tumors or the proliferation of roots at the site of infection (Nester *et al.*, 1984; Tempé and Casse-Delbart, 1989). The host range of *A. tumefaciens* is quite extensive, including most dicotyledonous plant species (De Cleene and De Ley, 1976). However, extensive differences in susceptibility have been shown when different strains of *Agrobacterium* were used to infect a single host plant species (Anderson and Moore, 1979). Certainly some of these differences can be accounted for by the different biotypes of *Agrobacterium,* some of which have a narrow host range. For example, group 3 infections are restricted to grapevine. Interestingly, almost all the current *Agrobacterium* vectors have been derived from group 1 bacteria. Experiments using different virulent strains of group 1 *Agrobacterium* and seedlings of various woody plants (almond, apple, apricot, cherry, citrus, Douglas-fir, grape, peach, and walnut) have shown wide ranges and variability in susceptibility to group 1 *Agrobacterium* (Martin, 1987). Walnut, grape, and apricot are the most susceptible in this study and Douglas-fir and citrus are the least, the rest being intermediate in response (Martin, 1987). Part of this variability could be explained by the variability of the genetic background of the seedling plant material used in the experiments.

Micropropagated plants are an excellent source material for transformation since they are genetically identical. Therefore, it is possible to perform a number of experiments using woody plants under controlled conditions in the laboratory. In recent years, several wild-type and recombinant strains of *Agrobacterium* have been evaluated for their ability to infect different woody plant species. The transformed cells have been evaluated for the expression of genes present on the T-DNA as well as those engineered into T-DNA (Table 7.1). In most instances (Table 7.1) the genes involved in the synthesis of opines, that is, nopaline, octopine, and agropine synthase (dehydrogenase) have been used because of the availability of sensitive assays for their detection (Otten and Schilperoort, 1978; Petit *et al.*, 1983). Also, opines are made only in transformed tissues; therefore, their presence is good evidence that transformation has occurred.

TABLE 7.1
Foreign Gene Expression in Tissue or Cells of Perennial Tree Crops

Plant	Method of gene transfer	Foreign gene[a]	Reference
Almond	*Agrobacterium*-mediated	*nos,ocs*	Martin (1987)
Alnus	*Agrobacterium*-mediated	*kan*	Mackay *et al.* (1988)
Apple	*Agrobacterium*-mediated	*gus*	Dandekar *et al.* (1990)
	Agrobacterium-mediated	*gus,agr*	Martin *et al.* (1990)
Apricot	*Agrobacterium*-mediated	*nos,ocs*	Martin (1987)
Birch	*Agrobacterium*-mediated	*kan*	Mackay *et al.* (1988)
Cherry	*Agrobacterium*-mediated	*nos*	Martin (1987)
Citrus	Free DNA uptake in protoplasts	*kan*	Kobayashi and Uchimiya (1989)
Douglas-fir	*Agrobacterium*-mediated	*kan,ocs*	Dandekar *et al.* (1987)
	Electroporation	*luc*	Gupta *et al.* (1988)
Grapevine	*Agrobacterium*-mediated	*kan*	Baribault *et al.* (1989)
		gus	Baribault *et al.* (1990)
	Agrobacterium-mediated	*gus*	Colby *et al.* (1991)
	Agrobacterium-mediated	*agr*	Guellec *et al.* (1990a)
	Agrobacterium-mediated	*gcmvCp*	Guellec *et al.* (1990b)
	Agrobacterium-mediated	*ocs,nos,kan*	Martin *et al.* (1989)
Larch	*Agrobacterium*-mediated	*kan*	Karnosky *et al.* (1988)
Loblolly pine	Electroporation	*luc*	Gupta *et al.* (1988)
Mango	*Agrobacterium*-mediated	*kan*	Litz *et al.* (1990)
Papaya	*Agrobacterium*-mediated	*nos*	Pang and Sanford (1988)
Peach	*Agrobacterium*-mediated	*ocs*	Hammerschlag *et al.* (1989)
	Agrobacterium-mediated	*nos,ocs*	Martin (1987)
	Agrobacterium-mediated	*gus*	Martin *et al.* (1990)
	Agrobacterium-mediated	*kan*	Scorza *et al.* (1990)
Pecan	*Agrobacterium*-mediated	*ocs*	Martin (1987)
	Agrobacterium-mediated	*kan,gus*	G. H. McGranahan (unpublished observations)
Pine	*Agrobacterium*-mediated	*ocs,nos,agr*	Sederoff *et al.* (1986)
	Microprojectile bombardment	*gus*	Stomp *et al.* (1990)
Spruce	*Agrobacterium*-mediated	*nos,agr*	Ellis *et al.* (1989)
Strawberry	*Agrobacterium*-mediated	*ocs,nos,agr*	Uratsu *et al.* (1991)
Walnut	*Agrobacterium*-mediated	*ocs,nos,kan*	Dandekar *et al.* (1988)
Willow	*Agrobacterium*-mediated	*kan*	Vahala *et al.* (1989)

[a] *kan*, Kanamycin phosphotransferase; *ocs*, octopine synthase; *nos*, nopaline synthase; *agr*, agropine synthase; *gus*, β-glucuronidase; *gcmvCp*, coat protein for grapevine chrome mosaic virus; *luc*, firefly luciferase.

7.2.3 *AGROBACTERIUM* VIRULENCE AND TRANSFORMATION EFFICIENCY

Significant variation in virulence has been reported with *Agrobacterium* strains carrying different Ti plasmids in some plant species (Owens and Cress, 1985; Dandekar *et al.*, 1987,1988,1990; Martin, 1987; Ellis *et al.*, 1989; Uratsu *et al.*, 1991). The *vir* regulon of a Ti plasmid can be divided into six complementation groups, based on mutational and DNA sequence analysis, that have been designated *virA, B, C, D, E,* and *G* (Stachel and Nester, 1986; Rogowsky *et al.*, 1990). The induction of transcription of the individual genes within the *vir* regulon are mediated through a two-component regulatory system involving *virA* and *virG* in responses to plant signal molecules (Stachel and Zambryski, 1986; Winans *et al.*, 1986; Leroux *et al.*, 1987). Genetic differences exist in virulence among *Agrobacterium* strains carrying different Ti plasmids. For example, there are differences in the induction of virulence and the structure of the *virA* gene in limited host range *Agrobacterium* compared with wide host range strains (Yanofsky *et al.*, 1985; Yanofsky and Nester, 1986; Leroux *et al.*, 1987). The induction of virulence is also responsible for the increased virulence and host range of *A. tumefaciens* strains carrying the Ti plasmid pTiBo542 (Hood *et al.*, 1986; Jin *et al.*, 1987; Pythoud *et al.*, 1987; Dandekar *et al.*, 1990; Raineri *et al.*, 1990). The genetic basis for the observed increased virulence is the higher level of expression of *virG* in strains carrying pTiBo542 compared with strains carrying other plasmids, for example, pTiA6 (Jin *et al.*, 1987). This observation has been established further in experiments in which increased virulence has been obtained, for example, in tobacco, petunia, potato, tomato, soybean, carrot, kalanchoe (Jin *et al.*, 1987), apple (Dandekar *et al.*, 1990), poplar (Pythoud *et al.*, 1987), and rice (Raineri *et al.*, 1990), through the introduction of additional copies of *virG* in *Agrobacterium* strains carrying pTiBo542.

Apart from the genetic factors, there are physiological and biochemical factors that influence virulence gene expression. The most notable is the induction mediated by plant signal molecules, that is, plant phenolic compounds (Bolton *et al.*, 1986) such as acetosyringone and α-hydroxyacetosyringone. Other factors have been reported to influence the induction of virulence genes, for example, sucrose, pH, temperature (Alt-Mörbe *et al.*, 1988,1989), opines (Veluthambi *et al.*, 1989), sugars (Ankenbauer and Nester, 1990; Cangelosi *et al.*, 1990), the osmoprotectant compound glycine betaine, and Ca^{2+} ions (Vernade *et al.*, 1988). These factors could be pertinent to the growth of *Agrobacterium* strains prior to cocultivation with target plant material in gene transfer experiments.

7.2.4 DIRECT DNA TRANSFER

In addition to *Agrobacterium,* the most promising delivery systems for transformation of woody plants are electroporation (Fromm *et al.*, 1985,1986), free

DNA uptake (Paszkowski *et al.*, 1984; Hain *et al.*, 1985), and DNA-coated microprojectiles (Klein *et al.*, 1987; McCabe *et al.*, 1988). For electroporation to be successful, one of the major requirements is a functional protoplast-to-plant regeneration system. With the few exceptions that have been published so far, this approach would certainly be difficult. However, mesophyll protoplasts have been isolated from a number of different woody plants (Revilla *et al.*, 1987). Regeneration of such protoplasts is not always possible. Examples of woody plant genera in which regeneration from protoplasts has been successful include *Citrus* (Vardi *et al.*, 1982), *Prunus* (Ochatt *et al.*, 1987; Ochatt and Power, 1988), *Malus* (Patat-Ochatt *et al.*, 1988; Wallin and Johansson, 1989), *Pyrus* (Ochatt and Caso, 1986), *Broussonetia* (Oka and Ohyama, 1985), *Populus* (Russel and McCown, 1986,1988), Douglas-fir, and loblolly pine (Gupta *et al.*, 1988). Transient gene expression of the firefly luciferase gene was reported after electroporation of protoplasts of Douglas-fir and loblolly pine (Gupta *et al.*, 1988). Electroporation could be an efficient means of introducing DNA into plant cells, especially in those species not susceptible to *Agrobacterium* infection. One of the challenges that faces anyone attempting this approach is the successful regeneration of plants from protoplasts. Interestingly, electroporation has been shown to improve the culture and regenerability of some protoplasts, possibly by the enhanced uptake of plant growth substances or essential nutrients (Rech *et al.*, 1987). Some of the major problems with transformation of protoplast-to-plant regeneration are the possibility of somaclonal variation among regenerants, the presence of multiple copies of inserted DNA, rearrangements and instability of inserted DNA, and inability to define accurately all the foreign sequences in regenerated plants (often a requirement for regulatory purposes).

7.3 Regeneration of Plants

Woody plants generally are considered to be recalcitrant in culture and are difficult to regenerate (Haissig *et al.*, 1987). However, as always, there are exceptions that serve as models for other woody plants. Two modes of regeneration have been observed: indirect and direct. Indirect regeneration involves an intermediate callus phase (unorganized proliferation of cells) whereas direct regeneration does not. In an indirect system, cells of the callus differentiate to produce adventitious meristems or somatic embryos. In a direct system, somatic embryos are produced directly from explant cells without an intervening callus phase.

7.3.1 ADVENTITIOUS BUD FORMATION

Adventitious meristem formation in plants can be induced directly or indirectly from callus (Williams and Maheswaran, 1986). The critical question is whether the bud is formed from a single cell or more than one cell. Several lines of evidence point to a multicellular origin (D'Amato, 1985; Monacelli *et al.*, 1988; Colby *et al.*, 1991). If the origin is a single cell, then transformation should occur at a high frequency as long as the cells undergoing adventitous bud formation are sensitive to *Agrobacterium*. If the origin involves more than one cell type, chimeras of transformed and nontransformed tissues could occur. In the direct pathway, to avoid chimeras it would be necessary to transform all the cells involved in adventitious bud formation rather than some or one of the cells. One possibility for obtaining fully transformed shoots from adventitious buds originating from more than one cell would be through a callus phase. The callus can be selected, allowing the proliferation of only the transformed cells, not the untransformed cells. Fully transformed plants then can be regenerated from these selected calluses, even if more than one cell was involved. Somaclonal variation could be a problem in regeneration events that go through a callus phase (Larkin and Scowcroft, 1983).

7.3.2 SOMATIC EMBRYOGENESIS

The process of somatic embryo formation has been reviewed extensively (Ammirato, 1983,1987; Williams and Maheswaran, 1986; Tulecke, 1987; Wann, 1988) and has been shown to occur in a number of woody plant species (Tulecke, 1987; Wann, 1988). Somatic embryogenesis is a process in which somatic plant cells undergo differentiation to form embryos. Somatic embryogenesis occurs from somatic tissue through an asexual process of differentiation that is not understood entirely but can be distinguished from normal zygotic embryos that are formed as a result of sexual hybridization. However, like zygotic embryos, somatic embryos can be germinated to form plants and can multiply to produce many more somatic embryos through a process referred to as secondary or repetitive embryogenesis, an important property of somatic embryos that has been exploited for the transformation of walnuts. In walnut, secondary somatic embryos are formed from single epidermal cells on existing somatic embryos in culture (Polito *et al.*, 1989). These epidermal cells in walnut somatic embryos are probably the sites of *Agrobacterium* infection and, when allowed to proliferate, give rise to transformed embryos that can, in turn, be germinated to give transformed plants (McGranahan *et al.*, 1988,1990; Dandekar *et al.*, 1989).

7.4 Genetic Selection of Transformed Cells

Chimeric genes constructed by fusing the structural sequences of genes encoding antibiotic resistance with promoter sequences expressed in plants have proved to be very valuable in selecting transformed cell populations. Although a variety of different genes have been used as selectable markers in the transformation of herbaceous species, the gene encoding resistance to the antibiotic kanamycin has been most widely used for the transformation of woody plant species. Toxicity of the selective agent to untransformed plant tissue is critical when evaluating each selective agent. Typically, the optimal concentration of the selective agent should be one that prevents regeneration from untransformed cells without being toxic to the target explant. In grapevine, for example, it has been found that adventitious shoot formation is highly sensitive to kanamycin, exceeding the sensitivity reported in any other species, and has been reported to hinder the recovery of transgenic plants (Colby and Meredith, 1990). Conversely, other plant tissues, such as walnut, were found to be quite resistant to kanamycin, possibly because of endogenous nonspecific kinases (Dandekar *et al.*, 1988).

The genetic selection of transformed cells in walnut was made more efficient with the use of a scorable marker gene in addition to the selectable marker (McGranahan *et al.*, 1990). The gene used was the *gus* gene that encodes a β-glucuronidase enzyme that is an excellent reporter gene (Jefferson, 1987; Jefferson *et al.*, 1987). The transformed embryos were separated after visual identification of *gus* expression in the transformed cells.

7.5 Transgenic Woody Plants

Regeneration of transgenic plants from transformed cells is a major challenge for most plant species. In woody plants, the problem is exacerbated by the lack of reliable tissue culture and regeneration methodology for many species. Transgenic woody plants have been obtained using both *Agrobacterium*-mediated and direct DNA transfer procedures; these are listed in Table 7.2 and some are summarized in subsequent sections.

7.5.1 APPLE (*MALUS PUMILA*)

In apple, a wide variety of tissue culture procedures has been developed successfully, including micropropagation, (Jones, 1976; Zimmerman, 1983), adventitious bud formation (James, 1987), and somatic embryogenesis from leaf explants (James *et al.*, 1984). The first transgenic apple plants were obtained after transformation of leaf disks of the apple cultivar 'Greensleeves' with *Agrobacterium* containing the binary vector pBin6 (James *et al.*, 1989). Other apple cultivars that are good targets for obtaining transgenic plants are those in which regeneration from leaf disks already has been reported. These include 'McIntosh,' 'Triple Red Delicious' (Fasolo *et al.*, 1989), 'M26,' 'M25,' 'M9,' and 'Greensleeves' (James, 1987; James *et al.*, 1988). A potential problem is the low frequency of transformation observed in apple (James *et al.*, 1989; James and Dandekar, 1991). Studies indicate that several factors could influence the

TABLE 7.2
Perennial Crops in which Transgenic Plants Have Been Recovered

Tree crop	Gene transfer procedure	Genes[a]	Reference
Allocasurina	*Agrobacterium rhizogenes*-mediated	*ocs*	Phelep *et al.* (1990)
Apple	*Agrobacterium*-mediated	*kan*	James *et al.* (1989)
		CpTI	D. J. James (unpublished observations)
Birch	*Agrobacterium*-mediated	*kan,lac*	Mehto *et al.* (1990)
Citrus	Free DNA uptake	*kan*	Vardi *et al.* (1990)
Cranberry	Microprojectile bombardment	*kan*	Serres *et al.* (1990)
Eucalyptus	Electroporation	*kan,cry*	Kawazu *et al.* (1990)
Grapevine	*Agrobacterium*-mediated	*kan,gus*	Mullins *et al.* (1990)
Neem	*Agrobacterium*-mediated	*ocs*	Naina *et al.* (1989)
Papaya	DNA-coated microprojectile	*kan,gus*	Manshardt *et al.* (1990)
	DNA-coated microprojectile	*prvCp*	Fitch *et al.* (1990)
Peach	*Agrobacterium*-mediated	*ipt*	Smigocki and Hammerschlag (1990)
Plum	*Agrobacterium*-mediated	*kan,gus*	Mante *et al.* (1990)
Poplar	*Agrobacterium*-mediated	*kan,aro*	Fillatti *et al.* (1987)
	Agrobacterium-mediated	*kan,bar*	DeBlock (1990)
Strawberry	*Agrobacterium*-mediated	*kan,ipt*	James *et al.* (1990a)
Walnut	*Agrobacterium*-mediated	*kan*	McGranahan *et al.* (1988)
		kan,gus	McGranahan *et al.* (1990)
		*cry*IA(c)	A. M. Dandekar (unpublished observations)

[a] *prvCp,* Papaya ring spot virus coat protein; *ipt,* isopentyltransferase gene (cytokinin) of *Agrobacterium*; *cry,* insecticidal crystal protein of *Bacillus thuringiensis*; *CpTI,* cowpea trypsin inhibitor; *lac,* β-galactosidase. For additional abbreviations, see footnote to Table 7.1.

transformation frequency of apple, including the type and physiology of explant tissue used, the strain of *Agrobacterium,* and the design of plasmid vectors (James *et al.,* 1988; Dandekar *et al.,* 1990; James and Dandekar, 1991). *Agrobacterium* strains containing pTiBo542 or pTiC58 and the derivatives thereof were found to be most virulent on apple (Dandekar *et al.,* 1990). Genetic alterations in an *Agrobacterium* strain by introducing additional copies of the virulence genes *virA, B,* and *G* greatly stimulated virulence on apple (Dandekar *et al.,* 1990). Detailed methods for the *Agrobacterium*-mediated transformation of apple have been published (James and Dandekar, 1991). Apart from *Agrobacterium*-mediated transformation, it should be possible to try direct DNA transfer methods since reports show that apple plants can be regenerated from protoplasts (Patat-Ochatt *et al.*, 1988; Wallin and Johansson, 1989), although the disadvantages of using this approach have been mentioned already.

7.5.2 GRAPEVINE (*VITIS VINIFERA*)

Agrobacterium-mediated transformation was used successfully to obtain transgenic grapevines after the regeneration of adventitious buds from hypocotyl explants obtained from somatic embryos of the rootstock *Vitis rupestris* (Rupestris St. George) (Mullins *et al.*, 1990). Transformed buds were obtained also from the cultivars 'Cabernet Sauvignon' and 'Chardonnay' (*Vitis vinifera*) after cocultivation of petiole segments that were obtained from expanding leaves of *in vitro* grown cultures. A binary plasmid vector pCGN7320 was used that contained chimeric *gus* and *kan* genes. The T-DNA on this plasmid was introduced into grapevine using the disarmed host *Agrobacterium* LBA4404. The binary vector plasmid pCGN7320 is very similar to pCGN7314 that was used to transform walnut (McGranahan *et al.*, 1990). The ability to select transformed grapevine was dependent on the level of the antibiotic kanamycin used for genetic selection. No transformants were detected at kanamycin concentrations below 10 μg/ml whereas, at 15 μg/ml, all regenerating shoots were transformed.

7.5.3 NEEM (*AZADIRACHTA INDICA*)

Transgenic neem plants were obtained by Naina *et al.* (1989) when shoots regenerated from tumor tissue obtained after infection with recombinant tumor-inducing *Agrobacterium* strains K12 × 562E and K12 × 167. These tumor-inducing *Agrobacterium* strains have been described previously (Dandekar *et al.*, 1987) and were used to mobilize two different recombinant T-DNAs into cells of aseptic neem seedlings. The recombinant T-DNAs contained a chimeric *kan* gene encoding the Tn*5* kanamycin phosphotransferase enzyme [APH(3′)II], conferring resistance to the antibiotic kanamycin, and octopine synthase, for the synthesis of the opine compound octopine. The seedlings were wounded by one of three methods and a suspension of the bacteria was applied subsequently. Infections with strain K12 × 562E gave rise to tumors which then produced shoots. Strain K12 × 167 produced no tumors, but did give rise to shoots with teratomas at the base. Both types of shoots were rooted and transferred to soil. Opine assays revealed the presence of octopine in shoots derived from infections with both strains, but no Southern analysis data were presented.

7.5.4 PLUM (*PRUNUS DOMESTICA*)

An abstract describing the transformation of plum has been presented (Mante *et al.*, 1990), that includes a description of a regeneration/transformation system that uses plum hypocotyl segments. A binary vector was used, the T-DNA of which carried chimeric *kan* and *gus* genes. Regenerants were selected on the basis of root formation on kanamycin-containing medium, and were assayed for APH(3′)II and GUS activity. Transformed plants were detected. Southern analyses, indicating insertion of the T-DNA into the genome of the regenerated plants, were reported also.

7.5.5 POPLAR (*POPULUS ALBA × GRANDIDENTATA*)

Transgenic poplar plants were first obtained by Fillatti *et al.* (1987). Leaf explants of this hybrid poplar were cocultivated simultaneously with two strains of *Agrobacterium tumefaciens,* one a disarmed strain (LBA4404; Hoekema *et al.*, 1983) containing the binary plasmid pPMG85/587 and the other a tumor-forming strain containing the Ti plasmid pTiC58. The binary plasmid pPMG85/587 contained a recombinant T-DNA that included two chimeric *kan* genes and one chimeric *aroA* gene that encodes the bacterial enzyme 5-enolpyruvylshikimate-3-phosphate (EPSP) synthase, conferring resistance to the herbicide glyphosate. The explants were selected on kanamycin and resistant shoots were obtained from adventitious meristems. Western blots revealed that the introduced *aroA* gene was expressed in the transformed plants. An interesting feature of this study was that no transformants were obtained from those explants cocultivated with only the disarmed strain. Only explants that were exposed to the combination of armed and disarmed plasmids gave rise to transformants.

7.5.6 STRAWBERRY (*FRAGARIA × ANANNASSA*)

James *et al.* (1990a,b) have reported transformation of the commercial and octoploid strawberry cultivar ‘Rapella’ using a disarmed binary vector system. APH (kanamycin resistance) was used as a scorable marker and nopaline synthase (NOS, nopaline dehydrogenase) was used as a scorable marker. The short generation time of strawberry, compared with other woody plants, was advantageous for the purpose of carrying out inheritance studies. Seeds from the selfed transformants were germinated and the tissue from these plants was tested either for nopaline synthase activity or for the ability to produce callus on a medium containing kanamycin. All nopaline-positive plants were also kanamycin resistant, as expected. Nopaline synthase was found to segregate in the F_2 generation in a 3:1 ratio (3 nopaline-positive to 1 nopaline-negative). Additionally, strawberry plants were transformed with the T-DNA-encoded isopentyltransferase (*ipt*) gene, the product of which catalyzes the production of the cytokinin *trans*-zeatin. Such plants were characterized by a lack of apical dominance, hormone autotrophy, and an inability to form roots.

Another report of strawberry transformation is that of Nehra *et al.* (1990) of the cultivar 'Redcoat.' *Agrobacterium* was used to introduce a recombinant T-DNA carrying three genes: *kan, nos,* and *gus*. Assays for APH(3′)II and GUS activity and Southern blot analysis all indicated the insertion of the T-DNA and marker genes into the regenerated plants.

Diploid strawberry (*Fragaria vesca;* $2n = 2x = 14$) has been suggested as a model system for woody plant transformation studies (P. Schuerman, unpublished results). *Fragaria vesca* has a generation time of 7 wk, is highly inbred (isogenic), is propagated easily and has a small genome, approximately 22 times that of *Escherichia coli* (Ahmadi *et al.*, 1988). It is also regenerable from various tissues (P. Schuerman and A. M. Dandekar, unpublished results) and infectable by wild-type strains of *Agrobacterium* (Uratsu *et al.*, 1991).

7.5.7 WALNUT (*JUGLANS REGIA*)

In walnut, adventitious bud formation has not, as yet, been reported to occur from tissue segments, making it impossible to adapt procedures such as the leaf disk transformation system described by Horsch *et al.* (1985). However, like a number of woody species, walnut undergoes somatic embryogenesis (Tulecke and McGranahan, 1985). Walnut somatic embryos proliferate in culture, producing numerous secondary embryos that originate from single cells in the epidermal layer of the embryo (Tulecke and McGranahan, 1985; Polito *et al.*, 1989). These epidermal cells appear to be susceptible to infection with wild-type *Agrobacterium,* producing tumors on the surface of the embryo (McGranahan *et al.*, 1988). Transformation of somatic embryos with recombinant disarmed strains of *A. tumefaciens* resulted in the regeneration of transgenic plants (McGranahan *et al.*, 1988; Dandekar *et al.*, 1989). Using a single selectable marker (*kan*), transgenic plants containing the marker gene were selected at a low frequency (McGranahan *et al.*, 1988). It has been shown that the frequency of transformation is as high as 30% with the use of both selectable and scorable markers (McGranahan *et al.*, 1990). This increased efficiency of transformation was accomplished using binary plasmid vectors containing two marker genes, *gus,* a scorable marker gene, and *kan,* a selectable marker gene (Dandekar *et al.*, 1989; McGranahan *et al.*, 1990).

7.6 Molecular Breeding: Useful Traits

Desirable traits for woody tree crops include pest resistance, disease resistance, drought and cold tolerance, herbicide tolerance, improved fruit quality, rootability of mature tissue, reduced juvenility, and alteration of tree form and architecture (e.g., dwarfing, semidwarfing, or compact/columnar habit). A major obstacle is the lack of useful genes since very few of these traits have been characterized in sufficient detail (i.e., to the DNA sequence level) to be incorporated into vectors for gene transfer. Although traits such as disease and pest resistance can be found in plant germ plasm and, in some instances, have been shown to be controlled by a single locus, no disease-resistance gene has been cloned at present. However, transgenic plants with resistance to disease, herbicides, and insect damage have been produced using genes from both plant and nonplant sources.

7.6.1 DISEASE AND PEST TOLERANCE

In cultivated crops, disease resistance has been approached through viral cross-protection using chimeric genes specifying viral coat protein as a way of protecting plants from viral disease (Bevan *et al.*, 1985; Powell *et al.*, 1986). This approach now has been used successfully against a number of different viruses, for example, tobacco mosaic virus (TMV; Bevan *et al.*, 1985; Powell *et al.*, 1986; Nelson *et al.*, 1988), alfalfa mosaic virus (AMV; Tumer *et al.*, 1987; Van Dun *et al.*, 1987), cucumber mosaic virus (CMV; Cuozzo *et al.*, 1988), and potato virus X (PVX; Hemenway *et al.*, 1988). This could be a viable approach for many woody plant viral diseases, for example, plum pox, *Citrus* tristeza virus, and blackline disease of walnuts.

In plants, resistance to insect pests has been achieved by employing two types of gene products; (1) protease inhibitors and (2) insecticidal crystalline proteins (ICPs) from *Bacillus thuringiensis*. Protease inhibitors are synthesized in a variety of plants in response to insect predation and have been found to be toxic to diverse groups of insect species (Green and Ryan, 1972; Gatehouse and Boulter, 1983; Ryan *et al.*, 1986). The gene encoding the cowpea trypsin inhibitor (CpTI) was transferred and expressed in tobacco plants and shown to protect the plant against *Heliothis virescens* (Hilder *et al.*, 1987). Transgenic apple plants containing CpTI have been produced (D. J. James, unpublished results) using the identical construct reported by Hilder *et al.* (1987).

The ICPs are a class of proteins that are highly toxic to a specific group of insects, especially lepidopteran species (Aronson *et al.*, 1986; Whiteley and Schnepf, 1986). These proteins are encoded by a diverse set of genes in the microbe *B. thuringiensis* that can be separated into 13 groups based on differences in DNA sequence (Höfte and Whiteley, 1989). Many of the commercially significant insect pests of woody plants belong to Lepidoptera and may be sensitive to *B. thuringiensis*-encoded ICPs. Transgenic tobacco (Vaeck *et al.*, 1987), tomato (Fischhoff *et al.*, 1987), and cotton (Perlak *et al.*, 1990) plants expressing ICPs have been obtained using two genes, *cryIA*(*c*) and *cryIA*(*b*). These transgenic plants expressed sufficient quantities of ICP to protect the plants from damage caused by the feeding larvae of tobacco and tomato hornworm, cabbage looper, and beet armyworm, all voracious pests. The *cryIA*(*c*) gene has been transferred into walnut using the approach described by Vaeck *et al.* (1987); the transgenic plants are being field tested (A. M. Dandekar, unpublished results).

Important considerations accompany this approach. The ICP must be made in sufficient quantities to kill the insect pest and must be stable in the plant tissue and not degrade rapidly. The production of the protein must not interfere with the metabolism of the plant, and must not compromise the quality of the produce in any way.

7.6.2 HERBICIDE TOLERANCE

Herbicide tolerance has been achieved through the identification of genes in bacteria that (1) are insensitive to the herbicide (Comai *et al.*, 1983) or (2) detoxify the herbicide (DeBlock *et al.*, 1987; Stalker *et al.*, 1988). Bacterial genes have been used successfully to confer tolerance to the popular broad spectrum herbicide glyphosate ("Roundup") in plants (Comai *et al.*, 1985; Shah *et al.*, 1986). The gene encoding resistance to glyphosate has been transferred into poplar (Filatti *et al.*, 1987).

7.6.3 FRUIT QUALITY

Trends in the fruit markets in the United States and in other countries around the world show that there is a greater emphasis on taste and innovation in marketing and distribution (Cook, 1992). Fruit are subject to the demands of the consumer, and scientific trends may often be in conflict. There is an increase in the demand of both fresh and processed fruit, especially among the middle age group (The Food Institute Report, 1989). This increase has been attributed to an increase in yield and quality of fruit after World War II (Cook, 1992). Another important factor has been the improvement in storage and transport. The improvement in maritime transportation with container lines and palatized boats has made it possible to export highly perishable fruit (e.g., kiwi) over long distances. The emergence of large worldwide markets makes these technologies a feasible proposition. The export of fruit from the southern hemisphere, for example, Australia, New Zealand, and South America, has provided fruit during seasons when none are growing in the northern hemisphere. It is also clear, in view of the advances in shipping, that seasons will be less important since fruit would be available all season long, making it no longer necessary to develop early varieties. However, as the demand becomes more sophisticated, innovation would be the main source to stimulate demand—modern distribution, good appearance, good shelflife, consistent production, optimum size, reasonable price, and, above all, a greater emphasis on food safety (Zind, 1990). The per capita consumption of tree fruit and nuts is dampened by smaller crops and higher prices;

therefore it will be important to sustain both yield and quality (Fruit and Tree Nuts Situation and Outlook Report Yearbook, 1990). As the use of various chemicals is banned, there will be a need to enhance fruit color and to prevent softening and decay through genetic modification. Currently, "antisense" constructs of ripening-specific genes are being tested in transgenic plants to slow ripening and to prevent the softening of tomato fruit (Sheehy *et al.*, 1988; Smith *et al.*, 1988). Similar studies need to be done with other fruit crops.

7.6.4 PLANT FORM AND ARCHITECTURE

At the orchard level, at least two important trends in tree form and architecture exist: (1) reducing tree stature and (2) reducing orchard cycle times. The modification of tree stature would lead to a reduction in production costs, that is, pruning, thinning, spraying, and harvesting. Genes conferring dwarf (Hanche, 1986a) and columnar habit (Lapins and Watkins, 1973) have been used in breeding programs and are encoded by traits that are inherited simply. The *co* gene is a dominant gene in apple that controls the compact and columnar habit observed in the apple mutant 'McIntosh Wijcik' (Lapins and Watkins, 1973). The control of tree stature through the breeding of genetic dwarfs or dwarfing rootstocks would allow high density plantings and an increased mechanization of orchard management. The second important trend is the shortening of orchard lifespan and establishment period, resulting in a faster turnover of new cultivars. This can only be accomplished through the identification of genes controlling juvenility. The juvenile trait in *Prunus* appears to be a simply inherited trait (Hanche, 1986b). A reduction in the juvenile phase in a fruit or nut tree crop would result in trees that mature earlier. However, in the case of forest trees it may be desirable to lengthen the juvenile phase, thus allowing the trees to grow faster. Identification of these traits at the molecular level is a challenge that will have to be met through genome mapping projects in these tree crops.

7.7 Environmental Impact of Transgenic Woody Plants and Potential Resistance Development

The most compelling arguments against the release of genetically modified organisms into the environment are ecological ones (Gould, 1988; Pimentel *et al.*, 1989). These arguments stress that important and thorough consideration should be given to the possible impact of placing a transformed plant in the environment. Typically cited are ecological disasters that have occurred in the past when newly introduced organisms were allowed to infiltrate and were found subsequently to dominate existing ecosystems. Woody plants, once introduced, will survive for decades; therefore it is vital that long-term research be carried out to address each ecological issue specifically and to evaluate the risk to the environment. Transgenic woody plants compared with annual plants provide a good experimental system for such studies.

The release of plants that are resistant to insects, for instance, will place pressure on insect populations to overcome this resistance. Viral populations will respond in a similar fashion when plants with viral resistance are released. The likelihood of resistance development in these populations must be estimated in properly constructed field trials before the true risk to the environment can be assessed. Extreme reliance on a single resistant cultivar could be devastating in the event that the resistance was overcome. It is important that long-term research be conducted to estimate the probability of resistance development and to develop alternative strategies. Woody plants expressing ICPs are good candidates for such studies because of their longevity compared with annual plants that can be rotated each year.

Gene regulation in an environment may be the most critical factor in the success or failure of a genetic engineering program in woody plants. Gene regulation can be manipulated on a population level, on a temporal level, or on an anatomical level, and can be accomplished using a combination of these manipulations. Examples would be the incorporation of an insect resistance gene into only a proportion of the trees in an orchard (population level), expression of the foreign gene only in some of the tissues of the plant (anatomical level), or expression only when the tissues are immature (temporal level). This approach would reduce the intensity of the selection pressure on the insect population and would maintain a susceptible population that would not damage the produce of the transgenic trees.

Ecological and genetic studies will doubtless be of assistance in risk evaluation. With respect to insect resistance genes, many studies already have been undertaken to estimate the likelihood of susceptibility loss in insect populations in the laboratory (McGaughey, 1985; McGaughey and Johnson, 1987; Stone *et al.*, 1989; Van Rie *et al.*, 1990). By learning from ecological mistakes committed in the past, and with a commitment to the type of responsible management that any worthwhile endeavor deserves, the risks of transgenic plant release can be minimized.

Perhaps the least helpful response to transgenic release comes from those individuals who confuse the issue by portraying manmade taxonomic divisions as sacred barriers or complaining that genetic engineering is simply too dangerous to be done in the first place. Of more assistance are authors who offer guidelines for the evaluation of risk associated with transgenic tree release (Raffa, 1989).

References

Ahmadi, H., Bringhurst, R. S., Uratsu, S. L., and Dandekar, A. M. (1988). *Genome* **30**(*Suppl. 1*).

Alt-Mörbe, J., Neddermann, P., von Lintig, J., Weiler, E. W., and Schröder, J. (1988). *Mol. Gen. Genet.* **213,** 1–8.

Alt-Mörbe, J., Kühlmann, H., and Schröder, J. (1989). *Mol. Plant–Microbe Inter.* **2,** 301–308.

Ammirato, P. V. (1983). *In* "Handbook of Plant Cell Culture" (D. A. Evans, W. R. Sharp, P. V. Ammirato, and Y. Yamada, eds.), Vol. 1, pp. 82–123. Macmillan, New York.

Ammirato, P. V. (1987). *In* "Plant Tissue and Cell Culture" (C. E. Green, D. A. Somers, W. P. Hackett, and D. D. Biesboer, eds.), pp. 57–81. Liss, New York.

Anderson, A. R., and Moore, L. (1979). *Phytopathol.* **69,** 320–323.

Ankenbauer, R. G., and Nester, E. W. (1990). *J. Bacteriol.* **172,** 6442–6446.

Aronson, A. I., Beckman, W., and Dunn, P. (1986). *Microbiol. Rev.* **50,** 1–24.

Baribault, T. J., Skene, K. G. M., and Scott, N. S. (1989). *Plant Cell Rep.* **8,** 137–140.

Baribault, T. J., Skene, K. G. M., Cain, P. A., and Scott, N. S. (1990). *J. Exp. Bot.* **41,** 1045–1049.

Bevan, M. W., Mason, S. E., and Goelet, P. (1985). *EMBO J.* **4,** 1921–1926.

Bolton, G. W., Nester, E. W., and Gordon, M. P. (1986). *Science* **232,** 983–985.

Cangelosi, G. A., Ankenbauer, R. G., and Nester, E. W. (1990). *Proc. Natl. Acad. Sci. U.S.A.* **87,** 6708–6712.

Colby, S. M., and Meredith, C. P. (1990). *Plant Cell Rep.* **9,** 237–240.

Colby, S. M., Juncosa, A. M., and Meredith, C. P. (1991). *J. Am. Soc. Hort. Sci.* **116,** 356–361.

Comai, L., Sen, L., and Stalker, D. (1983). *Science* **221,** 370–371.

Comai, L., Facciotti, D., Hiatt, W. R., Thompson, G., Rose, R. E., and Stalker, D. E. (1985). *Nature (London)* **317,** 741–744.

Cook, R. L. (1990). *In* "Postharvest Technology of Horticultural Crops" (A. A. Kader, ed.). University of California Division of Agriculture and Natural Resources. pp. 3–13.

Cuozzo, M., O'Connell, K. M., Kaniewski, W., Fang, R.-X., Chua, N.-H., and Tumer, N. E. (1988). *Bio/Technology* **6,** 549–558.

D'Amato, F. (1985). *CRC Crit. Rev. Plant Sci.* **3,** 73–112.

Dandekar, A. M. (1992). *In* "Biotechnologies of Perennial Fruit Crops" (F. Hammerschlag and R. Litz, eds.). C.A.B. International. (In press.)

Dandekar, A. M., Gupta, P. K., Durzan, D. J., and Knauf, V. (1987). *Bio/Technology* **5,** 587–590.

Dandekar, A. M., Martin, L. A., and McGranahan, G. H. (1988). *J. Am. Soc. Hort. Sci.* **113,** 945–949.

Dandekar, A. M., McGranahan, G. H., Leslie, C. A., and Uratsu, S. L. (1989). *J. Tissue Cult. Meth.* **12,** 145–150.

Dandekar, A. M., Uratsu, S. L., and Matsuta, N. (1990). *Acta Hort.* **280,** 483–494.

DeBlock, M. (1990). "Abstracts of the VIIth International Congress on Plant Tissue and Cell Culture," p. 49.

DeBlock, M., Botterman, J., Vanderwiele, M., Dockx, J., Thoen, C., Gossele, V., Rao Movva, N., Thompson, C., Van Montagu, M., and Lee Mans, J. (1987). *EMBO J.* **6,** 2513–2518.

De Cleene, M., and De Ley, J. (1976). *Bot. Rev.* **42,** 389–466.

De Cleene, M., and De Ley, J. (1981). *Bot. Rev.* **47,** 147–194.

Ellis, D., Roberts, D., Sutton, B., Lazaroff, W., Webb, D., and Flinn, B. (1989). *Plant Cell Rep.* **8,** 16–20.

Fasolo, F., Zimmerman, R. H., and Fordham, I. (1989). *Plant Cell Tissue Organ Cult.* **16,** 75–87.

Fillatti, J. J., Sellmer, J., McCown, B., Haissig, B., and Comai, L. (1987). *Mol. Gen. Genet.* **206,** 192–199.

Fischhoff, D. A., Bowdish, K. S., Perlak, F. J., Marrone, P. G., McCormick, S. M., Niedermeyer, J. G., Dean, D. A., Kusano-Kretzmer, K., Mayer, E. J., Rochester, D. E., Rogers, S. G., and Fraley, R. T. (1987). *Bio/Technology* **5,** 807–813.

Fitch, M. M. M., Manshardt, R. M., Gonsalves, D., Slightom, J. L., and Sanford, J. C. (1990). "Abstracts of the VIIth International Congress on Plant Tissue and Cell Culture," p. 55.

Fraley, R. T., Rogers, S. G., and Horsch, R. B. (1986). *CRC Rev. Plant Sci.* **4,** 1–46.

Fromm, M., Taylor, L. P., and Walbot, V. (1985). *Proc. Natl. Acad. Sci. U.S.A.* **82,** 5824–5828.

Fromm, M., Taylor, L. P., and Walbot, V. (1986). *Nature (London)* **319,** 791–793.

Fruit and Tree Nuts Situation and Outlook Report Yearbook (1990). USDA Economic Research Service, Washington, D.C.

Gasser, C., and Fraley, R. T. (1989). *Science* **244,** 1293–1299.

Gatehouse, A. M. R., and Boulter, D. (1983). *J. Sci. Food Agric.* **34,** 345–350.

Gheysen, G., Dhaese, P., Van Montagu, M., and Schell, J. (1985). *In* Plant Gene Research: Genetic flux in plants. (B. Hohn and E. S. Dennis, eds.), pp. 11–47. Springer-Verlag, Wien.

Gould, F. (1988). *BioScience* **38,** 26–33.

Green, T. R., and Ryan, C. A. (1972). *Science* **175,** 776–777.

Guellec, V., David, C., Branchard, M., and Tempé, J. (1990a). *Plant Cell Tissue Organ Cult.* **20,** 211–215.

Guellec, V., Netzer, M. H., Robin, O., David, C., Mauro, M. C., Deloire, A., and Tempé, J. (1990b). "Abstracts of the VIIth International Congress on Plant Tissue and Cell Culture," p. 59.

Gupta, P. K., Dandekar, A. M., and Durzan, D. J. (1988). *Plant Sci.* **58,** 85–92.

Hain, R., Stable, P., Czernilofsky, A. P., Steinbib, H. H., Herrera-Estrella, L., and Schell, J. (1985). *Mol. Gen. Genet.* **199,** 161–168.

Haissig, B. E., Nelson, N. D., and Kidd, G. H. (1987). *Bio/Technology* **5,** 52–59.

Hammerschlag, F. A., Owens, L. D., and Smigocki, A. C. (1989). *J. Am. Soc. Hort. Sci.* **114,** 508–510.

Hanche, P. E. (1986a). *Hort. Sci.* **21,** 1193–1195.

Hanche, P. E. (1986b). *Hort. Sci.* **21,** 1197–1198.

Hemenway, C., Fang, R.-X., Kaniewski, W. K., Chua, N.-H., and Tumer, N. E. (1988). *EMBO J.* **7,** 1273–1280.

Hilder, V. A., Gatehouse, A. M. R., Sheerman, S. E., Barker, R. F., and Boulter, D. (1987). *Nature (London)* **330,** 160–163.

Höfte, H., and Whiteley, H. R. (1989). *Microbiol. Rev.* **53,** 242–255.

Hoekema, A., Hooykaas, P. J. J., and Schilperoort, R. A. (1983). *Nature (London)* **303,** 179–180.

Hood, E. A., Helmer, G. L., Fraley, R. T., and Chilton, M.-D. (1986). *J. Bacteriol.* **168,** 1291–1301.

Horsch, R. B., Fry, J. E., Hoffmann, N. L., Eichholtz, D., Rogers, S. G., and Fraley, R. T. (1985). *Science* **227,** 1229–1231.

James, D. J. (1987). *In* "Biotechnology and Genetic Engineering Reviews" (G. E. Russell, ed.), Vol. 5, pp. 33–79. Intercept, London.

James, D. J., and Dandekar, A. M. (1991). *In* "Plant Tissue Culture Manual: Fundamentals and Applications" (K. Lindsey, ed.), pp. B8/1–B8/18. Kluwer, The Netherlands.

James, D. J., Passey, A. J., and Deeming, D. C. (1984). *J. Plant Physiol.* **115,** 217–229.

James, D. J., Passey, A. J., and Rugini, E. (1988). *J. Plant Physiol.* **132,** 148–154.

James, D. J., Passey, A. J., Barbara, D. J., and Bevan, M. W. (1989). *Plant Cell Rep.* **7,** 658–661.

James, D. J., Passey, A. J., and Barbara, D. J. (1990a). *Plant Sci.* **69,** 79–94.

James, D. J., Passey, A. J., and Barbara, D. J. (1990b). *In* "Genetic Engineering of Crop Plants" (G. Lycett and D. Grierson, eds.), pp. 239–248. Butterworths, London.

Jefferson, R. A. (1987). *Plant Mol. Biol. Rep.* **5,** 387–405.

Jefferson, R. A., Kavanagh, T. A., and Bevan, M. W. (1987). *EMBO J.* **6,** 3901–3907.

Jin, S., Komari, T., Gordon, M. P., and Nester, E. W. (1987). *J. Bacteriol.* **169,** 4417–4425.

Jones, O. P. (1976). *Nature (London)* **262,** 392–393.

Karnosky, D. F., Diner, A. M., and Barnes, W. M. (1988). *In* "Somatic Cell Genetics of Woody Plants" (M. R. Ahuja, ed.), pp. 55–63. Kluwer, London.

Kawazu, T., Doi, K., Ohta, T., Shinohara, Y., Ito, K., and Shibata, M. (1990). "Abstracts of the VIIth International Congress on Plant Tissue and Cell Culture," p. 64.

Keane, P. J., Kerr, A., and New, P. B. (1970). *Aust. J. Biol. Sci.* **23,** 585–595.

Kerr, A., and Panagopoulos, C. G. (1977). *Phytopathol.* **90,** 172–179.

Klein, T. M., Wolf, E. D., Wu, R., and Sanford, J. C. (1987). *Nature (London)* **327,** 70–73.

Kobayashi, S., and Uchimiya, H. (1989). *Jpn. J. Genet.* **64,** 91–99.

Lapins, K. O., and Watkins, R. (1973). "Report of East Malling Research Station for 1972," pp. 136.

Larkin, P. J., and Scowcroft, W. R. (1983). *In* "Genetic Engineering of Plants: An Agricultural Perspective" (T. Kosuge, C. P. Meredith, and A. Hollaender, eds.), pp. 289–314. Plenum Press, New York.

Leroux, B., Yanofsky, M. F., Winans, S. C., Ward, J. E., Ziegler, S. F., and Nester, E. W. (1987). *EMBO J.* **6,** 849–856.

Lichtenstein, C. P., and Fuller, S. L. (1987). *Genet. Eng.* **6,** 103–183.

Litz, R. E., Mathews, H., Bharathan, N., and Narayanan, K. R. (1990). "Abstracts of the VIIth International Congress on Plant Tissue and Cell Culture," p. 67.

McCabe, D. E., Swain, W. F., Martinell, B. J., and Christou, P. (1988). *Bio/Technology* **6,** 923–926.

McGaughey, W. H. (1985). *Science* **229,** 193–195.

McGaughey, W. H., and Johnson, D. E. (1987). *J. Econ. Entomol.* **80,** 1122–1126.

McGranahan, G. H., Leslie, C. A., Uratsu, S. L., Martin, L. A., and Dandekar, A. M. (1988). *Bio/Technology* **6,** 800–804.

McGranahan, G. H., Leslie, C. A., Uratsu, S. L., and Dandekar, A. M. (1990). *Plant Cell Rep.* **8,** 512–516.

Mackay, J., Seguin, A., and Lalonde, M. (1988). *Plant Cell Rep.* **7,** 229–232.

Manshardt, R., Fitch, M., Gonsalves, D., Slightom, J., and Sanford, J. (1990). "Abstracts of contributed papers at XXIII International Horticultural Congress," p. 132.

Mante, S., Cornell, U., Morgens, P., Scorza, R., Cordts, J., and Callahan, A. (1990). *In Vitro* **26,** 44A.

Martin, G. C., Miller, A. N., Castle, L. A., Morris, J. W., Morris, R. O., and Dandekar, A. M. (1990). *J. Am. Soc. Hort.* **115,** 686–691.

Martin, L. (1987). Genetic transformation and foreign gene expression in tissue of different woody species. Master's Thesis, University of California, Davis.

Martin, L. A., Stamp, J. A., Meredith, C. P., and Dandekar, A. M. (1989). *In* "Proceedings of the AgBiotech'89 International Conference and Exposition," pp. 172–185.

Mehto, M., Tujunen, K., Honkanen, J., and Seppanen, P. (1990). "Abstracts of the VIIth International Congress on Plant Tissue and Cell Culture," p. 62.

Monacelli, B., Altamura, M. M., Pasqua, G., Biasini, M. G., and Sala, F. (1988). *Protoplasma* **142,** 156–163.

Mullins, M. G., Tang, F. C. A., and Facciotti, D. (1990). *Bio/Technology* **8,** 1041–1045.

Naina, N. S., Gupta, P. K., and Mascarenhas, A. F. (1989). *Curr. Sci.* **58,** 184–187.

Nehra, N. S., Chibbar, R. N., Kartha, K. K., Datla, R. S. S., Crosby, W. L., and Stushnoff, C. (1990). *Plant Cell Rep.* **9,** 10–13.

Nelson, R. S., McCormic, S. M., Delanney, X., Dube, P., Layton, J., Anderson, E. J., Kaniewska, M., Proksch, R. K., Horsch, R. B., Rogers, S. G., Fraley, R. T., and Beachy, R. N. (1988). *Bio/Technology* **6,** 403–409.

Nester, E. W., Gordon, M. P., Amasino, R. M., and Yanofsky, M. F. (1984). *Ann. Rev. Plant Physiol.* **35,** 387–413.

Ochatt, S. J., and Caso, O. H. (1986). *J. Plant Physiol.* **122,** 243–249.

Ochatt, S. J., and Power, J. B. (1988). *Plant Sci.* **56,** 75–79.

Ochatt, S. J., Cocking, E. C., and Power, J. B. (1987). *Plant Sci.* **50,** 139–143.

Oka, S., and Ohyama, K. (1985). *J. Plant Physiol.* **119,** 455–460.

Otten, L. A. M. B., and Schilperoort, R. A. (1978). *Biochim. Biophys. Acta* **527,** 497–500.

Owens, L. D., and Cress, D. E. (1985). *Plant Physiol.* **77,** 87–94.

Pang, S. Z., and Sanford, J. C. (1988). *J. Am. Soc. Hort. Sci.* **113,** 287–291.

Paszkowski, J., Shillito, R. D., Saul, M., Mandak, V., Hohn, T., Hohn, B., and Potrykus, I. (1984). *EMBO J.* **3,** 2717–2722.

Patat-Ochatt, E. M., Ochatt, S. J., and Power, J. B. (1988). *J. Plant Physiol.* **133,** 460–465.

Perlak, F. J., Deaton, R. W., Armstrong, T. A., Fuchs, R. L., Sims, S. R., Greenplate, J. T., and Fischhoff, D. A. (1990). *Bio/Technology* **8,** 939–943.

Petit, A., David, C., Dahl, G. A., Ellis, J. G., Guyon, P., Casse-Delbart, F., and Tempé, J. (1983). *Mol. Gen. Genet.* **190,** 204–214.

Phelep, M., Petit, A., Allioux, B., Dulioux, E., and Tempé, J. (1990). "Abstracts of the VIIth International Congress on Plant Tissue and Cell Culture," p. 54.

Pimentel, D., Hunter, M. S., LaGro, J. A., Efroymson, R. A., Landers, J. C., Mervis, F. T., McCarthy, C. A., and Boyd, A. E. (1989). *BioScience* **39,** 606–614.

Polito, V. S., McGranahan, G. H., Pinney, K., and Leslie, C. (1989). *Plant Cell Rep.* **8,** 219–221.

Powell, A. P., Nelson, R. S., De, B., Hoffmann, N., Rogers, S. G., Fraley, R. T., and Beachy, R. N. (1986). *Science* **232,** 738–743.

Pythoud, F., Sinkar, V. P., Nester, E. W., and Gordon, M. P. (1987). *Bio/Technology* **5,** 1323–1327.

Raffa, K. F. (1989). *BioScience* **39,** 524–534.

Raineri, D. M., Bottino, P., Gordon, M. P., and Nester, E. W. (1990). *Bio/Technology* **8,** 33–38.

Rech, E. L., Ochatt, S. J., Chand, P. K., Power, J. B., and Davey, M. R. (1987). *Protoplasma* **141,** 169–176.

Revilla, M. A., Ochatt, S. J., Doughty, S., and Power, J. B. (1987). *Plant Sci.* **50,** 133–137.

Rogowsky, P. M., Powell, B. S., Shirasu, K., Lin, T.-S., Morel, P., Zyprian, E. M., Steck, T. R., and Kado, C. I. (1990). *Plasmid* **23,** 85–106.

Russel, J. A., and McCown, B. H. (1986). *Plant Sci.* **46,** 133–142.

Russel, J. A., and McCown, B. H. (1988). *Plant Cell Rep.* **7,** 59–62.

Ryan, C. A., Bishop, P. D., Graham, J. S., Broadway, R. M., and Duffey, S. S. (1986). *J. Chem. Ecol.* **12,** 1025–1035.

Schuerman, P., and Dandekar, A. M. (1990). *In* "Plant Genetic Engineering: Subcellular Biochemistry" (B. B. Biswas and J. R. Harris, eds.), Vol. 17, pp. 81–105. Plenum Press, London.

Scorza, R., Morgens, P. H., Cordts, J. M., Mante, S., and Callahan, A. M. (1990). *In Vitro Cell. Dev. Biol.* **26,** 829–834.

Sederoff, R., Stomp, A. M., Chilton, W., and Moore, L. (1986). *Bio/Technology* **4,** 647–649.

Serres, R. A., McCown, B. H., Zeldin, E., McCabe, D. E., and Stang, E. J. (1990). "Abstracts of contributed papers at XXIII International Horticultural Congress," p. 127.

Shah, D. M., Horsch, R. B., Klee, H. J., Kishore, G. M., Winter, J. A., Tumer, N. E., Hironaka, C. M., Sanders, P. R., Gasser, C. S., Aykent, S., Siegel, N. R., Rogers, S. G., and Fraley, R. T. (1986). *Science* **233,** 478–481.

Sheehy, R. E., Kramer, M., and Hiatt, W. R. (1988). *Proc. Natl. Acad. Sci. U.S.A.* **85,** 8805–8809.

Smigocki, A. C., and Hammerschlag, F. A. (1990). "Abstracts of the VIIth International Congress on Plant Tissue and Cell Culture," p. 77.

Smith, C. J. S., Watson, C. F., Ray, J., Bird, C. R., Morris, P. C., Schuch, W., and Gierson, D. (1988). *Nature (London)* **334,** 724–726.

Stachel, S. E., and Nester, E. W. (1986). *EMBO J.* **5,** 1445–1454.

Stachel, S., and Zambryski, P. (1986). *Cell* **46,** 325–333.

Stalker, D. M., McBride, K. E., and Malyj, L. D. (1988). *Science* **242,** 419–423.

Stomp, A.-M., Bergman, B. A., Weissinger, A., Sederoff, R., and Haggman, H. (1990). "Abstracts of the VIIth International Congress on Plant Tissue and Cell Culture," p. 48.

Stone, T. B., Sims, S. R., and Marrone, P. G. (1989). *J. Invert. Pathol.* **53,** 228–234.

Tempé, J., and Casse-Delbart, F. (1989). *In* "Cell Culture and Somatic Cell Genetics" (J. Schell and I. K. Vasil, eds.), Vol. 6, pp. 25–49. Academic Press, New York.

The Food Institute Report (1989). "Demographic Directions," Vol. I. American Institute of Food Distribution, Washington, D.C.

Tulecke, W. (1987). *In* "Cell and Tissue Culture in Forestry" (J. M. Bonga and D. J. Durzan, eds.), Vol. 2, pp. 61–91. Martinus Nijhoff, Boston.

Tulecke, W., and McGranahan, G. H. (1985). *Plant Sci.* **40,** 53–67.

Tumer, N. E., O'Connell, K. M., Nelson, R. S., Sanders, P. R., Beachy, R. N., Fraley, R. T., and Shah, D. M. (1987). *EMBO J.* **6,** 1181–1189.

Uratsu, S. L., Ahmadi, H., Bringhurst, R. S., and Dandekar, A. M. (1991). *Hort. Sci.* **26,** 196–199.

Vaeck, M., Reynaerts, A., Hofte, H., Jansens, S., DeBeuckeleer, M., Dean, C., Zabeau, M., Van Montagu, M., and Leemans, J. (1987). *Nature (London)* **328,** 33–37.

Vahala, T., Stabel, T., and Eriksson, T. (1989). *Plant Cell Rep.* **8,** 55–88.

Van Dun, C. M. P., Bol, J. F., and van Vloten-Doting, L. (1987). *Virology* **159,** 299–305.

Van Rie, J., McGaughey, W. H., Johnson, D. E., Barnett, B. D., and Van Mellaert, H. (1990). *Science* **247,** 72–74.

Vardi, A., Bleichman, S., and Aviv, D. (1990). "Abstracts of the VIIth International Congress on Plant Tissue and Cell Culture," p. 80.

Vardi, A., Spiegel-Roy, P., and Galun, E. (1982). *Theor. Appl. Genet.* **62,** 171–176.

Veluthambi, K., Krishnan, M., Gould, J. H., Smith, R. H., and Gelvin, S. B. (1989). *J. Bacteriol.* **171,** 3969–3703.

Vernade, D., Herrera-Estrella, A., Wang, K., and Van Montagu, M. (1988). *J. Bacteriol.* **170,** 5822–5829.

Wallin, A., and Johansson, L. (1989). *J. Plant Physiol.* **135,** 565–570.

Wann, S. R. (1988). *In* "Horticultural Reviews" (J. Janick, ed.), Vol. 10, pp. 153–181. Timber Press, Portland, Oregon.

Whiteley, H. R., and Schnepf, H. E. (1986). *Ann. Rev. Microbiol.* **40,** 549–576.

Williams, E. G., and Maheswaran, G. (1986). *Ann. Bot.* **57,** 443–462.

Winans, S., Ebert, P., Stachel, S., Gordon, M., and Nester, E. W. (1986). *Proc. Natl. Acad. U.S.A.* **83,** 8278–8282.

Yanofsky, M. F., and Nester, E. W. (1986). *J. Bacteriol.* **168,** 244–250.

Yanofsky, M., Lowe, B., Montoya, A., Rubin, R., Krul, W., Gordon, M., and Nester, E. W. (1985). *Mol. Gen. Genet.* **201,** 237–246.

Zambryski, P., Tempé, J., and Schell, J. (1989). *Cell* **56,** 193–201.

Zimmerman, R. H. (1983). *In* "Methods in Fruit Breeding" (J. N. Moore and J. Janick, eds.), pp. 124–135. Purdue University Press, West Lafayette, Indiana.

Zind, T. (1990). *In* "Fresh Trends '90," pp. 37–68. Market Facts Inc. and Vance Research Series, Prairie View, Illinois. *The Packer Focus 1989–90.*

8

Transgenic Cotton

Elizabeth E. Murray*

McArdle Laboratory for Cancer Research
University of Wisconsin
Madison, Wisconsin 53705

David L. DeBoer

Monsanto Corporation
St. Louis, Missouri, 63198

Ebrahim Firoozabady

DNA Plant Technology Corporation
Oakland, California, 94608

In the past decade, many breakthroughs have occurred in the genetic transformation and regeneration of crop species. Cotton is an important crop worldwide, grown for its fiber, oil, and by-products. Cotton breeders have developed many cultivars through quantitative selection for traits such as fiber quality, earliness, resistance to pests, disease, and stress, and other desirable morphological traits. Although cotton has been a recalcitrant plant to regenerate after tissue culture, much progress has been made toward regeneration of fertile transgenic plants from several cultivars. In this chapter, we will review methods for regeneration and genetic transformation of cotton and discuss applications of genetic engineering for cotton improvement.

8.1 Economic Importance of Cotton as World Crop

The genus *Gossypium* L. includes 39 species, 4 of which are domesticated. Wild cotton grows in the arid and semiarid tropics and subtropics of Africa, Central and South America, and Australia. Cotton was domesticated independently in

*Present Address: Promega Corporation, 2800 Woods Hollow Road, Madison, Wisconsin 53711

both the Old and New World (Fryxell, 1979). The two cultivated cotton species native to the Old World, *Gossypium arboreum* L. and *G. herbaceum* L., are diploid. Because of varying fiber quality and low yields, these diploid species account for less than 2% of world cotton production (Lee, 1984). Two other cotton species were domesticated in the New World, *G. hirsutum* L. (upland cotton) and *G. barbadense* L. (extra long-staple Egyptian and pima cotton). These allotetraploids arose adventitiously in South America through chromosome doubling after hybridization of African and South American diploid species (Cherry *et al.*, 1970). About 90% of the cotton grown worldwide is *G. hirsutum,* whereas *G. barbadense* accounts for the remaining 8% (Lee, 1984).

Cotton is an important crop grown worldwide where cultivation requirements for length of growing season, temperature, and sunlight are met. In 1987, cotton was grown in over 90 countries on 32.6 million hectares. Developing countries account for about 75% of the area grown, but only 62% of total production. In 1987, the major producers, accounting for 75% of production, were China, the United States, the Soviet Republics, Pakistan, and India (in order of contribution) (FAO, 1987). Annually, about 16 million tons of cotton fiber and about 33 million tons of cotton seed are produced worldwide. The United States is a major cotton producing country, accounting for 14% of world production in 1987 (Cotton:World Statistics, 1987). It is the fourth largest crop in the United States, ranking behind corn, soybeans, and wheat. The value of United States production was over $2.4 billion in 1986 and $3.6 billion in 1985 (USDA-National Agricultural Statistical Services, 1987).

Cotton is grown primarily for its lint, which provides much of the high quality fiber for the textile industry. Additional short fibers called linters are removed from cotton seed and used in cellulose production. Cotton seed is also a valuable commodity, producing oil, meal, and seed hulls. Cotton oil is the most valuable seed product and the second largest source of edible oil produced worldwide (FAO, 1987). After oil is extracted from the seeds, the remaining meal is used for animal feed. Although this meal contains high quality protein, its use as animal feed is restricted to ruminants because of the presence of gossypol, a toxic yellow polyphenolic pigment. When gossypol is extracted, the meal also can be used in poultry and swine feed. Several cultivars have seeds that lack gossypol glands; this meal is used for human as well as animal consumption.

8.2 Genetic Engineering

8.2.1 AGRONOMIC POTENTIAL OF GENETIC ENGINEERING IN COTTON

Extensive cultivation of cotton since the 18th century has resulted in selection for increased yield and improved fiber quality. Cotton breeders have continually sought to improve the crop through selection for cultivar types that suit mechanized cultivation and harvest, a wide range of environmental conditions, and the pressures of pests, weeds, and disease. Genetic engineering allows researchers to introduce novel genes for insect, bacterial, nematode, virus, and herbicide resistance into already existing cotton varieties. Some of these genes have been identified; their potential for cotton improvement is discussed in the following sections.

8.2.2 GENETIC ENGINEERING FOR INSECT-RESISTANT COTTON

Cotton growers use almost half the insecticides applied to crops in the United States (Adkisson *et al.*, 1982). Three-fourths of the damage to cotton in the United States is caused by the boll weevil (*Anthonomous grandis*) and the cotton bollworms (*Heliothis zea* and *Heliothis virescens*) (Pendergrass, 1989). Other damaging pests include the tarnished plant bug (*Lygus lineolaris*), the cotton fleahopper (*Pseudatomoscelis serriatus*), cotton leafworms (*Spodoptera littoralis*), and the pink bollworm (*Pectinophora gossypiella*). A number of other pests sporadically cause serious losses.

During the past 40 years, growers have sought to control cotton pests through application of synthetic insecticides. Three major classes of insecticide have been introduced sequentially: organochlorines, organophosphates, and, most recently, synthetic pyrethroids. These three classes of insecticides are broad-spectrum, controlling virtually all species of insect pests. However, these insecticides also interfere with the insect enemies (parasites and predators) of cotton pests and can reduce populations of honeybees, an important cotton pollinator. The use of these insecticides has followed a similar cycle: rapid expansion, with limited use providing excellent insect control and improved crop yields; "crisis," when pests become resistant, and many applications are required to provide control; followed by "disaster" in some areas, when the product becomes unusable (Adkisson *et al.*, 1982; Morton and Collins, 1989). Reports of pyrethroid-resistant insects have begun to appear (Morton and Collins, 1989), signaling the approach of another "crisis" in cotton pest management.

In response to the problems of insecticide use, cotton growers in the United States and other countries began to devise strategies of integrated pest management (IPM) in the 1970s. IPM involves limited use of insecticides, altering cultivation practices to reduce pest populations, planting rapid-fruiting short-season varieties of cotton that are harvested before insect predation reaches a maximum, and biological control of pests through natural plant pathogens (Adkisson *et al.*, 1982; Frisbie *et al.*, 1989; Green and Lyon, 1989).

Bacillus thuringiensis (*B. thuringiensis*) is a gram-positive soil bacterium that is one such natural plant pathogen (reviewed in Hofte and Whitely, 1989; see

Chapter 14 in Volume I of this treatise). *Bacillus thuringiensis* produces crystalline protein inclusions (ICP) during sporulation that are specifically toxic to a variety of insect hosts. Whereas most *B. thuringiensis* strains described are active against lepidopteran pests, including cotton bollworms, a few strains toxic to dipteran or coleopteran species have been described. Although *B. thuringiensis* spore/crystal mixtures are toxic to some cotton pests and have been used commercially as a microbial pesticide on cotton, problems associated with limited field stability and narrow host ranges have limited their use.

The limitations of ICPs as insecticides could be overcome by cloning two or more ICP genes and expressing them in cotton plants under the control of plant promoters and 3′ regulatory sequences. Several genetic engineering companies have reported on the generation of transgenic tobacco and tomato plants using a number of lepidopteran-toxic ICP genes (Adang *et al.*, 1987,1989; Barton *et al.*, 1987; Fischhoff *et al.*, 1987; Vaeck *et al.*, 1987); experiments are underway to introduce ICP genes toxic to cotton bollworms into transgenic cotton. To date, the levels of ICP gene expression reported in transgenic plants have been low, possibly because of mRNA instability in plant cells (Murray *et al.*, 1991). Efforts are underway in several laboratories (M. Adang, personal communication; K. Barton, personal communication) to rebuild ICP genes for high expression in plants.

Insects resistant to *B. thuringiensis* have not been reported in the field to date (Hofte and Whitely, 1989), possibly because of its rapid turnover in the environment. If transgenic cotton plants expressing high levels of ICPs became widely used, several insect generations per year would be exposed continuously to the ICPs, providing strong selective pressure for the evolution of resistant insects. McGaughey (1985) reported that the Indian mealworm (*Plodia interpunctella*) developed resistance to a commercial preparation of *B. thuringiensis* var. *kurstaki* HD-1. He reared a colony of Indian mealworms on a diet containing *B. thuringiensis* var. *kurstaki* HD-1 at a level predicted to produce 70–90% larval mortality. After two generations of selection, the larvae were approximately 30-fold more resistant, and after 15 generations, resistance had plateaued at a level 100-fold higher than the control level. The resistance persisted after selection was terminated and was inherited as a recessive trait. McGaughey and Johnson (1987) observed that other *B. thuringiensis* strains were still toxic to the resistant larvae. Hofte and Whitely (1989) suggest that simultaneous expression in a transgenic plant of two or more ICP genes that act independently against the same insect could reduce the probability of evolution of resistant insect populations. A second strategy to reduce the development of ICP-resistant insect populations is to express the ICP genes in plants at levels high enough to insure a profitable cotton crop, but not so high that they provide strong selective pressure for resistance. This could be achieved through the use of appropriate cotton promoters. Hopefully, plant biotechnology companies seeking to engineer insect-resistant cotton have learned respect for the high potential for evolution of resistance from the history of insecticide use on cotton pests.

A number of other genes have the potential to confer insect resistance in transgenic cotton plants. Cowpea (*Vigna unguiculata*) seeds contain trypsin inhibitors (CpTIs), small polypeptides of 80 amino acids belonging to the Bowman–Birk type of double-headed serine protease inhibitor (Hilder *et al.*, 1987; see Chapter 13 in Volume 1 of this treatise). Feeding trials with purified CpTI in artificial insect diet showed these to be toxic to a wide range of insects, including *H. virescens* and *S. littoralis*. Transgenic tobacco plants containing cloned CpTI under the control of the cauliflower mosaic virus (CaMV) 35S promoter were tested in an insect bioassay with *H. virescens* larvae (Hilder *et al.*, 1987). Levels of insect resistance were correlated positively with high levels of CpTI protein. Other insecticidal protein genes could be introduced to cotton from insect viruses, bacterial pathogens, or predators such as scorpions.

8.2.3 USE OF HERBICIDE RESISTANCE GENES IN TRANSGENIC COTTON

Cotton losses to weeds can be severe because of competition for water, nutrients, and sunlight. In some cases, allopathic weed–crop interactions may inhibit cotton growth, reducing yields and lint quality (Frans and Chandler, 1989). Weeds also can serve as hosts for numerous pests, so weed control can reduce losses due to insects and diseases (Ridgway *et al.*, 1984). Although herbicides have proved valuable in improving cotton yields, they are costly and can affect the cotton plants adversely. Introduction of genes for herbicide resistance into cotton could reduce the physiological impact of these nonselective herbicides.

Glyphosate [*N*-(phosphonomethyl)glycine] is a broad-spectrum herbicide widely used for both pre- and postemergence weed control in cotton. However, it is reported to stunt cotton and increase abcission after direct contact (Ridgway *et al.*, 1984; Frans and Chandler, 1989). Glyphosate inhibits 5-enolpyruvylshikimate-3-phosphate (EPSP) synthase, a key enzyme in the biosynthesis of aromatic amino acids in plants and bacteria (Botterman and Leemans, 1988). Glyphosate resistance has been conferred on transgenic petunias through introduction of multiple copies of the wild-type precursor EPSP synthase under the control of the strong CaMV 35S promoter (Shah *et al.*, 1986). A glyphosate-tolerant gene for EPSP synthase has been isolated from *Salmonella typhimurium* that, when introduced to transgenic tobacco, tomato, and petunia (Gasser and Fraley, 1989), increased glyphosate tolerance. Thus, EPSP synthase is an excellent candidate gene for introduction to transgenic cotton. Resistance genes for several other broad-spectrum herbicides have been introduced to transgenic plants (see Chapter 9 in Volume 1 of this treatise). Such genes may confer herbicide in cotton and expand the number of useful cotton herbicides.

8.3 Cotton Regeneration

A prerequisite to transformation of cotton is the development of methods to regenerate healthy fertile plants from tissue culture efficiently. Tissue culture methods for cotton have lagged behind those for other major crops, with the exception of soybean (*Glycine max*). A number of successful procedures for regeneration of cotton plants from tissue culture have been reported in recent years. We will review these methods and discuss their application for the different transformation procedures used for cotton.

8.3.1 SOMATIC EMBRYOGENESIS AND PLANT REGENERATION IN COTTON

Ammirato (1983) provides a general discussion of somatic embryogenesis as a method of plant regeneration. The basic somatic embryogenesis protocol involves use of an explant of the desired species to establish callus tissue on a primary medium providing nutrients, vitamins, and an auxin source. The callus proliferates and dedifferentiates, becoming a mass of unorganized cells and cell clusters. Transfer to a second medium devoid of growth regulators prompts initiation of embryo development or maturation of embryos initiated on the primary medium. Embryogenesis protocols typically are highly specific for the type of plant and explant source regenerated. Somatic embryos have been regenerated both from cotton callus directly or from embryogenic callus placed in suspension culture.

The first successful regeneration of *G. hirsutum* L. cv. 'Coker 310' was reported by Davidonis and Hamilton (1983). They initially established callus from seedling cotyledon explants on modified LS medium (Linsmaier and Skoog, 1965) containing 30 g/liter glucose, 2 mg/liter α-naphthaleneacetic acid (NAA), and 1 mg/liter kinetin. After 3 months, a slow-growing gray friable callus was established. The callus then was subcultured on LS medium containing 30 g/liter glucose, 1 mg/liter NAA, and 0.5 mg/liter kinetin over a 2-yr period. An embryogenic callus formed after transfer to modified LS medium lacking growth regulators. Development of the somatic embryos was enhanced by transfer to modified Murashige and Skoog (MS) media, lacking NH_4NO_3 but containing twice the standard KNO_3 concentration and 0.1 mg/liter gibberellic acid (GA). Root initiation and growth was promoted by reducing glucose concentration to 5 g/liter. Although plants were generated following this procedure, the long induction period and low efficiency of embryo formation were undesirable.

Shoemaker *et al.* (1986) evaluated 17 *G. hirsutum* cultivars for somatic embryogenesis and plant regeneration using three initiation and three induction media. Callus cultures were initiated from hypocotyl sections of aseptically germinated seeds. A wide variety of callus initiation and proliferation responses was observed among the different cultivars tested on all three initiation media; only cv. 'Acala 1517' showed no response (Table 8.1). All three media resulted in callus proliferation within 3–5 days, with 2I resulting in approximately 30% more callus mass than 1I or 3I, although this varied among cultivars. Calluses from each initiation regime were subcultured onto the three maintenance media. Results were variable but, generally, 3M showed the least vigorous growth and 1M the most. Only cultivars 'Coker 201' and '315' developed embryogenic calluses under this regime. Emergence of the calluses was not correlated with any of the specific induction/maintenance combinations. Mature embryos and, eventually, plants developed after sequential subculture of embryogenic calluses on SC1 and SC2 media. Transfer of embryogenic calluses from media containing 3% glucose to 3% sucrose was essential for maintenance of embryonic development. Approximately 40% of mature embryos underwent normal germination and produced plants. This procedure was simple, straightforward, and rapid, producing embryogenic calluses in 4–6 wk and normal-appearing transgenic plants.

Gawel *et al.* (1986) describe somatic embryogenesis from callus cultures of mature leaf and petiole explants from six cotton varieties using several different media. Using leaf disks, only media 3 containing LS salts and vitamins, 30 g/liter glucose, 0.5 mg/liter NAA, 1.0 mg/liter kinetin, 0.1 mg/liter GA, and 0.4 mg/liter thiamin produced embryos, and then only with 'Coker 312.' Embryos developed on callus formed from petiole tissue in all three media used, but only for varieties 'Coker 312' and 'T25.' The investigators conclude that cotton embryogenesis is influenced strongly by genotype.

Two protocols for regeneration of cotton plants from embryogenic suspen-

TABLE 8.1
Media Used in Shoemaker *et al.* (1986)

Medium	Salts[a]	Glucose (G) or Sucrose (S) % (w/v)	NAA (mg/liter)	2iP (mg/liter)	Kinetin (mg/liter)	Adenine (mg/liter)	Glutamine (m*M*)
1I	MS	3 G	1	—	1	40	—
2I	MS	3 G	2	—	1	40	—
3I	LS	3 G	2	—	1	—	—
1M	MS	3 G	1	10	—	—	—
2M	MS	3 G	1	—	0.5	—	10
3M	LS	3 G	1	—	0.5	—	—
SC1	MS	3 G	2	—	1	—	—
SC2	MS	3 S	2	—	1	—	—

[a]MS, Murashige and Skoog (1962); LS, Linsmaier and Skoog (1965).

sion cultures have been described. The first was developed by Trolinder and Goodin (1987,1988a,b). Embryogenic calluses were initiated in eight cultivars using MS media supplemented by 100 mg/liter *myo*-inositol, B_5 vitamins, 30 g/liter glucose, and four different hormone regimes (Trolinder and Goodin, 1987). These were 0.5 mg/liter 2,4-D; 0.1 mg/liter 2,4-D; 0.1 mg/liter 2,4-D and 0.1 mg/liter kinetin; and 0.1 mg/liter 2,4-D and 0.5 mg/liter kinetin. 'Coker 312' developed normal-appearing embryogenic tissue under all regimes tested, although all varieties tested developed some embryos in more than one of the regimes tested. 'Coker 312' hypocotyl explants also were evaluated for embryogenic potential on MS initiation medium containing two levels (0.1 mg/liter or 0.5 mg/liter) of the cytokinin kinetin but lacking the auxin 2,4-D. Both regimes produced embryogenic calluses. A higher percentage of calluses was embryogenic with 0.5 mg/liter kinetin medium, but this tissue was brown and less friable.

Subsequently suspension cultures were established from 'Coker 312' embryogenic calluses initiated on medium containing 0.1 mg/liter 2,4-D and 0.1 mg/liter kinetin. These calluses had been maintained in culture for 4 wk to 1 yr prior to establishing suspension cultures; 6-wk cultures were optimal. Suspensions were initiated by sieving calluses through consecutively finer sieves and inoculating liquid callus initiation media (lacking 2,4-D or kinetin) with 100 mg callus tissue/10 ml media. Liquid cultures were rotated at 120 rpm under light and subcultured after sieving every 3–6 wk. Development of embryos was induced by plating on solid induction media without hormones but with an additional 1.9 g/liter KNO_3. Embryos developed about 6 wk after plating; embryos greater than 5 mm were germinated successfully in vermiculite soaked in SH (Stewart and Hsu, 1977) germination medium containing 0.1 mg/liter indole-3-acetic acid (IAA). Of the regenerated plants that flowered, only 15.4% set seed, possibly because of the length of time the initial callus was cultured prior to suspension and plating.

In a further study, Trolinder and Goodin (1988a) compared somatic embryogenesis in 'Coker 312' using a wide range of explant sources and hormone regimes. All explants from 3-day-old seedlings readily formed embryogenic callus under all treatment regimes. Mature tissues formed embryos under a more limited number of treatments and were difficult to sterilize. Treatments containing 2,4-D as an auxin source produced callus quickly and readily. Callus formed more slowly after NAA treatment, but frequently contained later stages of embryos. Generally, (2-isopentyl)adenine (2i)P and kinetin were better cytokinins than benzyladenine (BA).

Finer (1988) also described a procedure for plant regeneration from somatic embryogenic suspension cultures established from cotyledons of 'Coker 310.' Embryogenic calluses were induced by placing explants on media containing MS salts, B_5 vitamins, and 3% glucose containing 2 mg/liter NAA, 1 mg/liter kinetin, and 8% agar (pH 5.7). After four weekly subcultures on this media, calluses were transferred to the induction medium, containing sucrose in place of glucose, for 1 month. 250 mg callus was place in 50 ml suspension culture induction medium containing MS salts, B_5 vitamins, 2% sucrose, and either 0.1 mg/liter 2,4-D or 0.5 mg/liter picloram and agitated at 150 rpm. Embryogenic tissue detected after 4–8 wk was transferred to embryo proliferation medium substituting 5 mg/liter 2,4-D as an auxin. Initiation of suspension cultures with a low inoculum produced very prolific embryogenic suspensions.

Somatic embryos developed when the suspension was plated to a solidified hormone-free medium. However, embryos also developed successfully in hormone-free liquid medium containing 15 m*M* glutamine. Early embryogenesis was fairly synchronous, and large numbers were produced. Replacement of culture medium after 1 wk appeared to stimulate embryo development. Embryos were germinated on solidified medium containing modified MS salts (no NH_4NO_3 and $2 \times KNO_3$), B_5 vitamins, 1% sucrose, and 0.2% Gel-rite (pH 5.7). After root and shoot elongation (approximately 2–3 months), the plantlets were transferred to soil. Regenerated plants were fertile.

8.3.2 OTHER APPLICATIONS OF COTTON TISSUE CULTURE

A number of the wild species of *Gossypium* have characteristics, including disease resistance, insect-pest resistance, drought resistance, and superior fiber qualities, that are lacking in the cultivated varieties (Bajaj and Gill, 1986). *Gossypium arboreum,* a cultivated diploid cotton, is more tolerant of environmental stress and more insect resistant than the widely grown tetraploid *G. hirsutum* (Gill and Bajaj, 1987). Cotton breeders would like to overcome barriers to hybridization, and use these species as reservoirs for new genetic variation. After sexually induced hybridization, tissue culture techniques can nurture the otherwise abortion-prone embryo through cell divisions until an F_1 plant regenerates. Embryo rescue techniques have led to transfer of novel disease resistance from wild to cultivated rice species (Cohen *et al.*, 1988). Gill and Bajaj (1987) have reported some success in regenerating F_1 plants from *G. arboreum* × *G. hirsutum,* but fertility barriers in these F_1 offspring must now be overcome. The investigators also were able to introduce traits from the wild species *G. anomalum* to *G. arboreum* following F_1 embryo rescue and backcross (Bajaj and Gill, 1986).

Pollen mother cells isolated from anthers can be isolated and cultured to produce haploid plants that can be screened rapidly for desirable traits. A number of new plant varieties have been developed following establishment of homozygous lines for hybrid seed production using haploid tobacco, rice, and rape plants (Morrison and Evans, 1988). Bajaj and Gill (1986) report that anther cultures from *G. arboreum* proliferated and underwent pollen embryogenesis as well as callus formation.

8.4 Systems of Plant Genetic Engineering

Several methods have been developed for introducing foreign genes into plants (reviewed in Gasser and Fraley, 1989; see Chapter 4 in Volume 1 of this treatise). *Agrobacterium tumefaciens* provides a natural gene-transfer mechanism for transfer of specific DNA sequences into the genomes of many dicotyledonous plants. Several direct DNA transfer methods also have been developed, including microinjection, electroporation, and particle gun technology. Application of these DNA transformation systems requires optimization of selectable markers such as antibiotic resistance in regenerating plants. We will discuss these systems and their application in cotton transformation and regeneration.

8.4.1 *AGROBACTERIUM*-MEDIATED GENE TRANSFER TO COTTON

Agrobacterium gene transfer has several advantages in cotton transformation (Gasser and Fraley, 1989). (1) Whole plant tissues can be transformed directly, bypassing the need for a regenerable protoplast system. (2) T-DNA integration generally results in transfer of intact genes contained within T-DNA borders. These genes are expressed stably and inherited with Mendelian segregation ratios; chromosomes into which they are inserted do not appear to be rearranged grossly.

Two groups (Firoozabady *et al.*, 1987; Umbeck *et al.*, 1987) have reported regeneration of transgenic cotton using *Agrobacterium* vectors. Umbeck *et al.* described regeneration of 'Coker' varieties 310, 312, and 5110 using hypocotyl sections inoculated with *Agrobacterium* strain LBA4404 containing plasmid CMC1204. This plasmid contained the marker enzymes neomycin phosphotransferase (NPTII) and chloramphenicol acetyltransferase (CAT), each flanked by the promoter and 3′ polyadenylation region of the nopaline synthase (*nos;* nopaline dehydrogenase) gene of pTiT37. Hypocotyl sections were inoculated with 10 μl of an overnight culture of *Agrobacterium* (10^8 cells per ml) and maintained for 3 days without selection on SH medium (Stewart and Hsu, 1977). The explants were placed on induction media [MS salts with B_5 vitamins, 2.4-D (0.1 mg/ml), kinetin (0.1 mg/ml), Gel-rite (1.6 g/liter), and $MgCl_2$ (0.75 g/liter)]. The antibiotics cefotaximine (50–100 mg/liter) and carbenicillin (400–500 mg/liter) were added to control *Agrobacterium* growth and kanamycin sulfite (5–50 mg/liter) was added as a selection agent. Subcultures were made every 4–6 wk. After tissue amplification, embryogenic tissues were transferred to Trolinder

TABLE 8.2
Media Used in Firoozabady *et al.* (1987)[a]

Medium	Salts	Glucose % (w/v)	NAA (mg/liter)	2iP (mg/liter)	Zeatin (mg/liter)	Gibberellic acid (mg/liter)	Ascorbic acid (mg/liter)
1/2G0	1/2MS	1.5	—	—	—	—	—
1/2G0AA	1/2MS	1.5	—	—	—	—	50
1/2G0iAA	1/2MS	1.5	—	0.1	—	—	50
G0	MS	3	—	—	—	—	—
G2	MS	3	0.1	5	—	—	—
G3	MS	3	5	0.1	—	—	—
GRMgn	SH	0.5	0.01	—	—	0.1	—

[a]MS, Murashige and Skoog (1962); SH, Stewart and Hsu (1977).

and Goodin (1987) regeneration medium. When embryos were 4 mm long, they were transferred to vermiculite saturated with SH medium plus IAA, GA, and kinetin (0.1 mg/liter each). Three transgenic plants were identified that expressed NPTII and CAT. Southern blot analysis of a single plant indicated that portions of the modified T-DNA region were deleted in this regenerant. A time frame for regeneration of transgenic cotton was not given.

Firoozabady *et al.* (1987) published a different transformation and regeneration protocol, transforming cotyledon pieces from aseptically germinated 'Coker 201' seedlings with *Agrobacterium* strain LBA4404 containing plasmid pH575. Plasmid pH575 was a "micro Ti" binary vector containing the gene for NPTII under the control of the CaMV 19S promoter, and octopine synthase (*ocs*), with its own regulatory sequences. Media used are summarized in Table 8.2. Cotyledon sections (0.5 cm^2) from 10- to 14-day-old seedlings were dipped in *Agrobacterium* scraped off agar medium and resuspended in liquid G_2 medium (10^8 cells per ml) so all cut edges contacted bacterial solution. Explants were blotted dry and placed to insure flat contact on Whatman (Clifton, New Jersey) No. 1 filter paper on solidified G_2 medium without antibiotics for a 3-day cocultivation period. Use of filter paper reduced bacterial overgrowth on the explants. Tissues were incubated at 25°C with a 16-hr photoperiod (90 $\mu E\ m^{-2}\ sec^{-1}$). The explants were placed on G_2 induction medium containing carbenicillin (500 mg/liter) to control *Agrobacterium* growth and kanamycin (15–35 mg/liter) for selection of transformed tissues. Tissues were incubated at 30°C with a 16-hr photoperiod (90 $\mu E\ m^{-2}\ sec^{-1}$). Microcalluses (2–4 mm) began to appear in 2–4 wk (Fig. 8.1 A,B). They were excised and transferred to fresh medium. After 2–3 more wk, calluses were placed on G_0 medium under kanamycin selection (25 mg/liter) under lower light intensity (10 $\mu E\ m^{-2}\ sec^{-1}$) until friable embryogenic callus formed (Fig. 8.1 C). Mature somatic embryos

Figure 8.1 Regeneration of transgenic cotton plants. (A) Kanamycin-resistant microcallus (arrow) developed at wound site 10 days after coculture. Bar: 7 mm. (B) Cotyledon section showing several calluses on 25 mg/liter kanamycin 3 weeks after coculture. Bar: 1 mm. (C) Embryogenic calluses developed 6–8 weeks after transfer to medium G_0. Bar: 2 mm. (D) Somatic embryos with different morphologies. Bar: 2 mm. (E) Regenerated transgenic plants after transfer to soil.

(5–10 mm in length; Fig. 8.1 D) were transferred to GRMgn medium and incubated at 30°C with a 16-hr photoperiod (90 μE m^{-2} sec^{-1}) to germinate and form plantlets. Plantlets were transferred to soil, maintained at 25°C night, 30°C day, 16-hr photoperiod (150 μE m^{-2} sec^{-1}), and gradually hardened off before transfer to the greenhouse (Fig. 8.1 E). Transgenic plants expressed the *nptII* and *ocs* genes at varying levels. DNAs from 15 plants were analyzed on Southern blots. All plants contained intact *nptII* genes, but the *ocs* gene had been lost or rearranged in some plants. The time from initiation of transformation to identification of transgenic plants was 6–8 months using this protocol.

Despite success in regenerating transgenic cotton from 'Coker' varieties (grown primarily in the eastern United States), regeneration and transformation of other varieties such as 'Acala,' 'Delta,' 'Plains,' and 'GSA' has not been reported. We (E. Firoozabady and D. L. DeBoer, unpublished results, 1988) have been successful in transformation and regeneration of some of these varieties using a modification of our original published protocol. We were successful using this protocol with a wide variety of explant sources including cotyledons, hypocotyls, roots, and petioles from mature and precociously germinated seeds, and embryos from immature seeds. Cotyledon tissue, described in our original protocol, was less preferable than hypocotyl or immature embryo tissue for transformation, since it tended to curl up, out of contact with antibiotics in the medium. Media used are summarized Table 8.2.

Using hypocotyls as an explant source, callus initiation of G_2 cultures continued until calluses were sufficiently large that auxin levels used were not toxic to the cells, generally 3–4 wk for 'Coker' varieties but 4–6 wk for other varieties ('Acala SJ-5,' 'GSC25,' and 'GSA75'). Tissues were maintained at 30°C, 16-hr photoperiod (90 μE m^{-2} sec^{-1}) for the first 2 wk followed by low light (10 μE m^{-2} sec^{-1}). Calluses were then transferred to G_3 (embryogenic callus induction medium with a high auxin to cytokinin ratio) and maintained with periodic subculture every 4 wk. When callus became embryogenic (characterized by production of proembryoids with globular and heart-shaped structures) it was transferred to G_0 medium (without kanamycin or carbenicillin) and maintained with periodic subculture every 2–3 wk. Typically, embryogenic callus developed 4.5–7.5 months after inoculation with *Agrobacterium*. Initially, 'GSC' varieties developed somatic embryos after 13 months in culture. However, when regenerated 'GSC' plants were used as an explant source, embryogenic callus developed after 4–5 months.

Some somatic embryos began to mature in 3–4 wk, and normal-appearing embryos with periodic subculture every 2–3 wk. Mature somatic embryos suitable for transfer to embryo germination medium (GRM_{gn}) were 3–5 mm in size, green in color, and had a pair of cotyledons. Embryos at this stage were maintained at 30°C, 16-hr photoperiod (90 μE m^{-2} sec^{-1}). These embryos were inspected weekly for development of root and top growth, top growth being most crucial. Approximately 30% of embryos on GRM_{gn} germinated normal roots and top growth, and were transferred to $\frac{1}{2}G_0AA$ medium in magenta boxes. Embryos that did not mature initially in 4 wk on GRM_{gn} were transferred to $\frac{1}{2}G_{0i}AA$ media, which induced further top growth in approximately 20% of the remaining embryos. Some embryos failed to mature on either media and were discarded. Developing plantlets were maintained in magenta cubes at 25°C with high light intensity (150 μE m^{-2} sec^{-1}) until their leaves reached the tops of the cubes. The lids were cracked gradually and the plants were hardened in growth chambers before being transferred to soil.

Most regenerated cotton plants using this protocol were fertile and normal in appearance. However, almost half the plants regenerated initially showed somaclonal variation and lack of fertility, possibly because of longer periods in culture. Trolinder and Goodin (1987) also observed that plants regenerated over long periods in tissue culture exhibited lower fertility. When regenerated plants were used as an explant source, they regenerated and transformed better than control plants, possibly reflecting genotype selection for regeneration.

8.4.2 TRANSFORMATION OF COTTON VIA DIRECT DNA TRANSFER

Cotton transformation by direct DNA transfer has been reported by Zhou *et al.* (1980) through injection of DNA obtained from *Gossypium barbadense* into embryos in immature *G. hirsutum* bolls. Transfer of DNA was suggested by differences in fiber length and boll size in the resultant plants. The absence of a readily identifiable discrete foreign gene precluded analysis by DNA hybridization to genomic Southern blots, so these results await molecular confirmation. However, Southern blots confirming direct DNA transfer by microinjection of rye have been reported (de la Peña *et al.*, 1987).

Particle bombardment has been used to transform plant cells (Klein *et al.*, 1987) and shoot tips (McCabe *et al.*, 1988). In this system, metal microprojectiles are coated with DNA and accelerated to speeds up to several hundred meters per second. The particles carry DNA through the layers of cells and allow transformation in tissue explants or somatic embryos. Finer and McMullen (1990) report regeneration of transgenic cotton after particle bombardment. Embryonic suspension cultures were prepared as described previously (Finer, 1988) except that 5 mg/liter 2,4-D was substituted for picloram in the cotton embryo proliferation media (CEPM). Initial optimization was monitored after bombardment with tungsten particles coated with a plasmid containing the *gus* gene flanked by a CaMV 35S promoter and a 3′ end. Using the optimized procedure, embryos were bombarded with pellets coated with aminoglycoside phosphotransferase type IV gene (conferring hygromycin resistance) flanked by a CaMV 35S promoter and a 3′ end. Transgenic somatic embryos were selected by cultivation in CEPM containing 100 mg/liter hygromycin. Hygromycin-resistant tissue was selected visually and removed for further culture (hygromycin-sensitive cotton tissue turned white). Transgenic plants were regenerated within 5 months of particle bombardment using previously established embryogenic suspension lines. Particle bombardment of soybean shoot tips led to regeneration of chimeric plants (McCabe *et al.*, 1988). To date, all cotton tissues analyzed by Southern hybridization have been transformed, frequently with multiple copies of plasmid DNA, although rearrangements were detected also.

8.5 New Directions for Transgenic Cotton

Although there has been much progress in transformation and regeneration of cotton, improvements need to be made in the following areas. (1) Transformation and regeneration protocols must be developed for use on cultivars other than the easily regenerated ‘Coker’ varieties, and on the other cotton species grown commercially. (2) The time required for cotton regeneration must be shortened so regenerated plants have less somaclonal variation and retain the desirable agronomic traits of the original cultivar. (3) Tissue-specific and developmentally regulated promoters for cotton should be characterized, as should promoters capable of regulating transcription at varying levels. (4) Additional genes suitable for transfer to cotton should be identified. For example, genes useful in controlling male sterility in cotton (Galau and Wilkins, 1989) could simplify development of hybrid cotton. It will be more difficult to identify and clone single genes for improvement of traits such as fiber quality and yield, plant morphology and physiology, drought resistance, oil composition, and elimination of gossypol from seeds. However, cotton breeders are beginning to use techniques of restriction fragment length polymorphism mapping to identify such traits, and some eventually may be identified and cloned (Tanksley *et al.*, 1989).

Acknowledgments

This research was supported by Agrigenetics Advanced Sciences Corporation, Madison, Wisconsin 53716, and by Public Health Service grant CA-22443. E. E. Murray was supported by Public Health Service Training grant CA-9075.

References

Adang, M. J., DeBoer, D., Endres, J., Firoozabady, E., and Klein, J., *et al.* (1989). *In* "Biotechnology, Biological Pesticides and Novel Plant-Pest Resistance Strategies for Insect Pest Management" (Roberts, D. W. and Granados, R. R., eds.) pp. 31–37. Boyce Thompson Institute for Plant Research, Ithaca, New York.

Adang, M. J., Firoozabady, E., Klein, J., DeBoer, D., and Sekar, V., *et al.* (1987). *In* "Molecular Strategies for Crop Protection" (Arntzen, C. J., and Ryan, C., eds.), pp. 345–353. Alan R. Liss, New York.

Adkisson, P. L., Niles, G. A., Walker, J. K., Bird, L. S., and Scott, H. B. (1982). *Science* **216,** 19–29.

Ammirato, P. V. (1983). *In* "Handbook of Plant Cell Culture" (P. V. Ammirato, D. A. Evans, W. R. Sharp, and Y. Yamada, eds.). Macmillan, New York.

Bajaj, Y. P. S., and Gill, M. S. (1986). *In* "Biotechnology of Plants and Microorganisms" (O. J. Crocomo, W. R. Sharp, D. A. Evans, J. E. Bravo, F. C. A. Tavares, and E. F. Paddock, eds.), pp. 118–151. Ohio State University Press, Columbus.

Barton, K., Whitely, H., and Yang, N.-S. (1987). *Plant Physiol.* **85,** 1103–1109.

Botterman, J., and Leemans, J. (1988). *Trends Genet.* **4,** 219–222.

Cherry, J. P., Katterman, F. R. H., and Endrizzi, J. E. (1970). *Evolution* **24,** 431–477.

Cohen, J. I., Plunckett, D. L., Smith, N. J. H., and Jones, K.-A. (1988). *Bio/Technology* **6,** 387–392.

Cotton: World Statistics (1987). *Bull. Internat. Cotton Adv. Comm.* **40,** 25–39.

Davidonis, G. H., and Hamilton, R. H. (1983). *Plant. Sci. Lett.* **32,** 89–93.

de la Peña, A., Lörz, H., and Schell, J. (1987). *Nature (London)* **325,** 274–276.

Food and Agriculture Organization (FAO) (ed.) (1987). "FAO Production Yearbook," Vol. 41.

Finer, J. J. (1988). *Plant Cell Rep.* **7,** 399–402.

Finer, J. J., and McMullen, M. D. (1990). *Plant Cell Rep.* **8,** 586–589.

Firoozabady, E., DeBoer, D. L., Merlo, D. J., Halk, E. L., Amerson, L. N., Rashka, K. E., and Murray, E. E. (1987). *Plant Mol. Biol.* **10,** 105–116.

Fischhoff, D. A., Bowdish, K. S., Perlak, F. J., Marrone, P. G., and McCormick, S. M., *et al.* (1987). *Bio/Technology* **5,** 807–813.

Frans, R. E., and Chandler, J. M. (1989). *In* "Integrated Pest Management Systems and Cotton Production," (R. E. Frisbie, K. M. El-Zik, and L. T. Wilson, eds.), pp. 327–360. John Wiley & Sons, New York.

Frisbie, R. E., El-Zik, K. M., and Wilson, L. T. (1989). "Integrated Pest Management Systems and Cotton Production." John Wiley & Sons, New York.

Fryxell, P. A. (1979). "The Natural History of the Cotton Tribe." Texas A&M University Press, College Station.

Galau, G. A., and Wilkins, T. A. (1989). *Theor. Appl. Genet.* **78,** 23–30.

Gasser, C. S. and Fraley, R. T. (1989). *Science* **244,** 1293–1299.

Gawel, N. J., Rao, A. P., and Robacker, C. D. (1986). *Plant Cell Rep.* **5,** 457–459.

Gill, M. S., and Bajaj, Y. P. S. (1987). *Euphytica* **36,** 625–630.

Green, M. B., and Lyon, D. J. deB. (1989). "Pest Management in Cotton." Ellis Horwood, Chichester.

Hilder, V. A., Gatehouse, A. M. R., Sheerman, S. E., Barker, R. F., and Boulter, D. (1987). *Nature (London)* **330,** 160–163.

Hofte, H., and Whitely, H. R. (1989). *Microbiol. Rev.* **53,** 242–255.

Klein, T. M., Wolf, E. D., Wu, R., Sanford, C. (1987). *Nature* (London) **327,** 70–72.

Lee, J. A. (1984). *In* "Cotton" (R. J. Kohel and C. F. Lewis, eds.), pp. 6–26. American Society of Agronomy, Madison, Wisconsin.

Linsmaier, F. M., and Skoog, F. (1965). *Physiol. Plant* **18,** 100–127.

McCabe, D. E., Swain, W. F., Martinell, B. J., and Christou, P. (1988). *Bio/Technology* **6,** 923–926.

McGaughey, W. H. (1985). *Science* **299,** 193–195.

McGaughey, W. H., and Johnson, D. E. (1987). *J. Econ. Entomol.* **80,** 1122–1126.

Morrison, R. A., and Evans, D. A. (1988). *Bio/Technology* **6,** 684–689.

Morton, N., and Collins, M. D. (1989). *In* "Pest Management in Cotton" (M. B. Green and D. J. DeB. Lyon, eds.), pp. 155–165. Ellis Horwood, Chichester.

Murashige, T., and Skoog, F. (1962). *Physiol. Plant* **15,** 473–497.

Murray, E. E., Rocheleau, T., Eberle, M., Stock, C., Sekar, V., and Adang, M. (1991). *Plant Mol. Biol.* **16,** 1035–1050.

Pendergrass, J. (1989). *In* "Pest Management in Cotton" (M. B. Green and D. J. deB. Lyon, eds.), pp. 11–15. Ellis Horwood, Chichester.

Ridgway, R. L., Bell, A. A., Veech, J. A., and Chandler, J. M. (1984). *In* "Cotton" (R. J. Kohel and C. F. Lewis, eds.), pp. 266–367. American Society of Agronomy, Madison, Wisconsin.

Shah, D. M., Horsch, R., Klee, H., Kishore, G., Winter, J., Turner, N., Hironaka, C., Sanders, P., Gasser, C., Aykent, S., Siegel, N., Rogers, S., and Fraley, R. (1986). *Science* **233,** 478–481.

Shoemaker, R. C., Couche, L. J., and Galbraith, D. W. (1986). *Plant Cell Rep.* **3,** 178–181.

Stewart, J. D. McD., and Hsu, C. L. (1977). *Planta* **137,** 113–117.

Tanksley, S., Young, N. D., Paterson, A. H., and Bonierbale, M. W. (1989). *Bio/Technology* **7,** 257–263.

Trolinder, N. L., and Goodin, J. R. (1987). *Plant Cell Rep.* **6,** 231–234.

Trolinder, N. L., and Goodin, J. R. (1988a). *Plant Cell Tissue Organ Cult.* **12,** 31–42.

Trolinder, N. L., and Goodin, J. R. (1988b). *Plant Cell Tissue Organ Cult.* **12,** 43–53.

Umbeck, P., Johnson, G., Barton, K., and Swain, W. (1987). *Bio/Technology* **5,** 263–266.

USDA-National Agricultural Statistical Service (ed.) (1987). "Crop Production," Vol. CrPr2-2 (5-87), pp. A-16–A-17.

Vaeck, M., Reynaerts, A., Hofte, H., Jansens, S., De Beuckeleer, M., Dean, C., Zabeau, M., Van Montegue, M., and Leemans, J. (1987). *Nature (London)* **328,** 33–37.

Zhou, G. Y., Weng, J., Zeng, Y., Huang, J., Qian, S., and Liu, G. (1983). *Meth. Enzymology* **101,** 433–481.

9

Transgenic Arabidopsis

Brigitte Damm
MOGEN International
2333 CB Leiden, The Netherlands

Ursula Halfter
The Rockefeller University
New York, New York 10021

Thomas Altmann and
Lothar Willmitzer
Institut für Genbiologische Forschung GmbH
1000 Berlin 33, Germany

Arabidopsis thaliana (L.) Heynh., a member of the Brassicaceae, has been used in classical plant genetics since the 1940s (Laibach, 1943) because of the small size of the mature plant, its short generation time, and the large number of offspring normally produced after self-pollination. Due to the small size of the seeds, large scale mutagenesis is possible with this plant; thus, many mutations have been identified, studied, and mapped (for reviews, see Redei, 1975; Koornneef *et al.*, 1983; Estelle and Somerville, 1986; Meinke, 1986; Meyerowitz, 1987; Bowman *et al.*, 1988). Most of these gene loci, however, have not been cloned. The mutations have been assembled into a classical genetic linkage map comprising about 80 loci (Koornneef *et al.*, 1983; Koornneef, 1987). This genetic map consists of five linkage groups, in concordance with the haploid chromosome number of five. Multiple marker strains with mutations on each of the chromosomes are available (Koornneef, 1983; Koornneef and van der Veen, 1983; Koornneef *et al.*, 1987). In addition, recently two RFLP (restriction fragment length polymorphism) maps became available. They contain 90 (Chang *et al.*, 1988) or 94 (Nam *et al.*, 1989) molecular markers that are distributed randomly over the five chromosomes. The maps are integrated only partly with each other at present, but this work should be completed soon.

During recent years, *A. thaliana* also has been adopted as a model system for studies on the molecular genetic level (for recent reviews, see Meyerowitz, 1987,1989; Bowman *et al.*, 1988). It has been shown that *A. thaliana* has a small nuclear genome with a very low level of dispersed repetitive DNA (Leutwiler *et al.*, 1984; Pruitt and Meyerowitz, 1986). On the basis of DNA reassociation kinetics, Leutwiler *et al.* (1984) have estimated that the haploid nuclear genome of *A. thaliana* contains approximately 70,000 kilobase pairs (kb) which is thus the smallest genome known for higher plants so far. The genome organization of *Arabidopsis* offers significant advantages when considering genetic experiments designed to identify and isolate plant genes by chromosome walking, insertional mutagenesis, or complementation.

The features of the *Arabidopsis* genome allow rapid and repetitive screening of genomic libraries with minimal effort. A lambda or cosmid library representing several genomes worth of *Arabidopsis* DNA can be plated on a single large petri dish. This characteristic, along with the near absence of interspersed repetitive DNA (Pruitt and Meyerowitz, 1986), makes *A. thaliana* uniquely suited to chromosome-walking experiments. Starting from a known chromosomal region, overlapping clones are isolated successively, proceeding to a region containing a desired gene. In addition, the availability of dense genetic maps of marker genes and RFLPs further facilitates gene cloning by chromosome walking (Bowman *et al.*, 1988; Meyerowitz, 1989). For example, Bleeker *et al.* (1990) reported preliminary evidence that cosmid clones carrying the dominant *etr* mutation from an ethylene insensitive mutant of *A. thaliana* have been isolated by walking along the chromosome starting at flanking RFLPs. Giraudat *et al.* (1990) and Hauge *et al.* (1990) presented RFLP mapping evidence indicating that cosmids covering the *abi3* (abscissic acid-insensitive) and *ga2* (gibberellin biosynthesis) loci have been identified. The construction of physical maps of the *Arabidopsis* genome cloned in cosmids (Hauge *et al.*, 1990) or yeast artificial chromosomes (YACs) with large inserts of *Arabidopsis* DNA (Somerville and Grill, 1989) will simplify gene cloning via chromosome walking further.

One obstacle in the development of gene-cloning strategies by insertional mutagenesis or complementation was the lack of rapid and efficient transformation and regeneration procedures for *A. thaliana*. In addition, simple methods for the introduction of natural or modified genes into the *Arabidopsis* genome are prerequisites for the successful use of *A. thaliana* in the study of factors involved in gene expression or influencing homologous recombination.

Several protocols have been developed for *Agrobacterium tumefaciens*-mediated gene transfer to *A. thaliana* involving either wounded tissue, that is, explants (Lloyd *et al.*, 1986; Sheikholeslam and Weeks, 1987; Schmidt and Willmitzer, 1988; Valvekens *et al.*, 1988) or entire seeds (Feldmann and Marks, 1987), and resulting in the generation of transgenic plants. Only recently have protoplast-plant regeneration systems been established for this species (Damm and Willmitzer, 1988; Ford, 1990), which are prerequisites for the genetic manipulation of *A. thaliana* on the protoplast level. In our laboratory, we have been using a highly efficient regeneration system from mesophyll protoplasts (Damm and Willmitzer, 1988) for experiments on direct gene transfer to *Arabidopsis* (Damm *et al.*, 1989) as an alternative to DNA transfer methods involving *A. tumefaciens*.

9.1 Transformation Methods for *Arabidopsis thaliana*

9.1.1 *AGROBACTERIUM TUMEFACIENS*-MEDIATED GENE TRANSFER

Agrobacterium tumefaciens provides a natural gene-transfer mechanism that can be used to transfer a defined DNA sequence into the genome of cells of many dicotyledonous plants. In the late 1970s it had been shown that *Arabidopsis thaliana* is amenable to *A. tumefaciens*-mediated transformation (Aerts *et al.*, 1979). On the basis of *in vitro* regeneration protocols from explants of various plant organs (Negrutiu *et al.*, 1975; Scholl *et al.*, 1981; Feldmann and Marks, 1986; Patton and Meinke, 1988), and by applying the Ti plasmid (tumor-inducing)-based transformation system developed for other dicotyledonous plants (Horsch *et al.*, 1985), several laboratories have established efficient transformation protocols resulting in the production of large numbers of transgenic *Arabidopsis* plants (Lloyd *et al.*, 1986; Sheikholeslam and Weeks, 1987; Schmidt and Willmitzer, 1988; Valvekens *et al.*, 1988).

As an alternative, Feldmann and Marks (1987) have developed an *A. tumefaciens*-based transformation system for germinating seeds. This method does not involve tissue culture steps.

9.1.1.1 Leaf Disc or Cotyledon Transformation

Several transformation procedures for *Arabidopsis thaliana* using leaf or cotyledon explants infected by *Agrobacterium tumefaciens* have been published (Lloyd *et al.*, 1986; Sheikholeslam and Weeks, 1987; Schmidt and Willmitzer, 1988).

The first protocol for the regeneration of fertile transgenic *Arabidopsis* plants after leaf disk transformation and selection with hygromycin was developed by Lloyd *et al.* (1986). Leaf explants of *A. thaliana* var. *Columbia* were incubated for 2 days on shoot-inducing medium before they were cocultivated for 2 days with an *A. tumefaciens* strain harboring a modified Ti plasmid containing the gene for hygromycin phosphotransferase (*hpt*), on the same medium with the addition of a feeder layer of tobacco cells. Selection for transformed cell lines and the regeneration of transformed shoots was performed on the same shoot-inducing medium without feeder cells. Selective conditions using hygromycin were applied directly after cocultivation. Approximately one-third of the leaf explants originally plated gave rise to hygromycin-resistant calluses, of which more than half regenerated shoots. Stable transformation was proven by Southern blot hybridization, demonstrating the integration of one to several copies of the T-DNA into the plant genome. The shortest time interval from inoculation of the leaf pieces to the collection of seeds from regenerated transformants was 4 months. Limited data on progeny analysis indicated Mendelian segregation of the *hpt* gene. According to Lloyd *et al.* (1986), use of the *hpt* gene as a selectable marker was absolutely essential.

A leaf disk transformation protocol with improved efficiencies was developed thereafter for *A. thaliana* var. *Wassilewskija* by Sheikholeslam and Weeks (1987). The neomycin phosphotransferase gene (*nptII*) was used as a dominant resistance marker; transformed cell lines and shoots were selected using geneticin (G418). The increased efficiency of this protocol was brought about by the use of acetosyringone. Acetosyringone is a naturally occurring wound-response molecule that is secreted by certain dicotyledonous plants and elicits the activation of *Agrobacterium* genes that are involved in promoting DNA transfer from the bacterium to the plant cell. The rate of transformation was as high as 60% when acetosyringone was added to an *A. tumefaciens* culture prior to incubation with leaf segments. Without an acetosyringone pretreatment, the transformation rate was as low as 2–3%. These data suggest that the production of wound-response molecules by *A. thaliana* is low. However, a stimulation of the virulence region of the Ti plasmid is achieved by pretreatment with synthetic acetosyringone, resulting in an increase in the rate at which the bacterium is able to transform *Arabidopsis* tissue. Leaf explants were cocultured with the bacteria for 3 days on a callus-inducing medium. After that, the explants were cultured further on callus-inducing medium without selection for 7 days before they were transferred to selective conditions. When G418-resistant calluses developed, shoot regeneration from these calluses was performed on a medium containing a high level of cytokinin. A second factor that may contribute to the increased transformation frequency over that seen using the protocol of Lloyd *et al.* (1986) is the culture condition of the explants during and after transformation. The cocultivation of the leaf explants on a callus-inducing medium, their subculture on the same medium after the transformation, the delayed start of selection, and the shoot-regeneration conditions after visible antibiotic-resistant calluses have been formed appear to be important for an efficient regeneration of transformed shoots. Sheikholeslam and Weeks (1987) proved that regenerated plants resistant to G418 were transformed by Southern blot hybridization. They demonstrated the presence of one to several copies of the T-DNA in the genome of their primary transformants, of which a small number was subjected to progeny analysis. The inheritance of the *nptII* gene by the F_1 progeny was shown.

A highly efficient and fast *A. tumefaciens*-mediated leaf disk transformation system for *A. thaliana* genotype *C24* was developed by Schmidt and Willmitzer (1988). In a very detailed study, several parameters were tested that might influence the transformation and regeneration efficiency. Thus, a 2-day preculture of the explants on feeder cells on callus-induction medium, the transfer of the explants after cocultivation on selective shoot-inducing medium 1, and a subsequent plating of calluses with small developing shoots on a second shoot-inducing medium 2 were observed to increase the transformation frequency drastically compared with previously published protocols. Under optimal conditions 70–100% of the explants gave rise to antibiotic-resistant calluses, of which again more than 80% regenerated at least one shoot. Most of these shoots were fertile. The molecular analysis of selected *A. thaliana* tissue demonstrated that it was transformed stably. Seeds could be harvested as early as 2 months after the start of selection. Segregation analysis showed that the introduced genes were transmitted to the progeny. This altered regeneration scheme enabled a more uniform and faster development of transformed tissue than was described previously (Lloyd *et al.*, 1986; Sheikholeslam and Weeks, 1987). The shorter regeneration procedure may reduce the amount of somaclonal variation (Larkin and Scowcroft, 1981). Chromosome analysis, however, revealed that approximately 50% of the transformants have altered chromosome numbers; of these, tetraploid plants were the most abundant (R. Schmidt, unpublished results, 1989). The protocol was amenable to other ecotypes of *A. thaliana* such as *Landsberg erecta* and *Wassilewskija*. The procedure was applicable to leaf explants as well as cotyledons. Hygromycin as well as kanamycin and G418 selection was established.

9.1.1.2 Root Transformation

Valvekens *et al.* (1988) have examined the regenerative response of *Arabidopsis thaliana* roots because, in their hands, the regeneration of transgenic plants after leaf or cotyledon disk transformation generally was not efficient and took more than 4 months. It was found that *Arabidopsis* root explants have a high potential for rapid shoot regeneration.

Valvekens *et al.* (1988) developed culture conditions that induce *A. thaliana* root cuttings to regenerate shoots rapidly and at 100% efficiency. Optimal shoot regeneration was achieved when the root explants were incubated for 4 days on a callus-inducing medium containing a strong auxin prior to shoot induction on a medium containing high levels of cytokinin. Of the shoots, 50% formed roots, and almost all the shoots, regardless of whether they had formed roots or not, produced viable seeds. Using this method, fertile plants were regenerated reproducibly within 2–3 months. Valvekens *et al.*, (1988) combined this regeneration method with *Agrobacterium tumefaciens* infection and established an efficient and rapid transformation procedure for *Arabidopsis* root explants based on kanamycin selection. By using *A. tumefaciens* strains containing vectors carrying *nptII* genes, transformed fertile plants were obtained with frequencies ranging from 20 to 80% (root explants forming seed-producing transformants per total number of infected root explants) within 3 months of gene transfer. The analysis of root transformants on the molecular level revealed that, on average, only one or two copies of the T-DNA were integrated into the genome. Only in a few cases were up to four copies of T-DNA found. F_1 progeny of most of the primary transformants exhibited Mendelian segregation of the kanamycin resistance trait, confirming the small number of independent T-DNA insertion loci.

The transformation method was optimized for *A. thaliana* genotype *C24* but was equally efficient when applied to the *Landsberg erecta* ecotype. Root explants of *A. thaliana* var. *Columbia* also could be transformed and induced to form green calluses. However, when compared with those of the other ecotypes, these calluses were both slower and less efficient in shoot formation.

The advantages of the root transformation protocol over leaf disk or cotyledon transformation procedures are as follows. The method is rapid, and therefore reduces somaclonal variation due to prolonged tissue culture conditions. Less than 10% of the transformants have alterations in the chromosome number (D. Valvekens, personal communication, 1989). To date, it represents the only efficient protocol for the transformation of *A. thaliana* var. *Landsberg erecta,* a genotype that is of special interest because most currently available mutants are in this genetic background. In addition, most transformants produced by this method contain only one or two functional T-DNA inserts, facilitating genetic analysis of the transformants.

9.1.1.3 Seed Transformation

Feldmann and Marks (1987) have developed an *Agrobacterium tumefaciens*-mediated transformation protocol for germinating seeds of *Arabidopsis thaliana* as an alternative to transformation procedures involving explants and the subsequent regeneration of plants in tissue culture. The advantage of this procedure, as pointed out by Feldmann and Marks (1987), is that it requires no tissue culture step. Even under optimal transformation and regeneration conditions, somaclonal variation (Larkin and Scowcroft, 1981), spontaneous mutagenesis associated with *in vitro* regeneration, changes in the chromosome number among regenerated transgenic plants, and the loss of fertility due to tissue culture conditions have been reported often for *A. thaliana* (Negrutiu *et al.*, 1975; Scholl *et al.*, 1981; Feldmann and Marks, 1986; Lloyd *et al.*, 1986; Patton and Meinke, 1988; Valvekens *et al.*, 1988). The method of Feldmann and Marks (1987) represents a nontissue culture approach to *Arabidopsis* transformation and should circumvent the described phenomena. However, when scoring transformants generated via the seed transformation protocol, Meinke *et al.* (1990) observed mutations that are most likely generated during the transformation process.

Feldmann and Marks cocultivated germinating seeds (T_1 generation) of *A. thaliana* ecotype *Wassilewskija* with an *A. tumefaciens* strain carrying a modified Ti plasmid containing the *nptII* gene, thus conferring kanamycin and G418 resistance to transformed tissue. The highest number of transformed progeny (T_2 generation) was obtained when the seeds were imbibed for 12 hr in a commonly used plant tissue culture medium prior to a 24-hr coculture with the bacteria in the same medium. After imbibition and cocultivation, the seeds (T_1) were sown and grown to maturity. Seeds harvested from the T_1 plants were germinated in the presence of kanamycin. After the treatment of more than 60,000 seeds, about 200 kanamycin-resistant T_2 seedlings were isolated, giving a total transformation frequency of approximately 0.3%. Some of these T_2 seedlings and the T_3 families derived from them were analyzed for the integration and genetic segregation of functional copies of the *nptII* gene. Thereby, Feldmann and Marks (1987) could show clearly that imbibing germinating seeds with *A. tumefaciens* resulted in the production of transgenic *Arabidopsis* plants that inherited the introduced resistance trait to their progeny. A limited number of T_4 families were analyzed, confirming that the introduced *nptII* gene was transmitted through several rounds of meiosis in *A. thaliana* as a stable Mendelian factor. Although Feldmann and Marks (1987) have presented conclusive evidence that cocultivation of germinating seeds of *A. thaliana* with *A. tumefaciens* leads to the transformation of some of the resulting progeny, the method raised some questions of "how" and "when" the transformation event actually occurred. Nevertheless, in the Feldmann laboratory, the seed transformation protocol yielded large numbers of transgenic *Arabidopsis* plants. To date, more than 8,000 transformed kanamycin-resistant T_2 lines have been generated (Feldmann, 1991).

9.1.2 DIRECT GENE TRANSFER

From the early 1980s to date, several methods for direct gene transfer to protoplasts [i.e., uptake of naked DNA mediated by Ca^{2+} and polyethylene glycol (PEG; Krens *et al.*, 1982), Ca^{2+} and polyvinyl alcohol (PVA; Hain *et al.*, 1985), electroporation (Fromm *et al.*, 1986), or microinjection (Crossway *et al.*, 1986)] have been developed for different plant species. In our laboratory, we have chosen a combined Ca^{2+}/PEG procedure that originally was developed for direct gene transfer to protoplasts of *Nicotiana tabacum* and *N. plumbaginifolia* (Negrutiu *et al.*, 1987) for the stable transformation of *Arabidopsis thaliana* genotype *C24* (Damm *et al.*, 1989). Transgenic *Arabidopsis* plants resistant to hygromycin were regenerated from mesophyll protoplasts treated with PEG and plasmid DNA carrying the *hpt* gene. The transformation procedure reproducibly yielded transformants at frequencies of approximately 1×10^{-4} (based on the number of protoplasts treated) or 5% (based on the number of regenerating calluses). DNA of plants regenerated from hygromycin-resistant colonies was analyzed by Southern blot hybridization, demonstrating that the foreign gene was integrated stably into the plant chromosome. The majority of the transformants contained 1–4 copies of the introduced gene. When primary transformants were subjected to chromosome analysis, about 50% of them were found to have changes in the chromosome number, tetraploids being the most obvious (B. Damm, unpublished results, 1990). These data are comparable to those obtained after leaf disk or cotyledon transformation (R. Schmidt, unpublished results, 1989). Genetic analysis of several hygromycin-resistant plants showed that the *hpt* gene was transmitted to the progeny. Most of the segregation data could be explained by assuming either one or several independent loci for the integration of the resistance trait. Transformation experiments involving vectors conferring kanamycin and G418 resistance yielded about the same transformation frequencies as previously obtained using an *hpt*-containing plasmid (B. Damm, 1989).

On the basis of an efficient regeneration system (Damm and Willmitzer, 1988), the major factors affecting the transformation frequency were the final PEG concentration and the presence of carrier DNA in the transformation experiment (Damm *et al.*, 1989). These factors as well as the regeneration conditions must be varicd for protoplasts of different ecotypes and the optimum transformation method must be derived empirically.

The transformation and regeneration protocol was optimized for mesophyll protoplasts of *A. thaliana* genotype *C24* and the use of plasmids conferring resistance to hygromycin or kanamycin and G418, respectively. However, after several modifications, the given protocol could be applied to mesophyll protoplasts of several ecotypes of *A. thaliana* such as *Landsberg erecta,* the *Landsberg* mutant ap-1 (Koornneef *et al.*, 1983), and *Estland,* leading to the regeneration of hygromycin- as well as G418-resistant transformants (B. Damm, unpublished results, 1990).

Transformation experiments performed with selectable and nonselectable genes located on separate plasmids resulted in cotransformation rates of functionally active copies in 25–60% of the transformants analyzed. Further, the nonselected gene was inherited stably to the F_1 and F_2 progeny (B. Damm, 1989). Hence, direct gene transfer can be used to introduce nonselectable genes into the *Arabidopsis* genome.

Compared with *Agrobacterium tumefaciens*-mediated transformation, direct gene transfer has the advantage that the regenerated transformants are, as a rule, not chimeric. Possible applications of protoplast-plant regeneration and transformation methods are numerous, the most obvious ones being the use of *A. thaliana* in somatic hybridization experiments, the creation of mutants by insertional mutagenesis, the molecular identification of mutated genes by virtue of complementation, and studies aimed at homologous recombination. In addition, protoplast transformation and regeneration systems also could be the basis for experiments performed to transfer large pieces of DNA such as cosmid or YAC clones to *A. thaliana*.

9.2 Factors Influencing Transformation Efficiency

9.2.1 INFLUENCE OF PRECULTURE OF EXPLANTS

Several transformation protocols involve a short preculture of *Arabidopsis thaliana* explants on a callus-inducing (Schmidt and Willmitzer, 1988; Valvekens *et al.*, 1988) or shoot-inducing medium (Lloyd *et al.*, 1986) prior to cocultivation with *Agrobacterium tumefaciens*. This preculture could be shown to result in an increased regeneration frequency of transformed tissue.

Lloyd *et al.* (1986) reported that inoculating leaf disks of *A. thaliana* with *A. tumefaciens* on shoot-regeneration medium containing a feeder layer of tobacco cells strongly reduced the damage of the leaf pieces by the bacteria, therefore maximizing the recovery of transformed plants. A slight increase in the transformation efficiency mediated by a preculture of leaf or cotyledon explants on tobacco feeder cells was observed by Schmidt and Willmitzer also (1988).

9.2.2 INFLUENCE OF ANTIBIOTICS

To develop efficient transformation methods for *Arabidopsis thaliana*, a number of antibiotics that are used currently as selectable markers in plant transformation procedures also were assayed for their action on various *Arabidopsis* explants. In general, it was found that 50–100 mg kanamycin, 10–25 mg G418, or 20–30 mg hygromycin per liter completely blocked regeneration from untransformed explants and cells, and therefore could be used to select for transformed cell lines (Lloyd *et al.*, 1986; Sheikholeslam and Weeks, 1987; Schmidt and Willmitzer, 1988; Valvekens *et al.*, 1988; Damm *et al.*, 1989).

However, the ease of selection for transformed cell lines after transformation using the various antibiotics seems to be determined by the nature of the explant and/or the *Arabidopsis* genotype. Whereas kanamycin selection was used successfully to develop a reliable transformation procedure for root as well as leaf explants of *A. thaliana* genotype *C24* (Schmidt und Willmitzer, 1988; Valvekens *et al.*, 1988), previous reports on leaf disk transformation involving the *Arabidopsis* varieties *Columbia* (Lloyd *et al.*, 1986) and *Wassilewskija* (Sheikholeslam and Weeks, 1987) claimed that direct selection for kanamycin resistance was not possible. Instead, Lloyd *et al.* (1986) applied hygromycin and Sheikholeslam and Weeks (1987) used G418 selection. Although Schmidt and Willmitzer (1988) were able to establish kanamycin selection in leaf disk transformation experiments, they preferably used G418 to select transformed tissue. Only G418 selection was possible after direct gene transfer to mesophyll protoplasts of the *A. thaliana* genotypes *C24* and *Landsberg erecta* involving vectors carrying the *nptII* gene. Applying kanamycin selection resulted in a retardation of the growth of nontransformed protoplast-derived colonies; therefore, a high background of nontransformed protoplast-derived colonies was observed that hindered scoring (B. Damm, 1989).

Progeny analysis of transformed *A. thaliana* plants is performed using hygromycin at 20–25 mg per liter or kanamycin at 50–100 mg per liter. After 2–3 wk, resistant and sensitive seedlings are easily distinguishable. Only resistant seedlings develop green primary leaves and an elongated root. Although G418 is used successfully for the selection of transformed leaf explants, for the following reasons it is, as a rule, not used for progeny analysis. When applying G418 selection, seed germination is often low, nonresistant seedlings die a gradual death, and resistant seedlings are retarded in their growth so scoring is often difficult (Sheikholeslam and Weeks, 1987).

Antibiotics such as cefotaxime, vancomycin, triacillin, or carbenicillin that are used commonly to kill *Agrobacterium tumefaciens* after cocultivation with the plant material also might influence the viability of the explants. For example, Valvekens *et al.* (1988) found that cefotaxime severely inhibited regeneration from root explants whereas vancomycin had no inhibitory effect on the shoot formation of root explants. Schmidt and Willmitzer (1988) used cefotaxime to select against *A. tumefaciens* after leaf disk transformation. However, vancomycin selection was applied in experiments involving cotyledons since they exhibited a reduced viability in the presence of cefotaxime.

9.2.3 INFLUENCE OF SELECTABLE MARKER GENES

Several chimeric marker genes have been used in developing transformation procedures for *Arabidopsis thaliana*. For selection of stably transformed cell lines, chimeric *nptII* genes conferring resistance to kanamycin, G418, and paromomycin, or chimeric *hpt* genes conferring resistance to hygromycin were used most commonly (Lloyd *et al.*, 1986; Sheikholeslam and Weeks, 1987; Schmidt and Willmitzer, 1988; Valvekens *et al.*, 1988; Damm *et al.*, 1989). Only recently chimeric bialaphos resistance genes (*bar*) conferring resistance to the herbicide bialaphos were used at the mature plant level to select for transformation (Valvekens *et al.*, 1988; T. Altmann, unpublished results, 1990). In addition to the possibility to select in sterile culture, simple herbicide spraying assays can be used to monitor expression of the *bar* gene. This screening procedure facilitates gene mapping. In addition, spraying of F_1 progeny can be used to follow segregation. Although kanamycin and hygromycin spraying assays could be used also to select transgenic *Arabidopsis* plants and their offspring in soil (R. Masterson and T. Altmann, unpublished results, 1990), by spraying bialaphos, a clear distinction between resistant and sensitive plant material can be seen as soon as 2–3 days after a single application of the herbicide. Additionally, a nondestructive assay for sensitivity can be carried out by painting a single leaf with a bialaphos solution, leading to the death of the treated leaf only (T. Altmann, unpublished results, 1990).

Several laboratories that currently are trying to establish efficient transposon tagging systems in *Arabidopsis* (see Section 9.4.1.2.) have used additional marker genes, for example, the genes encoding streptomycin and spectinomycin phosphotransferase, a mutated acetolactate synthase, or dihydrofolate reductase, to monitor transposition (Masterson *et al.,* 1989; Dean *et al.,* 1990; Honma *et al.,* 1990).

9.2.4 INFLUENCE OF GENOTYPE

It has been shown by a number of laboratories that the *Arabidopsis thaliana* ecotypes and the explants of the ecotype used in regeneration as well as transformation experiments have a strong influence on the efficiency of a given protocol (Damm and Willmitzer, 1988; Schmidt and Willmitzer, 1988; Valvekens *et al.,* 1988). For example, the root transformation procedure worked out by Valvekens *et al.* (1988) resulted in high frequencies when applied to the *Arabidopsis* genotypes *C24* or *Landsberg erecta.* The procedure also could be transferred to *A. thaliana* var. *Columbia,* but the overall efficiency was reduced significantly.

The observations made concerning genotype dependency led to the conclusion that it is not the amenability to *Agrobacterium* infection that limits *Arabidopsis* transformation but regeneration problems occurring with several genotypes or different explants after transformation (Valvekens *et al.,* 1988). These data once again confirm the importance of efficient regeneration systems as the main prerequisite for experiments aimed at the development of high-frequency transformation procedures for *A. thaliana.*

9.2.5 INFLUENCE OF *AGROBACTERIUM TUMEFACIENS* STRAINS AND VECTORS

Although many dicotyledonous plants are naturally susceptible to *Agrobacterium tumefaciens* infection (Anderson and Moore, 1979), most plant species exhibit different sensitivities to certain strains of *A. tumefaciens* (DeCleene and Deley, 1976). Also, for *Arabidopsis thaliana,* an influence of the *Agrobacterium* strains and vectors used in transformation experiments on transformation efficiencies was observed (Schmidt and Willmitzer, 1988; Valvekens *et al.,* 1988). Performing experiments on direct gene transfer using plasmids of different sizes (4–

15 kb), harboring various chimeric resistance genes, we did not detect any influence of the vector on the transformation frequency (B. Damm, unpublished results, 1990).

9.3 Unusual Segregation of Transferred Genes

Feldmann and Marks (1987) were the first to report that a foreign gene (*nptII*) introduced into *Arabidopsis thaliana* was not inherited in a Mendelian fashion. They examined 124 T_3 families derived from kanamycin-resistant T_2 plants that had been produced by the seed transformation method. Of these, 111 families showed a Mendelian segregation, indicating that they contained 1–4 functional inserts or were homozygous for the inserted *nptII* gene. The 13 families that did not exhibit Mendelian segregation fell into two classes, both having a deficiency of kanamycin-resistant seedlings. In one group (6 T_3 families) only 0–2% of the seedlings were resistant to kanamycin, whereas in the other group (7 T_3 families) 8–63% of the seedlings were resistant. Feldmann and Marks (1987) detected NPTII activity in all the kanamycin-resistant plants and Southern blot analysis revealed that these plants contained at least one copy of the T-DNA. Surprisingly, many of the T_4 families resulting from these resistant T_3 plants segregated for the kanamycin resistance trait in a normal Mendelian fashion.

Unusual segregation ratios for introduced transgenes in *A. thaliana* also have been reported by several other groups (Schmidt and Willmitzer, 1988; Damm *et al.*, 1989; Kilby and Furner, 1990; Mittelsten-Scheid *et al.*, 1991), independent of how the gene had been introduced. Preliminary evidence suggested that the genes were inactivated by methylation and that this inactivation is reversible (Kilby and Furner, 1990; Mittelsten-Scheid *et al.*, 1991).

Non-Mendelian segregation patterns have been observed also in transgenic tobacco (Budar *et al.*, 1986; Heberle-Bors *et al.*, 1988), petunia (Horsch *et al.*, 1985), and tomato (McCormick *et al.*, 1986) plants. In those cases, the prevalence of sensitive seedlings might be due to the instability of the inserted gene, the loss of expression of the gene, or its inactivation. From the number of reports that are coming up in the literature currently, one can conclude that the inactivation of foreign genes in transgenic plants is a common phenomenon.

9.4 Applications of Transformation Procedures

The transformation protocols described represent simple methods for the introduction of natural or modified genes into *Arabidopsis thaliana*. Thus, the random creation of insertion mutants, the directed complementation of mutants, the study of gene expression, or the specific alteration of genes via homologous recombination is now a feasible task in *A. thaliana*.

9.4.1 INSERTIONAL MUTAGENESIS

The development of efficient transformation techniques for *Arabidopsis thaliana* has resulted in new approaches to gene cloning as alternatives to chromosome-walking experiments. Possible applications are the inactivation of genes by the integration of foreign DNA sequences (i.e., T-DNA, transposable elements, or plasmid DNA) into the *Arabidopsis* genome. Such inactivations not only create useful insertion mutants but also facilitate the ultimate cloning of inactivated genes by virtue of their immediate proximity to the inserted known vector sequences. The identified and isolated mutated genes are employed as probes to screen wild-type genomic or cDNA (complementary DNA) libraries. Isolated clones of wild-type genes are being used to complement the mutant.

9.4.1.1 T-DNA Tagging

Using the seed transformation procedure (see Section 9.1.1.3), Feldmann and coworkers have obtained more than 8,000 transgenic *Arabidopsis* plants (Feldmann, 1991). From the 8,000 independent transformants analyzed so far, approximately 1,000 lines that segregated for any visible phenotypic alteration have been recovered. The isolated phenotypes fell in several classes: seedling or embryo lethals and size variants were the most frequent.

Several laboratories have started to analyze those visible phenotypes that might be due to random insertion(s) of the T-DNA into the *Arabidopsis* genome. The first mutant that was shown to be generated by T-DNA insertional mutagenesis was a recessive dwarf mutation (Feldmann *et al.*, 1989). For several other cases, it also has been shown that the altered phenotypes represent stable recessive mutations that cosegregate with the T-DNA of *Agrobacterium tumefaciens* and its associated kanamycin resistance trait (Herman and Marks, 1989; Marks and Feldmann, 1989; Errampalli *et al.*, 1991; Yanofski *et al.*, 1990).

Studying transformants produced by Feldmann, in two cases T-DNA-tagged genes have been isolated and analyzed. The first gene that has been isolated via T-DNA tagging is a mutant allele of the *gl1* (*Glabrous1*) locus (Marks and Feldmann, 1989). The *Glabrous* locus controls the presence of trichomes on the surface of various organs. A clone covering the mutation was used to isolate the corresponding wild-type gene, which was used thereafter to complement the *gl1* mutant (Herman and Marks, 1989). The second gene that was isolated after T-DNA tagging is an allele of the *agamous* (*ag*) locus, which encodes a homeotic gene involved in the control of flower formation. Yanofski *et al.* (1990) have used the tagged mutated *ag* locus to isolate the wild-type gene, which was introduced into the genome of a homozygous mutant. The mutation was complemented and the resulting flower displayed the wild-type phenotype.

Recent improvements in *A. thaliana* transformation using leaf explants also have resulted in gene isolation via T-DNA insertional mutagenesis. Koncz *et al.* (1989) have introduced T-DNA constructs into *A. thaliana* that allow both transcriptional and translational fusions to a promoterless *nptII* reporter gene linked to the right border of the T-DNA. Approximately 500 primary transformants were regenerated using hygromycin selection. Gene fusions were identified by screening for NPTII activity in callus, leaf, stem, and root tissues of hygromycin-resistant T_2 progeny of transgenic plants. Approximately 30% of the T-DNA inserts resulted in gene fusions, suggesting a preference for insertions into or near potentially active chromosome loci. One of the regulatory sequences of a transcriptional fusion obtained by Koncz *et al.* (1989) was shown to direct callus root-specific expression. In addition, a recessive *pale* mutation that cosegregates with the T-DNA and the encoded hygromycin-resistance phenotype was isolated. Plant DNA sequences from the boundaries of the T-DNA insert rescued from the *pale* mutant were used as probes for the isolation of genomic and cDNA clones of the wild-type gene (Koncz *et al.*, 1990). Transformation of the *pale* mutant with these clones resulted in a normal green phenotype, demonstrating positive complementation of the T-DNA-induced mutant. Koncz *et al.* (1990) observed visible mutations in about 1% of the T_2 families, in contrast with the frequency of about 20% reported by Feldmann *et al.* (1989).

9.4.1.2 Transposon Tagging

An alternative method of tagging genes in plants is the use of transposable elements as insertional mutagens. Although a family of defective retrotransposons (Voytas and Ausubel, 1988) and a different family of transposon-like elements (Peleman *et al.*, 1991) have been identified in the *Arabidopsis* genome, no active endogenous transposons are characterized. Therefore, alternative approaches were initiated using heterologous transposons. For example, members of the maize transposable element family *Activator* (*Ac*)/*Dissociation* (*Ds*) have been introduced into the *Arabidopsis* genome and shown to transpose (Van Sluys *et al.*, 1987; Masterson *et al.*, 1989; Schmidt and Willmitzer, 1989).

Presently many groups are putting effort into developing *Ac*-based gene-tagging system in *A. thaliana*. Chimeric constructs in which the *Ac* element is inserted in the 5′-untranslated leader of a marker gene, thus preventing its expression, have been transformed into *Arabidopsis* via *Agrobacterium tumefaciens* (Schmidt and Willmitzer, 1989; Dean *et al.*, 1992; Lawson *et al.*, 1990). The excision of *Ac* was monitored by assaying for the reconstitution of the marker gene activity. Somatic excisions of *Ac* were observed in 25–75% of the primary *Ac*-containing transformants (Van Sluys *et al.*, 1987; Schmidt and Willmitzer, 1989). Schmidt and Willmitzer (1989) further showed that excision of *Ac* was followed by reinsertion, suggesting that the system can be used for transposon tagging. However, in *Ac*-containing lines in which *Ac* had not moved during the transformation/regeneration procedure, germinal excisions (i.e., excisions that happen in germline cells and are transmitted to the next generation) were observed only at a very low frequency of 0.2–0.5%. These frequencies were obtained when analyzing F_1, F_2, and F_3 plants, indicating that the *Ac* element remains active through several generations in the heterologous host *A. thaliana* (Schmidt and Willmitzer, 1989). Since the frequency of transposition events transmitted to the next generation is the major determinant that decides the feasibility of a transposon-tagging approach, for *A. thaliana* efficient transposon-tagging systems still have to be established. This development might either be achieved by increasing the germinal transposition frequencies of *Ac* due to modifications of the element itself (Lawson *et al.*, 1990) or by developing systems that allow selection of very large populations for transposition events.

Mutations created by the autonomous *Ac* element are potentially unstable. To overcome this problem, several groups currently also are developing two-component transposable elements systems using the well-characterized maize transposon family *Ac/Ds* (Fedoroff, 1983; Masterson *et al.*, 1989; Altmann *et al.*, 1990; Coupland *et al.*, 1990; Honma *et al.*, 1990). In those systems, the mobilizable nonautonomous gene tag (*Ds*) is separated from the nonmobile source of *trans*-activating transposase (*Ac**). The systems are developed so *Ac** and *Ds* are located on separate T-DNA constructs and introduced to *A. thaliana* by *A. tumefaciens*-mediated transformation. Thereafter, genetic crosses of transformed plants are used to bring *Ac** and *Ds* together. *Ac** might activate *Ds,* which might transpose to other chromosomal sites. These new *Ds* insertions become fixed when *Ac** segregates away from them. To date, there is strong evidence that *Ds* elements can be transactivated by *Ac**. Somatic excision events have been identified by DNA analysis (Masterson *et al.*, 1989), as well as phenotypically, using the streptomycin resistance gene, a cell-autonomous marker gene. Green spots indicating cells that became resistant after excision of *Ds* were detected on the background of bleached (i.e., sensitive) cotyledon tissue (Balcells *et al.*, 1990).

We constructed a system composed of a *Ds* element, derived from *Ac* by deleting 2.8 kb of internal sequences, and an *Ac**, generated by precise deletion of one of the terminal 11-base pair (bp) inverted repeats of *Ac*. A chimeric *bar* gene (see Section 9.2.3.) and a bacterial *supF* gene were inserted into the *Ds* element. Hence, the construct carries a plant-selectable marker gene useful for genetic analysis and a bacterial gene that will allow recloning into plasmid as well as into λ vectors. Excision of *Ds* will reconstitute the activity of an *nptII* gene previously inactivated by the insertion of *Ds* into its untranslated leader region. Transposition of this *Ds* element promoted by *Ac** could be detected by G418 selection of regenerating leaf explants from plants carrying both elements that were combined previously by crossing. DNA analysis of these regenerants as well as their progeny, and the recloning of both the empty *Ds* donor site and a DNA fragment covering the 5′ region of *Ds* with its flanking sequence at the new insertion site confirmed that the *Ds* element had excised and reinserted into a different position (Altmann *et al.*, 1990).

As for the autonomous element *Ac*, further development of *Ac**/*Ds* systems is being done currently, including the introduction of different activating elements such as the *Ac* coding region expressed under the control of heterologous promoters (Balcells *et al.*, 1990) and the test of different *Ds* elements (Honma *et al.*, 1990; Masterson *et al.*, 1990).

Other transposons such as *Mu1* from *Zea mays* or *Tam3* from *Antirrhinum majus* also have been introduced into *A. thaliana*, but no clear evidence for transposition has been reported (Zhang and Somerville, 1987; Jofuku *et al.*, 1990).

9.4.2 COMPLEMENTATION OF MUTANTS

Gene rescue by shotgun cloning and complementation is another cloning strategy that eventually may be applicable for some genes in *Arabidopsis thaliana*. Gene isolation via shotgun cloning involves transformation of mutant plants with a complete random recombinant library derived from wild-type DNA, followed by the assay of individual transformed plants for complementation of the mutant phenotype. However, even in *Arabidopsis* with its small genome size, this method is probably feasible only for phenotypes that can be selected or screened for at the tissue culture stage, since it is presently not feasible to attempt the regeneration of tens of thousands of transformed plants. This alternative approach to gene cloning has not been performed successfully in *A. thaliana*.

The complementation of mutants with wild-type genes that have been isolated by chromosome walking or insertional mutagenesis has been used as final proof for the identification of mutated genes (Herman and Marks, 1989; Koncz *et al.*, 1990; Yanofski *et al.*, 1990).

The complementation of mutants can be accomplished either by *Agrobacterium tumefaciens*-mediated transformation or by direct gene transfer to protoplasts. Large DNA fragments such as cosmid or YAC clones might be transformed into *A. thaliana* using direct gene transfer to protoplasts, although this has not yet been demonstrated.

9.4.3 HOMOLOGOUS RECOMBINATION

One possibility to elucidate gene function is the application of homologous recombination, meaning the targeting of an exogenous sequence to its endogenous homologous counterpart. Targeted recombination may be used to modify the coding sequence of endogenous genes of unknown function; this may result in an altered phenotype, allowing a functional analysis of these genes. In higher eukaryotes, the main difficulty in this approach is the very low ratio of homologous to illegitimate recombination.

To establish such a system, it is necessary to use genes that enable selection for the homologous recombination event. Homologous recombination can be achieved in plants, as was shown for tobacco (Paszkowski *et al.*, 1988; Lee *et al.*, 1990; Offringa *et al.*, 1990). On one hand, Paszkowski *et al.* (1988) used transgenic tobacco plants carrying copies of a partial nonfunctional *nptII* gene. These plants were transformed via direct gene transfer using DNA molecules that provided the missing part of the *nptII* gene and additional overlapping sequences. Selection on kanamycin required a corrected intact gene, resulting in detection of recombinants with a relative frequency of up to 1×10^{-4}. On the other hand, *Agrobacterium tumefaciens*-mediated gene transfer into protoplasts was used to target exogenous DNA to an endogenous selectable gene: Lee *et al.* (1990) reported homologous recombination with a mutant acetolactate synthase gene conferring resistance to the sulfonylurea class of herbicides, and Offringa *et al.* (1990) described the restoration of a defective *nptII* gene. Again, the targeting frequency was about 1×10^{-4} to 1×10^{-5}.

In our laboratory *A. thaliana* genotype *C24* was used to establish homologous recombination. Protoplasts of transgenic plants harboring a defective *hpt* gene inactivated by a 19-bp deletion in the coding region were transformed with DNA fragments containing a promoterless intact coding region, using the protocol of Damm *et al.* (1989). Selection with hygromycin enriched for candidates that had undergone homologous recombination. To distinguish between homologous and nonhomologous recombination in which the exogenously added *hpt* coding region gave rise to hygromycin resistance as a result of transcriptional or translational fusions to endogenous promoters, polymerase chain reaction (PCR) technology was used. In total, 3.8×10^8 protoplasts have been transformed with the described substrate to obtain recombinants. Resistant calluses were selected; 3 of these selected putative recombinants have been shown to give a correct PCR signal and have resistant F_1 progeny. In addition, Southern blot analysis was performed to show that the hygromycin resistance phenotype was due to homologous recombination. The ratio of homologous events to random insertion events was calculated to be 1×10^{-4} in these experiments (Halfter *et al.*, 1992).

To apply this technique to endogenous nonselectable genes, an efficient system must be set up that enables selection or enrichment at lest 100-fold for the homologous recombination event.

9.5 Prospects

The most exciting prospect offered by *Arabidopsis thaliana* is the potential for identification, genetic mapping, and cloning of genes that, on the basis of the effects of their mutations, are likely to be involved closely in key aspects of plant growth and development. Among the genes that may be cloned and analyzed are those affecting particular enzyme activities, those with specific effects on the ability of *Arabidopsis* to synthesize and/or respond to plant hormones, and those in which mutations result in specific developmental abnormalities.

An advantage of insertional mutagenesis as well as chromosome walking is that it allows identification of genes on the molecular level that cannot be cloned by conventional methods due to the lack of knowledge about the biochemical activity of the proteins encoded by these genes.

Finally, the possibility of retransforming mutant genotypes and possibly targeting modified genes to the normal chromosomal location via homologous recombination will allow establishment of a detailed structure–function relationship for many plant genes.

References

Aerts, M., Jacobs, M., Hernalsteens, J.-P., Van Montagu, M., and Schell, J. (1979). *Plant Sci. Let.* **17**, 43–50.

Altmann, T., Schmidt, R., and Willmitzer, L. (1990). *In* "IVth International Conference on *Arabidopsis* Research," Abstracts. p. 26. Vienna.

Anderson, A. R., and Moore, L. (1979). *Phytopathol.* **69**, 320–323.

Balcells, L., Swinburne, J., Ingle, K., Scofield, S., Jones, J., and Coupland, G. (1990). *In* "IVth International Conference on *Arabidopsis* Research," Abstracts. p. 27. Vienna.

Bleeker, A. B., Chang, C., and Meyerowitz, E. M. (1990). *In* "IVth International Conference on *Arabidopsis* Research," Abstracts. p. 131. Vienna.

Bowman, J. L., Yanofski, M. F., and Meyerowitz, E. M. (1988). *Oxford Surv. Plant Mol. Cell. Biol.* **5**, 57–87.

Budar, F., Thia-Toong, L., Van Montagu, M., and Hernalsteens, J.-P. (1986). *Genetics* **114**, 303–313.

Chang, C., Bowman, J. L., DeJohn, A. W., Lander, E. S., and Meyerowitz, E. M. (1988). *Proc. Natl. Acad. Sci. U.S.A.* **83**, 1408–1412.

Coupland, G., Balcells, L., Ingle, K., Scofield, S., and Swinburne, J. (1990). *In* "IVth International Conference on *Arabidopsis* Research," Abstracts. p. 106. Vienna.

Crossway, A., Oakes, J. V., Irvine, J. M., Ward, B., Knauf, V. C., and Shewmaker, C. K. (1986). *Mol. Gen. Genet.* **202**, 179–185.

Damm, B. (1989). Ph.D. Thesis. Free University of Berlin.

Damm, B., and Willmitzer, L. (1988). *Mol. Gen. Genet.* **213**, 15–20.

Damm, B., Schmidt, R., and Willmitzer, L. (1989). *Mol. Gen. Genet.* **217**, 6–12.

Dean, C., Sjödin, C., Page, T., Jones, J., and Lister, C. (1992). *Plant J.* **2**, 69–81.

DeCleene, M., and Deley, J. (1976). *Bot. Rev.* **42**, 389–466.

Errampalli, D., Patton, D., Castle, L., Mickelson, L., Hansen, K., Schnall, J., Feldmann, K., and Meinke, D. (1991). *Plant Cell* **3**, 149–157.

Estelle, M. A., and Somerville, C. R. (1986). *Trends Genet.* **2**, 89–93.

Fedoroff, N. V. (1983). *In* "Mobile genetic elements." Academic Press. (J. A. Shapiro, ed.). New York.

Feldmann, K. A. (1991). *Plant J.* **1**, 71–82.

Feldmann, K. A., and Marks, M. D. (1986). *Plant Sci.* **47**, 63–69.

Feldmann, K. A., and Marks, M. D. (1987). *Mol. Gen. Genet.* **208**, 1–9.

Feldmann, K. A., Marks, M. D., Christianson, M. L., and Quatrano, R. S. (1989). *Science* **243,** 1351–1354.

Ford, K. (1990). *Plant Cell Rep.* **8,** 534–537.

Fromm, M. E., Taylor, L. P., and Walbot, V. (1986). *Nature (London)* **319,** 791–793.

Giraudat, J., Hauge, B., Smalle, J., and Goodman, H. M. (1990). *In* "IVth International Conference on *Arabidopsis* Research," Abstracts. p. 134. Vienna.

Halfter, U., Morris, P.-C., and Willmitzer, L. (1992). *Mol. Gen. Genet.* **231,** 186–193.

Hain, R., Stabel, P., Czernilofski, A. P., Steinbiβ, H. H., Herrera-Estrella, L., and Schell, J. (1985). *Mol. Gen. Genet.* **199,** 161–168.

Hauge, B. M., Giraudat, J., Hanley, S., Inwan, H., Hohchi, T., and Goodman, H. M. (1990). *In* "IVth International Conference on *Arabidopsis* Research," Abstracts. p. 11. Vienna.

Heberle-Bors, E., Charvat, B., Thompson, D., Schernthaler, J. P., Barta, A., Matzke, A. J. M., and Matzke, M. A. (1988). *Plant Cell Rep.* **7,** 571–574.

Herman, P. L., and Marks, M. D. (1989). *Plant Cell* **1,** 1051–1055.

Honma, M. A., Waddell, C. S., and Baker, B. (1990). *In* "IVth International Conference on *Arabidopsis* Research," Abstracts. p. 12. Vienna.

Horsch, R. B., Fry, J. E., Hoffmann, N. L., Eichholz, D., Rogers, S. G., and Fraley, R. T. (1985). *Science* **227,** 1229–1231.

Jofuku, K. D., Den Boer, B., Van Montagu, M., and Okamuro, J. K. (1990). *In* "IVth International Conference on *Arabidopsis* Research," Abstracts. p. 112. Vienna.

Kilby, N. J., and Furner, I. J. (1990). *In* "IVth International Conference on *Arabidopsis* Research," Abstracts. p. 35. Vienna.

Koncz, C., Martini, N., Mayerhofer, R., Konce-Kalman, Z., Korber, H., Redei, G. P., and Schell, J. (1989). *Proc. Natl. Acad. Sci. U.S.A.* **86,** 8467–8471.

Koncz, C., Mayerhofer, R., Koncz-Kalman, Z., Nawrath, C., Reiss, B., Redei, G. P., and Schell, J. (1990). *EMBO J.* **9,** 1337–1346.

Koornneef, M. (1983). *Genetica* **62,** 33–40.

Koornneef, M. (1987). *In* "Genetic Maps 1987: A Compilation of Linkage and Restriction Maps of Genetically Studied Organisms" (S. J. O'Brian,, ed.), pp. 742–745. Cold Spring Harbor Laboratory Press, Cold Spring Harbor, New York.

Koornneef, M., and van der Veen, J. H. (1983). *Genetica* **61,** 41–46.

Koornneef, M., Van Eden, J., Hanhart, C. J., Stam, P., Braaksma, F. J., and Feenstra, W. J. (1983). *J. Hered.* **74,** 265–272.

Koornneef, M., Hanhart, C. J., van Loenen-Martinet, E. P., and van der Veen, J. H. (1987). *Arabidopsis Inf. Serv.* **23,** 46–50.

Krens, F. A., Molendijk, L., Wullems, G. J., and Schilperoort, R. A. (1982). *Nature (London)* **296,** 72–74.

Laibach, F. (1943). *Bot. Arch.* **44,** 439–455.

Larkin, P. J., and Scowcroft, W. R. (1981). *Theor. Appl. Genet.* **60,** 197–214.

Lawson, E., Lister, C., and Dean, C. (1990). *In* "IVth International Conference on *Arabidopsis* Research," Abstracts. p. 14. Vienna.

Lee, K. Y., Lund, P., Lowe, K., and Dunsmuir, P. (1990). *Plant Cell* **2,** 415–425.

Leutwiler, L. S., Hough-Evans, B. R., and Meyerowitz, E. M. (1984). *Mol. Gen. Genet.* **194,** 15–23.

Lloyd, A., Barnason, A. R., Rogers, S. G., Byrne, M. C., Fraley, R. T., and Horsch, R. B. (1986). *Science* **234,** 464–468.

McCormick, S., Niedermeyer, J., Fry, J., Barnason, A., Horsch, R., and Fraley, R. (1986). *Plant Cell Rep.* **5,** 81–84.

Marks, D. M., and Feldmann, K. A. (1989). *Plant Cell* **1,** 1043–1050.

Masterson, R. V., Furtek, D. B., Grevelding, C., and Schell, J. (1989). *Mol. Gen. Genet.* **219,** 461–466.

Meinke, D. W. (1986). *Oxford Surv. Plant Mol. Cell. Biol.* **3,** 122–165.
Meyerowitz, E. M. (1987). *Ann. Rev. Genet.* **21,** 93–111.
Meyerowitz, E. M. (1989). *Cell* **56,** 263–269.
Mittelsten-Scheid, O., Paszkowski, J., and Potrykus, I. (1991). *Mol. Gen. Genet.* **228,** 104–112.
Nam, H.-G., Giraudat, J., den Boer, B., Moonan, F., Loos, W. D. B., Hauge, B. M., and Goodman, H. M. (1989). *Plant Cell* **1,** 699–705.
Negrutiu, I., Beeftink, F., and Jacobs, M. (1975). *Plant Sci. Lett.* **5,** 293–304.
Negrutiu, I., Shillito, R., Potrykus, I., Biasini, G., and Sala, F. (1987). *Plant Mol. Biol.* **8,** 363–373.
Offringa, R., De Groot, M. J. A., Haagsman, H. J., Does, M. P., Ven den Elzen, P. J. M., and Hooykaas, P. J. J. (1990). *EMBO J.* **9,** 3077–3084.
Paszkowski, J., Baur, M., Bogucki, A., and Potrykus, I. (1988). *EMBO J.* **7,** 4021–4026.
Patton, D. A., and Meinke, D. W. (1988). *Plant Cell Rep.* **7,** 233–237.
Peleman, J., Cottyn, B., van Montagu, M., and Inzé, D. (1991). *Proc. Natl. Acad. Sci. U. S. A.* **88,** 3618–3622.
Pruitt, R. E., and Meyerowitz, E. M. (1986). *J. Mol. Biol.* **187,** 169–183.
Redei, G. P. (1975). *Ann. Rev. Genet.* **9,** 111–127.
Schmidt, R., and Willmitzer, L. (1988). *Plant Cell Rep.* **7,** 583–586.
Schmidt, R., and Willmitzer, L. (1989). *Mol. Gen. Genet.* **220,** 17–24.
Scholl, R. L., Keathley, D. E., and Baribault, T. J. (1981). *Z. Pflanzenphysiol.* **104,** 225–231.
Sheikholeslam, S. N., and Weeks, D. P. (1987). *Plant Mol. Biol.* **8,** 291–298.
Somerville, C., and Grill, E. (1989). *In* "EMBO–EMBL Symposium 1989: Molecular Communication in Higher Plants," pp. 57–58. Heidelberg.
Valvekens, D., Van Montagu, M., and Van Lijsebettens, M. (1988). *Proc. Natl. Acad. Sci. U.S.A.* **85,** 5536–5540.
Van Sluys, M. A., Tempé, J., and Fedoroff, N. (1987). *EMBO J.* **6,** 3881–3889.
Voytas, D. F., and Ausubel, F. M. (1988). *Nature (London)* **336,** 242–244.
Yanofski, M. F., Ma, H., Bowman, J. L., Drews, G. N., Feldmann, K. A., and Meyerowitz, E. M. (1990). *Nature (London)* **346,** 35–39.
Zhang, H., and Somerville, C. R. (1987). *Plant Sci.* **48,** 165–173.

Part III

Social and Economic Impacts of Transgenic Plants

10

Federal Regulation of Transgenic Plants

Edward L. Korwek

Hogan & Hartson Law Offices
Washington, D.C., 20004

The Food and Drug Administration (FDA), the United States Department of Agriculture (USDA), and the Environmental Protection Agency (EPA) are the three primary federal organizations with substantial regulatory authority over transgenic plants, including products thereof. Other federal bodies, such as the Department of Commerce, also can exercise regulatory powers, for example, over exports. States, too, administer a variety of laws governing plants, particularly seeds. More recently, states such as North Carolina have enacted laws covering diverse activities relating generally to the topic of "environmental releases." Other areas of legal importance include the protection of intellectual property, a topic of considerable interest currently to the seed industry. The focus of this chapter is, generally, EPA, FDA, and USDA regulation, not intellectual property matters.

10.1 Regulatory Principles

The range of federal laws actually or potentially applicable to transgenic plants is as diverse as the numerous applications to plants of modern methods of biotechnology. A great deal of uncertainty still exists about when and how transgenic plants and products thereof will be regulated, and by whom. Table 10.1 describes a number of important federal statutes or laws and the areas they cover.

Characterizing or classifying an activity or product from a regulatory standpoint is the first important step toward ascertaining the regulatory hurdles and

TABLE 10.1
Principal Laws Applicable to Transgenic Plants

Laws	Relevant jurisdiction
Food, Drug, and Cosmetic Act	Animal and human drugs, human biologics, cosmetics, animal and human food and food ingredients including color additives, medical devices
Virus-Serum-Toxin Act	Animal biologics
Federal Meat Inspection Act, Poultry Products Inspection Act	Meat and poultry products used as human food
Federal Plant Pest Act, Plant Quarantine Act	Plant pests
Federal Insecticide, Fungicide, and Rodenticide Act	Pesticides
Federal Noxious Weed Act	Noxious weeds
Toxic Substances Control Act	Chemicals not regulated by FDA or as pesticides by EPA
Export Administration Act	Exports of commodities or technical data
National Environmental Policy Act	Activities that trigger any type of federal review usually require environmental analyses

costs involved in a particular activity, for example, conducting an environmental-release experiment or marketing a particular product. For example, whether an article is intended for uses in humans or in other animals, or is a food, drug, pesticide, or plant pest, helps determine the relevant regulatory requirements applicable to research and commercial activities.

An important point to remember is that the same activity or product can be regulated by different federal agencies in very different ways. This discrepancy is because most laws have been enacted for different purposes; thus, different agencies and regulatory requirements can apply to the same activity. This diversity of laws contributes to the complexity of the oversight system applicable not only to transgenic plants, but to all articles or activities marketed or undertaken in the United States.

The relevant laws usually are triggered by three factors: (1) the types of plants, (2) their intended uses, and (3) the techniques used to effect genetic changes. The transgenic nature of a plant does not necessarily trigger federal regulation. Table 10.2 lists a variety of traits or uses of transgenic and other plants and the federal regulatory bodies responsible for their oversight.

With respect to the first factor, the nature or types of plants, those that are noxious weeds or plant pests, or that harbor plant pests, are potentially regulable

TABLE 10.2
Federal Jurisdiction over Transgenic Plants

Plant trait or uses	Agency or office
Herbicide tolerance	—
Plant pest	USDA/APHIS[a]
Noxious weed	USDA/APHIS
Animal feed	FDA/CVM[b]
Human food	FDA/CFSAN[b], USDA/FSIS[a]
Human drugs	FDA/CDER[b]
Human biologic	FDA/CBER[b]
Animal drug	FDA/CVM
Animal biologic	USDA/APHIS
Pest tolerance	EPA/OPP[c]
Commodity chemical	EPA/OTS[c]

[a]The United States Department of Agriculture (USDA) is composed of, in relevant part, the Animal and Plant Health Inspection Service (APHIS) and the Food Safety and Inspection Service (FSIS).

[b]The Food and Drug Administration (FDA) is part of the Department of Health and Human Services. FDA is composed of, in relevant part, the Center for Veterinary Medicine (CVM), the Center for Food Safety and Applied Nutrition (CFSAN), the Center for Drug Evaluation and Research (CDER), and the Center for Biologics Evaluation and Research (CBER).

[c]The Environmental Protection Agency (EPA) is composed of, in relevant part, the Office of Toxic Substances (OTS) and the Office of Pesticide Products (OPP).

by the USDA, either before or after their deliberate genetic alteration. The current regulation of transgenic plants by the USDA is based primarily on their characterization as potential plant pests, a topic discussed more fully subsequently.

In addition to the nature or characteristics of a plant, another relevant regulatory trigger is its intended uses. Thus, agriculturally important plants such as corn and wheat potentially would be subject to different regulation than decorative plants such as roses. The use of transgenic plants for pharmaceuticals or as human food or animal feed, in the production of such items, or as components of each of these articles usually results in more stringent regulation than uses of plants for nonhealth or nonnutritive purposes, for example, for decorative uses.

Other current applications of modern methods of biotechnology triggering significant regulatory oversight include the creation of plants expressing various proteins that make the altered plants either toxic to insects (often called "pesticidal plants") or resistant to herbicides.

Finally, the techniques used to alter plants genetically can influence the types of laws that are triggered. The use of various vectors to introduce genetic information into plants typically involves federal oversight because plant pathogens such as *Agrobacterium tumefaciens* are used. Much of the regulatory debate thus far has focused on the scope of organisms, in this case plants, that should be subject to federal oversight if they are altered genetically (Korwek, 1990a). A recent document published by the federal government [Office of Science and Technology Policy (OSTP, 1992)] suggests that vascular plants altered by traditional techniques such as mutagenesis should not be subject to special regulation. Nevertheless, the debate continues over the orientation of federal or state regulatory initiatives around product- or process-based criteria and the extent to which organisms that result from "traditional" genetic techniques should be regulated (Korwek, 1990a).

10.2 Types of Oversight

Regulatory agencies usually have two types of general oversight capabilities. The first involves premarket clearance authority, which means that some type of "prior regulation" exists, usually in the form of required permits, licenses, regulations, premanufacture notification submissions, or product approvals. Although federal regulation often is thought to be confined to industrial activities, it exists also for research involving nonprofit motives, especially when funding is obtained at least in part by federal grants (Korwek, 1990b). For example, the National Institutes of Health (NIH) Guidelines for Recombinant DNA Research have been applicable to most federally funded research, but the NIH Guidelines also have become the standard for conducting non-government-funded recombinant DNA research. Similarly, the USDA is developing guidelines for conducting environmental-release experiments, including those involving genetically altered plants. Such guidelines also will likely become the standard for conducting environmental-release experiments, whether federally funded or not, regardless of the techniques of genetic alteration that are used.

The second type of oversight is postmarketing regulation. Once an article is marketed or an activity is undertaken, various regulatory agencies have numerous statutory powers, sometimes under emergency provisions, to remedy any health or environmental problems that may arise. Thus, because a federal agency does not have prior review powers does not mean that it has no authority over a particular activity or product.

On the other hand, prior review authority generally is viewed as more protective of health and environment, because exposure can be prevented or limited before an activity is undertaken. Nevertheless, this approach to oversight assumes that relevant risks can be ascertained and that their impact on health or the environment can be assessed precisely, which often is not the case. This chapter is confined to a general discussion of the prior review powers of the federal government, particularly since the costs and time involved in conducting a particular activity, such as marketing a product or conducting a field test, are related to such preclearance authority.

10.3 Regulatory Agencies and Legislation

10.3.1 FEDERAL FOOD, DRUG, AND COSMETIC ACT; MEAT INSPECTION ACT; AND POULTRY PRODUCTS INSPECTION ACT

The likely use of many transgenic plants as foods, drugs, or, perhaps, color additives and cosmetics, or as components of each of these products, would trigger FDA oversight at least under the Federal Food, Drug, and Cosmetic Act (FD&C Act). Each of these product categories is subject to varying degrees of regulatory oversight, including premarket clearances, although cosmetics usually are not subject to review prior to marketing.

Most FDA applications of transgenic plants probably involve premarketing clearances. For example, plant-derived pharmaceuticals usually would be subject to FDA clearances prior to their use in clinical settings or at the commercial level. The regulatory picture is more complicated for food products, since different kinds of food ingredients can be regulated differently; some require premarket clearances whereas others do not (Korwek, 1990c).

Ingredients of meat and poultry products typically are also regulated by the Food Safety Inspection Service (FSIS) under the Federal Meat Inspection Act (FMIA) and Poultry Products Inspection Act (PPIA). Although the FDA and the USDA both have regulatory authority over meat and poultry products, the authority of the FSIS is much broader because it includes prior clearances of all ingredients, labeling, and materials brought into federally inspected meat and poultry plants. FDA clearance under the FD&C Act usually is a prerequisite for obtaining USDA clearance of ingredients of meat and poultry products.

10.3.2 FEDERAL INSECTICIDE, FUNGICIDE, AND RODENTICIDE ACT

The Federal Insecticide, Fungicide, and Rodenticide Act (FIFRA) defines a pesticide as essentially any substance intended for preventing, destroying, repelling, or mitigating any pest. Pesticides typically are the subject of prior regulation in the form of product registrations. A pesticide product usually must undergo extensive tests before it can be registered by the EPA and legally marketed. Tolerances, which specify levels of pesticides that cannot be exceeded in order to prevent unnecessary exposure, usually exist for pesticides in or on raw agricultural commodities.

For transgenic plants, questions exist regarding the regulation of so-called "pesticidal plants," including plants containing not only genes expressing toxins potent for various organisms, but also genes providing tolerances to various viruses. Whether the FDA, as part of its jurisdiction over food under the FD&C Act, would have primary jurisdiction over such pesticidal plants intended for food use or whether this jurisdiction would fall to the EPA under the FIFRA scheme for pesticides, or to both the FDA and the EPA, remains unclear. The possibility likely exists now that both agencies may have some type of premarket clearance powers over certain types of pesticidal plants intended for food applications.

10.3.3 FEDERAL PLANT PEST ACT, PLANT QUARANTINE ACT, AND NOXIOUS WEED ACT

The Federal Plant Pest Act (FPPA), the Federal Plant Quarantine Act (FPQA), and the Federal Noxious Weed Act (FNWA) potentially apply to all transgenic and other plants, whether genetically altered or not. They provide the USDA with extensive authority to oversee a wide range of activities that potentially may result in plant injury; moreover, these laws apply regardless of whether the transgenic plant is intended for use as food or for other purposes. The laws are oriented environmentally, not toward insuring the safety of food. Their purpose is generally to insure that plants, whether genetically altered or not, are not exposed to disease-causing or damaging agents that could devastate plant resources.

The most relevant statute to federal regulation of transgenic plants thus far is the FPPA, which currently is being used to require USDA clearances for field testing of plants genetically altered as a result of the use of vectors involving plant pathogens such as *Agrobacterium tumafaciens* and cauliflower mosaic virus. Thus far, use of the FPPA has not been problematic in delaying field tests; the USDA has cleared numerous field tests of genetically altered plants.

Another environmental statute, perhaps of emerging significance to the oversight of transgenic plants, is the FNWA. This law governs noxious weeds and could, in the future, serve as a source of oversight of transgenic plants, particularly as the use of pathogens such as *Agrobacterium* becomes more familiar and perhaps subject to USDA exemption under the FPPA (Korwek, 1990c).

10.3.4 TOXIC SUBSTANCES CONTROL ACT

The Toxic Substances Control Act (TSCA) provides the EPA with extensive authority to regulate a wide variety of research and development or commercial activities involving a "chemical substance." This term is defined broadly in the TSCA to include any organic or inorganic substance of a particular molecular entity. Because such articles as foods, drugs, cosmetics, and pesticides are excluded from the definition of a chemical substance, application of the TSCA to transgenic plants is generally quite limited, especially since the TSCA currently is interpreted not to apply to whole plants.

One potential TSCA application to transgenic plants is the production, for example, of commodity chemicals or use of diagnostics for plant diseases. Such articles not subject to the FD&C Act or FIFRA could be deemed "new chemical substances" subject to TSCA prior regulation. This interpretation generally would mean that commercial production of such chemicals could not occur prior to submission of a premanufacture notification.

10.3.5 NATIONAL ENVIRONMENTAL POLICY ACT

The National Environmental Policy Act (NEPA) basically requires all federal agencies to assess the environmental impact of their actions. This is a separate requirement apart from the preclearance authority an agency may exercise. Generally, a NEPA assessment must be performed whenever a federal agency other than the EPA is involved in exercising its premarket clearance authority. NEPA requires an assessment of the potential environmental impact of a contemplated activity, perhaps of a field-release experiment. The effect of NEPA in most regulatory areas is becoming increasingly apparent. A NEPA assessment is conducted regularly for field tests involving genetically altered plants in which plant pests are used as vectors. NEPA requirements also must be satisfied whenever FDA clearances are sought for food, drug, or other applications of modern biotechnology.

10.4 Regulatory Perspective

Various legislative initiatives in Congress have been introduced in the past and currently are under consideration to change the scheme of regulation of environmental releases, including those involving transgenic plants. The formulation of a science-based uniform system for overseeing environmental releases has been difficult for a variety of reasons (Korwek, 1990a), including the consideration that different laws are aimed at achieving different purposes, an observation that results in a complex system of regulation because different types of triggers for oversight exist. Different triggers create different bases for regulation, making consistency of approach sometimes difficult to achieve, particularly given the variety of applications of transgenic plants. Further, each intended use for plants also can trigger different oversight schemes, some involving premarketing clearances, others not. Many laws engender a lack of uniformity because an element of judgment is involved in their administration, for example, in deciding what is a plant pest. Thus, different agencies can administer different laws in a variety of ways, using diverse criteria to trigger or exclude products from oversight. Modern methods of biotechnology thus raise many old issues or problems in a new context.

References

Korwek, E. L. (1990a). *Chem. Reg. Rep.* **13,** 1454–1458.
Korwek, E. L. (1990b). "1990–91 Biotechnology Regulations Handbook." Center for Energy and Environmental Management.
Korwek, E. L. (1990c). *Food Tech.* **44,** 76–78, 80.
Office of Science and Technology Policy (OSTP) (1992). *Fed. Reg.* **57,** 6753–6762.

11

Social and Economic Aspects of Transgenic Plants

Douglas D. Parker

Center for Public Issues in Biotechnology
Maryland Biotechnology Institute
University of Maryland
College Park, Maryland 20742

Darrell Hueth

Department of Agricultural and Resource Economics
University of Maryland
College Park, Maryland 20742

Genetic manipulation of plants began thousands of years ago when humans first stopped wandering and began to cultivate and select desirable plant crops. Initially, these early innovators relied on chance and, after some time, experience to create new successful plant varieties. Today, scientists have discovered ways to alter plant characteristics through the intentional manipulation of their basic genetic codes. This type of genetic manipulation requires extensive amounts of information on the structure of the genes within a plant. The rate of innovation is limited by the amount of information available. As this information accumulates, the rate of innovation will increase, perhaps exponentially. The field of transgenic plants may be in the early stages of an exponential growth pattern. Several new transgenic plants are now in the final testing stages. Once these early innovations become commercially available, the speed at which subsequent innovations are developed will increase. This rapid development of new transgenic plants already has produced some public opposition.

Sociologists have demonstrated that the willingness of the public to accept change is related inversely to the rate of change and the degree of change. Biotechnology ranks high in both of these areas. In general, societies seek systems that are stable, often reacting irrationally to destabilizing innovations. Therefore, in order to insure public confidence and acceptance, care must be taken to insure that new transgenic plants are safe and that the introduction of these new plants is carried out both openly and honestly. This type of approach will increase public confidence, necessary for the successful establishment of this industry.

Lack of public support for new technologies often has been interpreted as a lack of communication. Recognition that public concern may result from value differences among groups may be the first stage in gaining support. In the past, the point of view of many industries promoting new technologies has been that if the public, which was assumed to be ignorant, could be made to understand the science behind the innovation then everyone would support these new technologies. This assumption ignores the fact that certain segments of the population may have different goals than the industry. Further, by treating the public as inferior, this approach increases the probability of public opposition. Disrespect damages the credibility of an industry and decreases the willingness of the public to accept new innovations.

Some initial opposition to new technologies may be unavoidable. Current public perception is that the potential benefits of a new innovation are known whereas some uncertainties remain over the costs. This perception appears to be a result of the belief that industry focuses the majority of its investigation on the potential benefits of the new product, leaving the potential costs unknown. Further, if the public is satisfied with its current technological base, the costs of a new technology may be weighted greater than the benefits. When the new technology proves beneficial, this weighting of benefits versus costs may change, producing greater public acceptance. This initial skepticism to new technologies may be traced to past technological advances.

After World War II, the public believed in the promise of better living through chemistry. Because many scientists also believed this, many products were introduced without a thorough investigation of their structure and their potential to do harm. People were told of the benefits of new chemicals but not of the possible consequences of use. Further, when some chemicals finally proved hazardous, the industry and government failed to respond quickly and properly. Through thorough investigation, brought about by the environmental movement of the 1960s and early 1970s, many products were found to produce unacceptable levels of harm. Current regulations require that new products be evaluated for adverse effects. This evaluation will insure that new products are safe and beneficial. Whereas the creation of transgenic plants also promises an increased standard of living, the potential for unforeseen harm does exist. With increased awareness, especially in the field of ecology, the introduction of transgenic plants is proceeding cautiously to insure public and environmental safety. Responsible introduction and regulation of transgenic plants is necessary to insure public confidence.

Finally, the issue of perceived versus scientific risk must be addressed when estimating public response to a new transgenic plant. We use the term scientific risk rather than real risk because, to an individual, perceived risk is real. The public often views risk as something that is voluntary, for example, the risk of driving an automobile. Scientists measure risk as some function of the probability of an event and the potential harm of an event. The public is often more willing to accept a voluntary risk over an involuntary risk, although the scientific risk is the same. It is also important to remember that risks and benefits do not accrue equally to all groups. In field tests of transgenic plants, local populations may be forced to take an involuntary risk while some large corporation, perhaps thousands of miles away, receives the benefits. Failure to recognize these points has caused corporations and regulators to miscalculate public response to the introduction of new transgenic plants.

The remainder of this chapter focuses on three areas in which the introduction of transgenic plants may affect society. The first area includes a wide range of social issues. This section briefly touches on the possible contributions from transgenic plants in the areas of food safety, food quality and nutritional composition, the environment, and farm worker and public health. We then focus on some of the potential economic impacts of transgenic plants. Finally, we consider some of the international contributions that may be made by this new field of science.

11.1 Social Issues

11.1.1 FOOD SAFETY

The issue of food safety has become a major concern for consumers. Although public concern has focused on chemicals used in agriculture, many scientific experts believe that greater threats to food safety arise from other sources. The general scientific ranking of food safety concerns lists the number one hazard as contamination from pathogenic microorganisms. The other concerns, in order of importance, are nutritional imbalances, natural toxicants, industrial pollutants, agricultural pesticides, and food additives (Hotchkiss, 1990). The creation of transgenic plants can affect each of these health issues, often affecting several of them simultaneously.

For example, the introduction of pest-resistant plants may lessen the need for pesticides and, through reduced damage, lessen the possibility of microbial contamination. This indirect benefit arises because most microorganisms require open wounds on the plant for attachment.

Although many transgenic plants may reduce food safety concerns, care must be taken to insure that other hazards do not increase. New varieties of plants, whether created by transgenic or traditional technologies, require close supervision in early trials to detect unforeseen changes. For example, some transgenic plants may be more susceptible than their traditional relatives to toxins. Through transgenic manipulation, it may be possible to decrease (increase) microbial contamination in plants by decreasing (increasing) the level of certain substances that induce microbes to grow.

11.1.2 FOOD QUALITY AND NUTRITIONAL COMPOSITION

Transgenic plants have the potential to produce changes in the nutritional composition of our diets. For populations that eat a wide variety of food stuffs, for example, the population of the United States, nutritionists believe that modification of the nutritional composition of specific foods is unnecessary. For certain subgroups of our society, the introduction of transgenic plants with increased levels of specific nutrients may offer increased health benefits.

For the general population, the introduction of transgenic plants most likely will have negligible effects on current nutritional balances. The dramatic increase in the consumption of poultry and fish during the past few years is strong evidence of the awareness and desire of the public to improve their diet nutritionally. According to many nutritionists today, variety in the diet is the key to good nutrition. The creation of transgenic plants, which are then promoted as providing complete daily nutritional needs, should not be pursued. Most nutritionists believe that a single food source would not be desirable, and promotion of "complete foods" would be misleading. Although the chances of a particular food becoming so dominant that it upsets the current nutritional balance appears small, nutritionists should continue to monitor the introduction of all new plants for this type of major dietary change.

The introduction of transgenic plants probably will benefit public health most by reducing the levels of harmful nutrients in foods, for example, saturated fats. The biggest nutritional imbalances in the United States today involve the overconsumption of items such as saturated fats, proteins, and cholesterol. By altering the amounts of these items in foods, better nutrition may be obtained. Low saturated fat transgenic vegetable oils are an example of products that are being pursued currently.

11.1.3 ENVIRONMENTAL IMPACTS

The introduction of new transgenic plants may produce both positive and negative environmental impacts. A great deal of effort has been put forth by environmental organizations in opposition to herbicide-resistant transgenic plants. These organizations claim that this resistance encourages the use of greater amounts of herbicides and increases the dependence of agriculture on chemicals. Work by Just and Hueth (1990) demonstrates this tendency for agricultural chemical companies to produce products that promote dependence. Industry analysts claim that these plants will allow farmers to use more environmentally benign herbicides and, because more weeds can be controlled per application, a decrease in total herbicide use may result.

Other transgenic plants also will affect the environment, either directly or indirectly. The introduction of pest-resistant transgenic plants may reduce the level of pesticides used, which will benefit the environment directly. Some groups have raised concerns over the safety of plants with "built in" pesticides. The possibility of these genetic manipulations being transferred to other plant species must be addressed. Such a transfer of herbicide resistance or pest resistance could have significant ecological consequences. There is also some concern, especially with nonedible plants, over the disposal of the plants once this pest resistance is introduced. Although many of the issues are mere speculation based on incomplete facts, these issues must be addressed adequately by those who are involved in the creation of these plants.

Some transgenic plants may affect the environment only indirectly. The creation of a tomato with increased solids by Calgene (Davis, California) has the potential to reduce harvesting and marketing damages to the product. This reduction in damage may in turn decrease the need for fungicides. Nitrogen fixation also may improve environmental quality by leading to a reduction in nitrogen-based fertilizer use. It also may lead to unforeseen changes if crop rotations are altered. This type of indirect change will produce uncertainties with respect to pest populations and soil erosion.

Some work also is being done to create transgenic plants that remove toxic metals and other environmental pollutants from soil or water sources. This type of innovation would be useful when cleaning up industrial pollutants by concentrating them in a controllable space. These plants then would represent a source of environmental pollution. Although the issue of what to do with these potentially toxic plants after they have removed the pollutant also must be addressed, by concentrating the pollutant into a movable convenient source the number of options for disposal increases.

Discussion of the wide range of possible environmental effects reinforces the need to consider carefully all the possible consequences of the introduction of transgenic plants. Care should be taken to insure that new plants do not produce unforeseen negative effects on the environment. We also should be careful not to misrepresent products and promote uncertain effects. Although the introduction of many transgenic plants may produce great environmental benefits, misleading promises should not be made lest the credibility of the entire industry suffer.

11.1.4 FARM WORKER AND PUBLIC HEALTH

Many of the potential health issues for transgenic plants can be linked to environmental issues. For example, plants that decrease the need for pesticide applications also will decrease the risk of farm worker exposure to both applications and residues. The introduction of herbicide-resistant plants may allow the use of less toxic chemicals which may lessen health risks to farm personnel. Reducing the use of fungicides could reduce risks greatly. Fungicides are a leading cause of farm worker illness from the use of agricultural chemicals. They also affect others in the food industry, for example, workers at grain storage sites and those involved in transporting foods.

Indirect effects on public health also will result from transgenic plants that lessen the need for agricultural chemicals. Accidental exposure of populations near farms could decrease. Further, by lessening the use of chemicals, fewer chemicals will be produced and transported, further reducing the public risk of accidental exposure from spills. The decrease in demand also will decrease the amount of residual or waste products generated by the production of agricultural chemicals.

As with the introduction of any new technology, transgenic plants offer some positive and some negative contributions to society. The fact that negative impacts are possible does not imply that this new technology should be abandoned. We have the ability to examine new innovations to assess the expected impacts. This information should be used to insure that only those beneficial innovations that keep costs within an acceptable range are introduced. Economists address this issue of costs versus benefits in many different ways, depending on the type of innovation. The next section briefly reviews how the costs and benefits of transgenic plants may be quantified.

11.2 Economic Impacts

The purpose of any new technology should be to increase the well-being of society. The creation of transgenic plants and other biotechnological innovations present some new and unique choices for society. The evaluation of the potential benefits and costs of these innovations by economists will require the application of a wide range of methodologies. The majority of past technological innovations in plant production are ones that increase farm yields or reduce costs. For transgenic plants, changes are possible that affect product quality, consumer acceptance, producer levels of risk and uncertainty, output fluctuations, environmental quality, and the more traditional yields and costs.

The field of economics provides a methodology, known as welfare economics, that is useful in the evaluation of potential benefits and costs of new innovations. This framework divides the effects into consumer and producer surplus, or gains, the sum of which equals the net change in welfare for society. The immediate effect of a new innovation is on the producers or users of it. The actions of this group will create market effects that will, in turn, affect other producers and, ultimately, consumers.

For example, herbicide-resistant plants most likely will increase yields and/or decrease costs to users. If enough growers adopt this new technology, the total supply of the crop will increase, causing a decrease in the price of a crop. The change in profits to growers using the new transgenic plant can be estimated. This estimate must account for both increases in revenues from increasing yields and decreases in revenues from decreasing prices. This reduction in price affects all growers of the crop, not just those using the new resistant variety. For growers who do not adopt the new transgenic plant, yields remain unchanged while prices decline. Therefore, revenues will decrease for this group. The sum of the profit effects on these two groups determines the net change in producer surplus.

The net change in social welfare, defined as the change in consumer welfare plus the change in producer welfare, often is viewed as the value of a new innovation. Although a new innovation may produce positive net social benefits, the distributional consequences may require some mitigation before the innovation is acceptable to all. The degree of analysis of distributional consequences varies widely in economic studies. In welfare economics analysis, it is possible to differentiate between users and nonusers of a new innovation and between consumers and producers.

In the next two sections, examples of how supply-increasing and demand-increasing innovations might be analyzed are given. These sections show how producer and consumer gains may be calculated and summed to provide a value for the net change in social welfare. These two sections are followed by a discussion of possible aggregate effects of transgenic plants on the structure of input and output markets. Finally, conflicts of interest in research and development will be addressed using the welfare economics methodology presented here. In particular, the conflicting goals of public versus private investments will be examined along with their potential effects on cooperative research projects.

11.2.1 YIELD-INCREASING AND COST-REDUCING INNOVATIONS

The majority of past technological advances in agriculture have focused on increasing yields. Although new innovations in transgenic plants will continue to increase yields, there probably will be an increase in other types of advances. Recent attention in agricultural economics has focused on the effects of yield increases on producer, or grower, welfare. Many of these studies show that, although the net change in welfare for society is positive, for those crops for which price responds greatly to changes in quantity the change in producer welfare may be negative, because the decrease in price from an increase in quantity overshadows the benefits of that change in quantity.

The economic effects of yield-increasing innovations are demonstrated in Fig. 11.1. Assume that, before the new innovation is introduced, the supply curve is S_0 and the demand curve is D. The initial equilibrium price, P_0, and quantity, Q_0, are determined by the intersection of the supply and demand curves. For producers, the total revenue received from the crop is equal to area ($b + c + e + f + h$). The total variable cost to producers is equal to the area under the supply curve S_0 and to the left of output quantity Q_0, or area ($c + f + h$). These are costs that producers can control in the short-run, for example, labor and chemical inputs. Producer surplus is defined loosely to be equal to the difference between these areas or the triangle represented by the area ($b + e$). Changes in this area can be thought of as changes in profits and, if positive, represent a net gain to producers from producing the item.

The demand curve represents the price that consumers would be willing to pay for the commodity at different levels of quantities. This willingness to pay can be thought of as the value of the product to consumers. For those consumers that would have paid a price greater than P_0, those represented by points on the demand curve to the left of Q_0, a surplus (measured as income) is generated equal to the price they were willing to pay, represented by the point on the demand curve minus the actual price paid, P_0. This surplus income is known as consumer surplus and is defined to be the area below the demand curve but above the actual or equilibrium price. From Fig. 11.1, the level of consumer surplus before the new innovation is shown to be equal to area (a).

After the introduction of a yield-increasing innovation, for example, a trans-

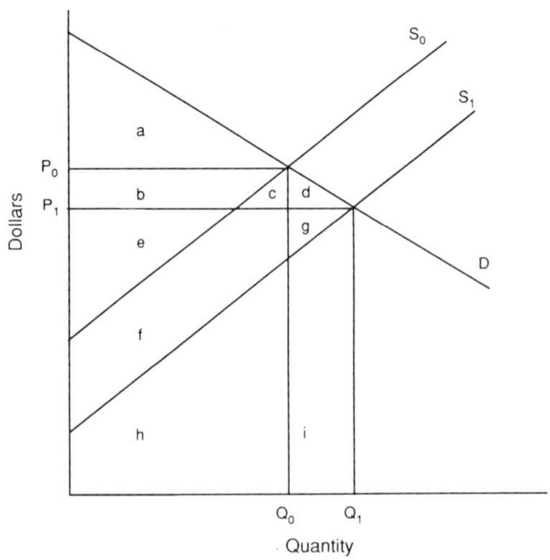

Figure 11.1 Welfare impacts of yield-increasing innovations.

genic plant, the supply curve is shifted outward to S_1. This shift represents the increase in output from the introduction of the new yield-increasing transgenic plant. This increase in supply does not affect the demand curve, which remains D. The shift in the supply curve to S_1 causes an increase in the equilibrium quantity to Q_1 and a decrease in the equilibrium price to P_1. This shift in equilibrium causes changes in both producer surplus and consumer surplus. The new level of producer revenue is equal to area $(e + f + g + h + i)$. The total costs now are represented by the area under the new supply curve, S_1, area $(h + i)$. The new level of producer surplus is equal to the new level of revenues minus the new level of costs, area $(e + f + g)$. The new level of consumer surplus is equal to the area below the demand curve and above the new equilibrium price P_1, or area $(a + b + c + d)$.

To identify the effects of the yield-increasing innovation, we must know what the changes in producer surplus, consumer surplus, and social welfare are. The change in producer surplus is equal to the new level of surplus, area $(e + f + g)$, minus the initial level of surplus, area $(b + e)$. Therefore, the change in producer surplus is equal to area $(f + g)$ − area b. Whether this term is positive or negative depends on the slope of the demand curve D. For very steep demand curves, this difference will be negative and producers lose from the new innovation. For these goods, demand is said to be inelastic, showing the increase in quantity to cause a relatively greater decrease in price. Unfortunately, this is the case for many agricultural commodities. For relatively flat, elastic demand curves the change in producer surplus will be positive. Therefore, the shape of the demand curve for the commodity determines the producer value of any new innovation that affects the supply of that item.

The change in consumer surplus from the yield-increasing innovation is somewhat easier to derive. The change in consumer surplus is equal to the new level of consumer surplus, area $(a + b + c + d)$, minus the initial level of surplus, area a. This is equal to area $(b + c + d)$, which is unambiguously positive. Although the size of the change in consumer surplus from a yield-increasing innovation depends on the slope of the demand curve, under normal conditions the change in consumer surplus will always be positive.

The net change in social welfare is equal to the sum of the changes in producer surplus, area $(f + g)$ − area b, and consumer surplus, area $(b + c + d)$, or area $(c + d + f + g)$. Therefore, we see that, while the effect on producers is uncertain, the positive effect on consumers insures that the yield-increasing innovation provides a net benefit to society. This net benefit does not include any type of measure on the social value of income distribution. If net transfers of wealth among subgroups is undesirable then the analysis should list changes in consumer and producer surplus separately.

As Schultz (1953) and others have noted, the demand for many food products is quite inelastic, causing their prices to fall substantially as quantities are increased. For example, United States demand elasticities for basic commodities (cereal, milk) are especially low, whereas demand elasticities for some specialty crops and fish tend to be high (George and King, 1971). Thus, transgenic plants that increase yields for commodities such as wheat or milk may harm producers whereas yield increases in certain fruit and vegetable crops as well as fish may benefit producers.

In the long run, the creation of transgenic plants that reduce costs will produce the same type of effects as yield-increasing innovations. Once a new equilibrium is established, the reduction in costs will have shifted the supply curve outward, as was shown in Fig. 11.1. Herbicide-resistant transgenic plants may fall into this category. By inducing resistance to a particular herbicide, growers will be able to reduce costs by applying a cheaper broader-range herbicide, which may reduce the number of treatments and the cost per treatment by eliminating the need for selective spraying. These new plants also may show increases in yields by reducing competition with weeds and by tolerance of herbicides. Although it is too early to predict exactly what the net effects will be, if the product is to be marketable either costs must decline or yields must increase. The net effect of herbicide-resistant plants will be an increase in supply. Whereas the net effect on social welfare is expected to be positive, an increase in supply will produce increases in producer surplus for crops with relatively high demand elasticities and losses in producer surplus for crops with relatively low demand elasticities. This analysis does not include effects on social welfare from changes in environmental quality, farm worker safety, or many other possible secondary effects. These effects, alluded to in the previous section, should be included in any analysis of a new innovation.

11.2.2 DEMAND-INCREASING INNOVATIONS

One of the unique attributes of biotechnology, compared with past technological advances, is its ability to alter the natural characteristics of plants. For example, biotechnology has been or could be used to alter the color of flowers, change the amino-acid content of oils, alter the processing and marketing characteristics of tomatoes, and alter the taste characteristics of fruits and vegetables.

Economic analysis of these types of innovations requires different techniques than those used for the more traditional innovations that increase yields. Quantification of changes in social welfare from new products is difficult. Economists may predict market conditions for a new product by analysis of conditions for similar products. For instance, the demand for a transgenic fruit that remains firm when ripe may be compared with the demand for other fruits that are tree ripened but must be marketed with special care to prevent damage.

Quantification of consumer response to changes in product characteristics often are extrapolated from direct surveys or from analysis of data that relates

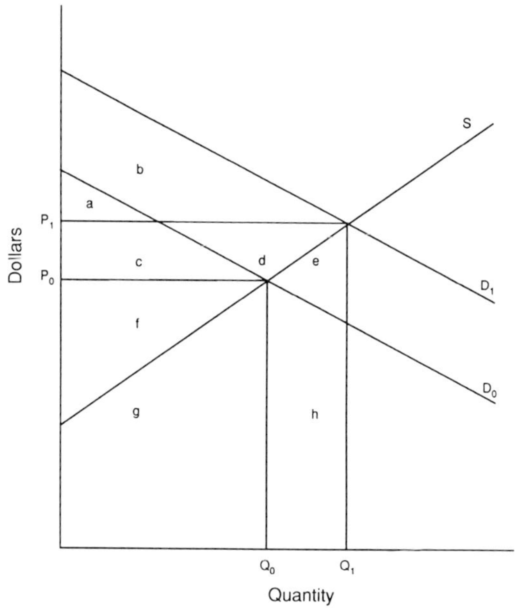

Figure 11.2 Welfare impacts of quality-increasing innovations.

price to quality characteristics (Waugh, 1928; LaFrance, 1983; Silberberg, 1985; Parker *et al.*, 1991). Valuation of characteristics is necessary to quantify shifts in the demand curve. The value of improved marketing or processing characteristics, which do not affect consumer demand, are obtained more easily via traditional cost-reducing loss-reducing types of analysis.

Once shifts in demand are quantified, the welfare effects of a change in quality characteristics can be modeled similarly to yield-increasing transgenic plants. This model is illustrated in Fig. 11.2. It is assumed that this new transgenic plant affects only the level of quality produced and not the yield or cost of production. Therefore, the supply curve will remain fixed at S. An increase in quality, while holding all else constant, would cause the demand curve to shift outward from D_0 to D_1. Following the analysis just presented for the welfare effects of a yield-increasing innovation, the initial level of producer surplus, when equilibrium price is P_0 and quantity is Q_0, is equal to area f. The introduction of the new transgenic plant shifts the demand curve for this product from D_0 to D_1, causing the new equilibrium price to be P_1 and quantity to be Q_1. The new level of producer surplus is therefore equal to area $(c + d + f)$. This leads to an unambiguous increase in producer welfare equal to area $(c + d)$. Note that this is in sharp contrast to the yield-increasing innovation which, for many agricultural products, will result in producer losses.

The effect on consumer surplus also can be shown. Initially, consumer surplus is equal to area $(a + c)$. With the introduction of the higher quality transgenic plant, the demand curve shifts outward to D_1. The new level of consumer surplus is equal to area $(a + b)$. The change in consumer welfare is therefore equal to area b − area c, which can be shown to be nonnegative. This is a case of a vertical supply curve, in which no adjustments in the quantity supplied are possible, that causes the net change to be zero. In all other cases, the net change in consumer welfare will be positive.

The net change in social welfare is equal to the change in producer welfare, area $(c + d)$, plus the change in consumer welfare, area b − area c, or area $(b + d)$. Therefore, holding all else constant, positive changes in characteristics lead to an increase in social welfare.

11.2.3 CHANGES IN STRUCTURE OF INPUT AND OUTPUT MARKETS

The creation of transgenic plants will cause changes in the input markets for agriculture and in output markets of agriculture. For example, the introduction of herbicide-resistant plants will increase the competitive advantage of chemical companies producing the herbicides for which tolerance is introduced. Further, changes in herbicide use may decrease the use of nonchemical alternatives. For instance, a decrease in manual weed control would lead to a decrease in the demand for labor. Therefore, farm workers might be expected to oppose the introduction of such technologies. Also, rural communities can be affected significantly by these types of shifts. The paths of changes from innovations are often long and unpredictable. For example, although fewer agricultural workers may be needed for weeding when herbicide-resistant crops are introduced, there may be an increase in the need for labor in the chemical industry. These types of shifts often result in a decrease in demand for unskilled labor and an increase in demand for skilled labor.

Transgenic plants also can be expected to cause shifts in the structure of output markets. The food processing industry has become a dominant part of our food system over the past several decades. By incorporating characteristics into transgenic plants that previously were added by food processors, this industry may change. For example, the production of transgenic broccoli or popping corn that has butter or cheese flavors has been proposed.

Plants that have better fresh characteristics also will have secondary market effects. Tomatoes that are firm when ripe will change the way in which tomatoes are handled, for example, decreasing the need for ethylene gassing of green tomatoes. Further, special packaging techniques may change to meet new needs. The creation of transgenic plants that increase the use of mechanical harvesting and packaging can have major effects on the packing industry.

Transgenic plants may have many such indirect effects on the entire food industry and its suppliers. Over time, these innovations will lead to new equilibriums in their respective economic markets. As with the introduction of any new innovation, some of the negative consequences may be mitigated through proper evaluation and planning.

11.2.4 CONFLICT OF INTEREST IN RESEARCH

In the welfare economics methodology presented earlier, it was shown that gains from innovations are not distributed evenly. This poses some problems for cooperative research projects in which public and private interests are combined. This conflict of interest will be examined using the demand-increasing analogy presented in Fig. 11.2.

The rate of return to an investment, I, is equal to the welfare gains divided by the amount of investment. This rate of return, R, can be separated into two categories: social (public) and private (producer). For example, from Fig. 11.2 these can be defined as

$$R^{\mathrm{s}} = \frac{\text{area}(b + d)}{I^{\mathrm{p}} + I^{\mathrm{s}}}$$

and

$$R^{\mathrm{p}} = \frac{\text{area}(c + d)}{I^{\mathrm{p}}}$$

where the superscripts s and p represent the social and private values.

The potential conflict of interest lies in the differences between the private rate of return, R^{p}, and the social or public rate of return, R^{s}. Under public financing only ($I^{\mathrm{p}} = 0$), the social rate of return is R^{s}. This shows that there should be an emphasis at universities and other public institutions on creating technologies that expand areas *b* and *d* of Fig. 11.2, that is, if universities are expected to maximize the social rate of return from public research funds, they should pursue research that increases the sum of producer gains plus consumer gains, paying no attention to the relative size of the gains to these two groups. For the private sector, only those technologies that increase areas *c* and *d* will be pursued. The conflict of interest for cooperative research results from an increased emphasis on expanding area *c* and a decreased emphasis on expanding area *b*.

Expansion of area *c* by the private sector does not create any net benefits to the enclosed system. What it does is transfer wealth from consumers to producers while using up resources, I^{p}.

It should be noted that consumers may benefit (or lose) from a particular biotechnology industry as the result of an improvement (or decline) in environmental quality. Because producers often do not receive payment (or pay costs) associated with these induced benefits (or losses), they do not include these as factors in their rate of return calculations. When maximizing true social welfare, public institutions such as universities should factor these indirect effects into their rate of return calculations.

This type of analysis demonstrates how economics may provide useful information for directing transgenic plant research. Determining optimal investment levels and strategies for public, private, and cooperative research is important. As we have illustrated, particular attention must be paid to the benefits and costs of public versus private research. In the field of biotechnology, the division of research interests is eroding particularly quickly. Analysis of the public versus private research issue must pay particular attention to the distributional effects of new innovations. Large shifts in wealth among consumers and producers may be undesirable. These changes can be analyzed using the welfare economics methodologies just presented.

Traditionally, funding for basic research has been provided by government or other public sources. This type of research is necessary to advance the general level of scientific knowledge. Through basic research, new techniques often are discovered that are used later in applied research. Further, it is often necessary to accumulate background information before applications of science may be made. Such fields as biotechnology require a great deal of basic information to progress. Information on techniques to analyze the genetic structure of a plant, and information on the genetic structure itself, must be available before any manipulation can be considered.

It may be necessary to carry out a great deal of research to gain a sufficient amount of information about the genetic makeup of an organism. Once this information is available, it may be possible to create several new innovations that become profitable once this initial investment in knowledge has been made. In this case, public research may take place at the university level to create the general knowledge. Once complete, this information may be used by private research institutions to create several new products. Although it may not be profitable for any one of the private parties to carry out the initial research, a net benefit to society can be realized from a combined effort that undertakes complete research. This type of information transfer has been important to the growth of our economy. Research and development of many new products may not be possible within either sector of our economy but, by joint research and information transfer, beneficial innovations can be achieved.

"Nonembodied innovations" represent another area in which the public sector may create new products beneficially, that otherwise may not be created in the private sector. These products contain new innovations that cannot be tied completely to the new product in order to insure exclusivity by the innovator. Examples of this include some new varieties of plants that seed propagate. Once released for use, control of the innovation is not held exclusively by the creator. The reason that these products will not be produced in the private sector is the inability to recover research and development costs.

The time it takes to develop and bring to market transgenic plants is estimated to be 5–6 yr. During this time, a company will spend $20–25 million researching, developing, and marketing the product (Fraley, 1990). If the company is unable to retain exclusive control over a new product for a sufficient period of time, the investment cannot be recovered.

This need of the private sector to recover their investment is met through legal protections such as patents. For some potentially beneficial innovations, legal protection may not be possible. In these cases, it may be beneficial for the public sector either to help fund the private investment in research and development or to carry out the research and development itself.

This type of government research and development may be especially important in less-developed countries. For these countries, the creation of transgenic plants may increase food supplies and nutritional composition for populations that are unable to pay the costs of research and development. Some of the potential international contributions of transgenic plants will be addressed in the next section.

11.3 International Effects

11.3.1 ALTERING COMPETITION IN WORLD AGRICULTURAL MARKETS

Transgenic plants, and biotechnology in general, will have significant effects on competitive agricultural markets of the world. The countries that excel in this area of research and development will gain, at least initially, additional competitive advantages in these markets. This will be the result of research that will be oriented toward domestic crops. The need to maintain a competitive advantage in agriculture often is used as a justification for public support of research into new technologies. Once these technologies are established, public support is phased out and the commercial sector takes control. This type of initial investment and support may increase the long-term competitiveness of the domestic agricultural industry.

11.3.2 ENHANCING WORLD FOOD SUPPLIES

The introduction of transgenic plants to the international community holds some particularly encouraging possibilities for developing an adequate world food supply. In the face of a rapidly increasing world population, this is essential.

Through the use of biotechnology, the food supply of the world can be increased in several different ways. Transgenic plants may be created that are environmentally tolerant. Drought resistance, salt resistance, and aluminum resistance are a few examples of attributes of transgenic plants that could increase the world food supply.

Other attributes through which transgenic plants may increase food supplies are pest and virus resistance, and enhancement of characteristics that lessen postharvest losses. The area of postharvest losses is just beginning to receive required attention. Most research in this field has been toward the creation of mechanical and marketing mechanisms to lessen postharvest losses. More research in the area of plant breeding and genetic engineering could help reduce this significant source of food loss.

Research should also be supported that will lead to increases in specific nutrients in foods for certain populations. World populations suffer from two sources of malnutrition: caloric deficiencies and nutrient deficiencies. Biotechnology can be instrumental in decreasing the instance of both of these problems.

As alluded to at the end of the previous section, who will pay for this type of research remains to be determined. Often such research is not profitable for the private sector in more-developed nations where these potential problems can be solved by alternative methods. Less-developed countries, where these types of innovations could create the greatest benefits, lack sufficient capital to support this type of research. Poor rural farmers in these countries will be unable to pay for expensive seeds that incorporate transgenic modifications. Under current economic conditions, the people who need these innovations the most will be unable to pay the research and development costs. There is a need for world food organizations and governments in more-developed nations to support the type of research that may benefit the entire world greatly.

The introduction of transgenic plants will have lasting effects on our economic and social systems. Through insightful monitoring and regulating of the industry, as is being performed already, we can insure that new innovations are not only beneficial but safe for humans and the environment. The fields of transgenic plants and biotechnology hold great promise for the improvement of our lives and of the environment in which we live.

References

Fraley, R. (1990). Presentation to the United States Department of Agriculture's Second National Conference on Federal and State Regulation of Biotechnology.

George, P. S., and King, G. A. (1971). "Consumer Demand for Food Commodities in the U.S. with Projections for 1980." Giannini Foundation Monograph 26. University of California.

Hotchkiss, J. H. (1990). Presentation to the National Agricultural Biotechnology Council's conference on Agricultural Biotechnology, Food Safety and Nutritional Quality for the Consumer.

Just, R. E., and Hueth, D. L. (1990). "The Role of Market Structure in Agricultural Biotechnology Investment and Production." University of Maryland, Department of Agricultural and Resource Economics working paper No. 90-29.

LaFrance, J. T. (1983). Economics of nutrients and consumer demand for food. Ph.D. Thesis. University of California, Berkeley.

Parker, D. D., Zilberman, D., and Moulton, K. (1991). *Calif. Agr.* **45,** 14–16.

Schultz, T. W. (1953). "The Economic Organization of Agriculture." McGraw-Hill, New York.

Silberberg, E. (1985). *J. Polit. Econ.* **93,** 881–900.

Waugh, F. V. (1928). *J. Farm Econ.* **10,** 185–196.

12

Social Implications of Transgenic Plants

John F. Gerber
Institute of Food and Agricultural Sciences
University of Florida
Gainesville, Florida 32601

Alvin L. Young
Office of Agricultural Biotechnology
United States Department of Agriculture
Washington, D.C. 20250

12.1 Plants and Society

An examination of the social implications of transgenic plants must begin with an understanding of the role of traditionally produced plants in our society. Moreover, since the genetic modifications that can be done with the new biology are novel, our experience with them is limited.

Plants and cultivation marked the beginning of the end of nomadic culture and laid the basis for farming and the organization of society into what is now civilization (van Raay and Lugo, 1974). Planting; cultivation; protection from insects, disease, and animal foraging; harvesting; and crop storage required husbandry in specific places at specific times.

12.1.1 TRADITIONAL PRACTICES

Successful plant husbandry required that the husbandman be able to select and grow superior plants—ones that survived drought and flood, that reliably produced a crop, and that had better seeds, grains, root and tuber sizes, fruits, or edible leaves. Traditional practice for centuries was to select seed or propagating parts from superior individual plants (Janick, 1988). Sometimes this was successful and sometimes it was not, but clearly and persistently the idea that superior plant parents produced superior plants was fixed in the human mind. Because certain events such as the weather greatly influenced successful husbandry, and other inexplicable events led to failed crops, it was assumed that magical and mystical powers were at work and that, for future success, placation of these forces was wise at both planting and harvest (Lehner and Lehner, 1960).

12.1.2 CELEBRATIONS

Cultural festivals and celebrations often use plants as the focal point. These rites are often thanksgiving festivals at harvest time not only for the good year, but also for the good seeds for next season, freedom from pests and pestilence, and freedom from hunger and disease. Decorations and ornamentation were commonly specimen plants and colorful wreaths and bouquets from the bountiful harvest. Having enough food to flaunt, even waste, and offer as a gift or sacrifice to those unknown and mysterious forces was a public declaration of the success of the culture. In western culture, both religious and secular holidays have plants associated with them (Lehner and Lehner, 1960; Metcalf, 1969): Christmas, coniferous evergreens, mistletoe, and poinsettias; Easter, lilies; Thanksgiving, pumpkins; and Mother's day, flowers. Most states have a state plant, tree, or flower.

12.1.3 PLANTS AND CULTURE

Plants have a multiplicity of roles that they perform in society. Aside from supplying food, shelter, and clothing, they have medicinal, asthetic, and cultural values (Swaninathan and Koleher, 1985). Woody plants have been used as fuel for cooking and heating (Smith, 1982). Wine, beer, and alcoholic spirits all are fermented plant products; often to the detriment of society, plants have been the source of hallucinogenic and addictive drugs (Mills, 1985). Although animal products are important food sources, animals can be viewed as bioprocessors of plant products (Pond, 1983).

Certain foods became marks of cultural groups. Rice was associated with the Orient, wheat with Europe, cassava and manioc with the tropics and Africa, and corn and potatoes with the Americas. Lentils, fruits, vegetables, and nuts were important additions to these dietary staples (Anderson, 1952; Swaninathan and Koleher, 1985). Although celebrations and festivals were built around certain seasonal foods or crops that signified success, good fortune, or portended a good future, other plants were seen as protectants, amulets, or powerful medicine. Adornment of dwellings with specific plants and even certain seeds, nuts, and bulbs, and tubers worn on the body were thought to ward off evil spirits and disease (Lehner and Lehner, 1960). Eating special foods made of certain plants at specific times of the season acquired both cultural and religious significance. Social concerns and trust in maintaining a favorable relation with "mother nature" and some of the expressed concern about using "unnatural" means to change plant genetics may be rooted in the implicit trust that society has invested in its symbiotic social and cultural relationship with plants.

12.1.4 ORNAMENTATION AND DECORATION

Flowers, trees, grass, gourds, unusually large and colorful seeds, fruits, and foliage all are used as ornaments and decorations. These are not as much a symbol of plenteous food, as of unusually skillful and successful husbandry. Color, texture, flavor, aroma, and taste are not elements of nutrition, but of the high degree of socialization of food and eating (Farb, 1980). Painters, photographers, and other artists portray foliage, fruits, and vegetables that are highly colored and free from blemishes as cultural symbols that represent the socialization of plants into the culture.

12.1.5 ECONOMICS AND PLANTS

Economics has been influenced greatly by plants and the ability to produce storable staple commodities that are the basis for trade on both local, regional, and continental scales (Almanac, 1990). Plants were grown and selected to produce marketable commodities (Hicks *et al.*, 1975). Infrastructure transportation and storage facilities helped avert famine, build fortunes, provide the underpinnings by which nations and governments waxed and waned, and provide employment for most of the population for most of the history of the world. Even today, when only a small percentage of the United States population is farmers, a large percentage of the work force is engaged in the food industry, albeit not in primary production (Almanac, 1990). The ability of an economy to produce food and food commodities without using most of the work force frees workers for other kinds of production and services and insures that food prices will be modest and supplies plentiful.

12.2 Social Implications of Transgenic Plants

What are the social implications of the ability and willingness to create transgenic plants? How will the introduction of such plants change social, cultural, and economic roles of traditional plants? What will the social impact be? There are really no answers to these questions. The questions may be framed another way. Who and what will drive the priorities that determine the characteristics to be incorporated into transgenic plants? How will potential social impacts be assessed? Some forces that will shape these changes are science, technology, the market place, and perceived social values. There is a public debate underway (Dayan *et al.*, 1988) on what values should guide the creation of transgenic plants or whether they should be created. The conflict between these values underlies this debate. Increasingly, perceived adverse social impacts are assuming more importance, often at the expense of the values of scientists and businessmen.

12.2.1 SOCIAL BENEFIT FROM TRANSGENIC PLANTS

Since the basic discoveries and inventions that led to molecular biology and the new biotechnology were paid for with public funds, is there an implied social contract to insure that public as well as private benefit results? If so, how will this be accomplished and who or what institutions or organizations will be responsible? The political arena and the marketplace provide the final arbiters (Hoban, 1989). If the market-place provides the arbiter of choice, can society depend on it to allocate the social impacts and benefits equitably and effectively?

12.2.2 UNANTICIPATED OR UNEXPECTED IMPACTS

When Henry Ford introduced mass-produced automobiles, he probably did not anticipate the social impacts his invention would have (Pound, 1962). Numerous other examples of such impact, such as radio, television (Williams, 1986), computers, and satellites, exist. Society has learned that the social impact of technological innovation can be huge and, through analysis, tries to anticipate it (Havik, 1987). Concern about environmental issues led to the National Environmental Policy Act, which sets a framework to assess impacts on the environment. This framework includes the human environment and social impacts of federal actions.

12.2.3 ASSESSING SOCIAL IMPACTS

In the popular book *Chaos,* James Gleick (1988) provides examples of attempts to explain behavior of natural and mechanical systems that failed either because of assumed linear behavior of nonlinear systems or because underlying complexity was perceived as randomness. Certainly social systems are nonlinear and complex, making analysis difficult. Can estimates of the degree of social impact of transgenic plants be made without knowledge or assumptions of what the impacts will be? Risks of all kinds evoke social concern. Generally, society is more concerned with undefined adverse risk than with well-defined risk. For example, health effects of smoking, obesity, and use of alcohol are well-known well-defined risks (Kris-Etherton *et al.*, 1988) about which only modest social concern exists. Unanticipated or unexpected impacts that are perceived as random results may be assumed to be unpredictable because they are poorly understood. One method of analysis is to hypothesize a group of scenarios of intuitively anticipated effects, analyze each, and make a case for it. This approach often uses similar past situations and assumes that they are congruent and may be used as predictors. When congruent cases are not found or when there is disagreement about the congruence of effects, the relative merits of competing or conflicting scenarios are debated and the resultant analysis presented as the "risk." Such risk may be based more on the basis of the skill of the debater than on the likelihood of the correctness of the scenario.

12.3 Risk Posed by Transgenic Plants

Risk is a perceived estimation of the chances that certain undesirable effects may occur. Once these effects occur, they are statistical data and not risk. Since individual perception varies greatly, there is not a good correlation between the probability of certain dangerous effects and the perception of risk. Consequently, well-known safety and health effects from certain actions, such as overeating, speeding, or driving while intoxicated, do not engender a social response commensurate with that to perceived risks that may be orders of magnitude less probable. Risk and risk analysis is based more on perception, constructed scenarios, and debating skill than on measuring the probability of an effect occurring or on ameliorating the effect or the probability of its occurrence. This type of analysis may become embedded in policy decisions. For example, when the Biotechnology Science Coordinating Committee attempted to determine which deliberate genetic modifications should require oversight, they encountered a very pragmatic problem. How could they provide oversight for "new" kinds of modification without including "traditional and accepted" modifications. Scenarios emerged. One, the oversight could be based, like the National Institutes of Health, on the "new" modifications. Two, the oversight should be blind to the method of modification and be based only on the "risk" inherent in the modified organism. These scenarios were elaborated to include experimental, historical, and, in some cases, even anecdotal evidence. The problem was and is the failure to recognize or admit that both scenarios are flawed and probably not applicable because risk is a perception, and therefore does not really serve as a predictor of which transgenic plants require oversight. The Biotechnology Science Coordinating Committee could not reach accord and left the decision up to the agencies.

All of the concerns mentioned are too broad to be answered globally, but discrete areas may be examined for specific issues for which social impacts can be anticipated.

12.3.1 ENVIRONMENT AND HEALTH

Perceived social concern about food safety, environmental effects, and health have affected the field testing and production of transgenic plants profoundly. Conversely, there is wide acceptance of the desirability of eliminating or reducing the need and dependence on "synthetic chemical pesticides." This acceptance is based on the implicit assumption (U.S. Department of Agriculture, 1989) that by eliminating the need for pesticides, pesticide residues will be eliminated, thereby improving the quality of the environment and human health. Plants that produce compounds that make them resistant or immune to insects and disease certainly would be socially and environmentally desirable, but would create an accompanying concern that the resistance may include compounds that may not be "safe" to consume from a health standpoint. Food modifications that lengthen the shelf-life or improve texture, aroma, or taste may encounter the same concern. Although scientists can demonstrate that components already are present in plants and are digested easily, and public health officials and regulatory agencies have tried to allay these concerns with rules, environmental assessments, and assurances of safety (Tolin and Vidaver, 1989), much of the public remains concerned.

12.3.2 WATER QUALITY

Plants that contain genes that make them resistant or immune to pests, or beneficial organisms that inhibit or prey on pests (biopesticides), can bring about dramatic reduction in the amounts of synthetic pesticides that enter ground and surface waters. When the plant produces its own "pesticide," none is released into the environment, and most decomposes with the plant residues after harvest (Fischoff *et al.*, 1987). Although the social impacts and benefits from reduced use of synthetic pesticides are large, the benefits to farmers will depend on effectiveness, cost, and reduced exposure to pesticides. Because of governmental mandates related to non-point-pollution and improved water quality, most farmers and manufacturers are seeking technologies that reduce the use of synthetic pesticides. However, market and technological forces are driving the development of plants tolerant to certain herbicides. Some of these herbicides are quite environmentally benign because they mimic natural compounds and easily decompose. Other herbicides are more persistent in the environment. Social concern about this issue is driven in part by the fear that there is a market strategy to broaden the use and market share of specific herbicides by development of economic transgenic plants tolerant to these herbicides. Some individuals visualize a scenario that would result in more, not less, water pollution and health risk (Goldberg *et al.*, 1990). There is an additional concern that outcrossing with native weeds could impart herbicide tolerance to weeds and increase the need and demand for herbicides.

A balancing concern is the continuing problem of cultivation that leads to erosion, with the attendant loss of valuable soil and silting and muddying of surface waters. Soil particles carry pesticide residues and fertilizer elements with them into the water (Khan, 1980). Wind erosion accounts for significant losses of soil and air pollution (Chepil, 1957). The USDA has estimated the extent of such erosion in the depletion of the soil resource, and has instituted programs such as the conservation reserve to set aside highly erodible soils that are planted with cover crops to reduce erosion greatly. Another approach is the use of minimum tillage, which can reduce such erosion substantially but depends on the use of herbicides at planting for control of weeds and competing plants (Schusky, 1989; Reganold *et al.*, 1990). Both short- and long-term benefit accrues to society from such programs.

12.3.3 AIR AND SOIL POLLUTION

Air and soil pollution from pesticides, irrigation water, waste dumps, and fertilizers all can be influenced by plants designed to eliminate the need for pesticides, by plants that are more drought and salt tolerant, and by plants that may fix nitrogen or are more efficient in using and removing nitrogen and other fertilizer elements. Landfills that contain garbage and other waste are special cases in which specially developed plants could be used to scavenge or concentrate elements or compounds that are deleterious to the environment or human health (Rochkind-Dubinsky *et al.*, 1987). Transgenic plants might be designed to exude or excrete enzymes or other compounds that could lead to the decomposition of persistent organic compounds when planted on or around landfill sites. Although this may be a socially desirable action, it is not clear that there is either a market or a technological motivation to invest in these technologies. Most economic plant scientists are more oriented to the changes that will protect the plant from pests and environmental stress or will produce higher yields, lower production costs, or better market qualities than to specialized social benefits.

12.3.4 FOOD SAFETY

New nonnative compounds formed by transgenic plants are of social concern. If genetic changes result in perceived social benefit, that is, better nutrition or lower fats and oils, it is anticipated that acceptance will be high. If the changes result in the production of a "foreign" compound or even in increased levels of "native" compounds resulting in resistance to pests, longer shelflife, and cosmetic properties, it is likely that concern will be expressed about approvals for commercial use (Hoban, 1989). Society wants assurance that foods that contain small residues of such compounds are "safe."

12.3.5 HEALTH AND NUTRITION

Developed countries are concerned about the relationship between obesity and health. This concern often goes beyond public health and engenders a desire to be "lithe" for asthetic and other reasons. This social concern has driven the commercial success of "diet" formulations and books that purport to reverse the accumulation of fat and to lower body fat levels. If plants could be altered to produce foods with all the desirable organoleptic qualities and significantly reduced caloric and fat content, there is little doubt that they would be an initial economic success. There is a market for "natural or organically" grown foods. Such foods are perceived as more healthful and safe because they were produced without synthetic pesticides or fertilizers. To the extent that plants could be altered to achieve these ends and still maintain their productive capacity, they could be accepted by this market.

Heart disease and obesity have become almost synonymous to society. Aside from the "lithe look" desired by younger members of society, middle-aged and elderly members of society find themselves at increasing health risk from heart disease (Ferguson, 1978; Kris-Etherton, 1988). The prescribed solution is to reduce the level of intake of fats and increase the intake of fiber and carbohydrates by eating more fruits and vegetables. Because cooking often adds fats, oils, and sweeteners, emphasis is on fresh fruits and vegetables. Making fresh fruits and vegetables tastier, more aromatic and flavorful, and less prone to softening from senescence should have beneficial social impacts in improved public health and quality of life.

Cancer has been demonstrated to be related to various chemicals in the environment (American Cancer Society, 1990). To some degree, this includes foods and food constituents. Certain toxins, such as mycotoxins, are produced by microorganisms that invade the edible portions of plants prior to or following harvest. These toxins also can become part of animal products if the animals are given feed containing the toxins. Transforming plants by making them resistant to infection by such organisms would benefit society, but might not be profitable or scientifically interesting. Other natural constituents of foods may have similar effects, many of which are not well defined. Altering food-producing plants to reduce such components would have high social impact and could become a major policy issue.

12.3.6 FLAVOR, TEXTURE, AND AROMA

Eating is a highly socialized activity. The pleasure derived from eating is often not only dependent on the food preparation but also on the properties of the raw food. Morel mushrooms and truffles are examples of highly prized gourmet food products. Organoleptic analysis can reveal those properties of food plants that impart desirable flavors, textures, and aroma and offer the possibility of transferring the genetic elements responsible to another species or plant. For example, it might be possible to create a potato that had the aroma and flavor of truffles, or even fish or fowl. Would such innovations be acceptable to society or does society want potatoes to taste like potatoes and truffles to taste like truffles (Anonymous, 1989)? Research has shown that color plays an important role in our perception of aroma and taste. The same fruits consumed in settings in which they are illuminated in different colors of light often are not perceived to taste alike. Certainly molecular biology makes it possible to tailor the flavor, texture, and aroma by enhancing the expression of or blocking enzymatic pathways that will lead to accumulation of specific characteristics. An example is the conversion of sugars to starch and starch to cellulose. If the conversion of sugar to starch is slowed, the product will taste sweeter and the sweet taste may persist longer without any increase in caloric content. If the conversion of starch to cellulose were accelerated, the digestible caloric content would be reduced and the traditional taste and texture might be maintained.

12.3.7 AESTHETIC VALUES OF PLANTS

Flowers, foliage, grass, trees, and shrubs are used for both practical and asthetic purposes (Swaninathan and Koleher, 1985), but the latter usually predominate. Mood, motivation, and activity level are influenced by plants (Grossman, 1979). Flowers that do not wilt, foliage that maintains color and texture, grass that remains green with little water or mowing, trees and shrubs that shade without excessive shedding or pruning, leaves that do not need to be raked, and plants that are evergreen all sound too good to be true (and are probably plastic), but plant breeding has brought great changes to ornamental plants and flowers. Indoor plantings grace homes, malls, and the lobbies of public buildings. Making flowers, trees, grass, and foliage plants resistant to pests does not pose the same social concerns as does manipulation of edible plants. Such modifications make plants easier to grow and maintain.

12.3.8 ROLE OF PLANTS IN GLOBAL CLIMATE CHANGE

It has been shown that trees are effective sinks of carbon dioxide (Farrell, 1987) and release oxygen. There is a scientific scenario that would encourage or mandate that marginal and tropical lands be planted with trees to slow the greenhouse effect by removing carbon dioxide from the atmosphere. Plants produce terpenes, volatile compounds thought to be responsible for the blue haze seen over forests (Selia *et al.*, 1982). The role these terpenes play in air quality is not well defined. Forest trees might be modified to reduce the production of terpenes if they are demonstrated to cause air quality problems, but the cost of modifications and the use of such trees would have to be guided by policy reasons because it is doubtful that it would add value to the trees for normal commercial uses. Use of such plants in urban settings that experience air pollution might be advantageous.

12.3.9 WILDLIFE HABITAT

Wildlife depends on plants to provide food and cover (Decker and Goff, 1989). Wildlife commonly forages from economic crops. Insects and small animals that feed on domestic plants are important in the food chain of wildlife. Woody species, weeds, and wild native plants provide most of the cover (Decker and Goff, 1989). Social concern for the success of wildlife, especially endangered or threatened species, has produced policy directions for their preservation. To the extent that transgenic plants are in conflict with these policies or will impact wildlife habitat adversely, the plants will be of social concern.

12.3.10 DRUGS AND PHARMACEUTICALS

Plants are a traditional source of drugs and medicinals (Mills, 1985). Homeopathy (Lehner and Lehner, 1960) has been practiced for centuries; much of it depended on the knowledge of plants and herbs and their special properties and uses. Legitimate drugs and illicit drugs can be produced from transgenic plants. Plant cell culture is being examined as a method of production of drugs and pharmaceuticals. Transgenic plants could become a source of illicit, recreational, and addictive drugs, which would become a major social issue. Conversely, plants that produced licit drugs and pharmaceuticals might lower costs and make the drugs available to people now denied them because of costs.

12.4 Conclusions

Plants have great traditional cultural and social significance in all societies. They are important hallmarks and parts of distinctive cultures. Concerns about social impacts and values will have significant impact on the production, testing, and commercialization of transgenic plants. Society has perceived satisfaction in the social contract with native or natural plants produced by classical techniques. There is concern about the direction and nature of "unnatural" changes that will be accomplished by the use of molecular genetic techniques. Transferring traditional properties among food plants and to plant foods poses interesting cultural and traditional questions. Social good may be achievable with transgenic plants that is not market, technologically, or scientifically driven. Production of plants with such properties may be driven by social and public health policy. Assurances by scientists, public health, and regulatory agencies that food produced by transgenic plants is "safe" may not be accepted. It is not possible to predict the social impact of transgenic plants, but it is possible to attempt to establish specific areas of possible social concerns. Concerns about social impact will influence the future development and adoption of transgenic plants.

References

Almanac (1990). Houghton Mifflin, Boston.

American Cancer Society (1990). "Cancer Facts and Figures—1990." Atlanta, Georgia.

Anderson, E. (1952). "Plants, Man, and Life." Little, Brown, Boston.

Anonymous (1989). *Sci. Am.* **260 (5),** 101.

Chepil, W. S. (1957). *In* "1957 Yearbook of Agriculture," pp. 308–314. U.S. Department of Agriculture, U.S. Government Printing Office, Washington, D.C.

Dayan, A. D., Campbell, P. N., and Jukes, T. H. (eds.) (1988). "Hazards of Biotechnology: Real or Imaginary?" Elsevier Applied Science, London.

Decker, D., and Goff, G. R. (1989). "Valuing Wildlife: Economic and Social Perspectives." Westview Press, Boulder, Colorado.

Farb, P. (1980). "Consuming Passions: The Anthropology of Eating." Houghton Mifflin, Boston.

Farrell, M. P. (ed.) (1987). "Master Index for the Carbon Dioxide Research State-of-the-Art Report Series." National Technical Information Service, Washington, D.C.

Ferguson, J. M. (1978). "A Change for Heart: Your Family and The Food You Eat." Bull Pub. Co., Palo Alto, California.

Fischoff, C. B., Bowdish, K. S., Perlak, F. J., Marrone, P. G., McCormick, S. M. Niedermeyer, J. G., Dean, D. A., Kusaon-Krettzmer, K., Mayer, E. J., Rochester, D. E., Goyers, S. G., and Fraley, R. T. (1987). *Bio/Technology* **5,** 807–813.

Gleick, J. (1988). "Chaos: Making a New Science." Penguin, New York.

Goldberg, R., Rissler, J., Shad, H., and Hassebrook, C. (1990). "Biotechnology's Bitter Harvest." The Tides Foundation.

Grossman, R. S. (1979). *HortScience* **14,** 690–691.

Havik, N. (1987). "Indicators of Social Change, Measure of the Quality of Life, Economic, Sociology, Complex Organizations and Futures Methodology: An Annotated Bibliography." Vance Bibliographies, Monticello, Illinois.

Hicks, J. R., Hayslip, N. C., and Showalter, R. K. (1975). *HortScience* **10,** 11–12.

Hoban, T. J., IV (1989). *Southern Rural Sociol.* **6,** 45–63.

Janick, J. (1988). *HortScience* **23,** 11–13.

Khan, S. U. (1980). "Pesticides in The Soil Environment." Elsevier Scientific, New York.

Kris-Etherton, P. M., Kummel, D., Dreon, D., Mackey, S., and Wood, P. (1988). *J. Am. Diet. Assoc.* **88,** 1373–1400.

Lehner, E., and Lehner, J. (1960). "Folklore and Symbolism of Flowers, Plants, and Trees." Tudor, New York.

Metcalf, E. (1969). "The Trees of Christmas." Abington Press, Nashville, Tennessee.

Mills, S. (1985). "The Dictionary of Modern Herbalism." Thomas, New York.

Pond, W. (1983). *Sci. Am.* **248,** 96–103.

Pound, A. (1962). "The Automobile and An American City." Wayne State University Press, Detroit.

Reganold, J. P., Papendick, R. J., and Parr, J. F. (1990). *Sci. Am.* **262,** 112–120.

Rochkind-Dubinsky, M. L., Saylor, G. S., and Blackburn, J. W. (1987). "Microbiological Decomposition of Chlorinated Aromatic Compounds." Dekker, New York.

Schusky, E. L. (1989). "Culture and Agriculture: An Introduction to Traditional and Modern Farming Systems." Bergin and Garvey, New York.

Selia, R. L., Arnts, R. R., Kuntz, R. L., Mowry, F. L., Knoerr, K. R., and Dudgeon, A. C. (1982). "Measurement of Loblolly Pine Terpene Emission, Research Triangle Park, North Carolina." EPA Environmental Sciences Research Laboratory, Cincinnati, Ohio.

Smith, W. R. (ed.) (1982). "Energy from Forest Biomass: XVII IVFRO World Congress Energy Group Proceedings." Academic Press, New York.

Swaninathan, M. S., and Koleher, S. L. (eds.) (1985). "Plants and Society." Macmillan, London.

Tolin, S., and Vidaver, A. K. (1989). *Ann. Rev. Phytopathol.* **27,** 551–581.

U.S. Department of Agriculture (1989). "Biological Control: Spreading the Benefits." APHIS, Plant Protection and Quarantine. Government Printing Office, Washington, D.C.

van Raay, H. G. T., and Lugo, A. E. (eds.) (1974). "Man and Environment Limited." Rotterdam University Press, The Netherlands.

Williams, T. M. (1986). "The Impact of Television: A Natural Experiment in Three Communities." Academic Press, Orlando, Florida.

13

Industrial Aspects of Transgenic Plant Products

Ralph W. F. Hardy

Boyce Thompson Institute for Plant Research
Ithaca, New York 14853

Transgenic plants can be and are being made of many agronomic, horticultural, and even tree crops. These transgenic plants are expected to have commercial potential by meeting customer needs for improved productivity, improved value-in-use, expanded nonfood and nonfeed uses, and environmentally neutral crop production practices. Several examples are in development, and the usefulness of a few has been demonstrated in field trials. Commercialization of the most advanced transgenic plants is expected by the mid-1990s, with rapid development thereafter.

Favorable changes in the 1980s in the policy of the U.S. Patent and Trademark Office (PTO) with respect to the patentability of novel and useful living organisms is providing proprietariness for transgenic bacteria, plants, and animals equivalent to that which has existed for over 200 years for novel and useful inanimate materials. The technology requirements for commercialization of transgenic plant products include state-of-the art germ plasm, adequate yield transformation processes, and genes with commercial utility. The small number of known genes is the largest limitation.

The agribusiness input industry of fertilizers, agrichemicals, and seeds is changing in response to maturation of the fertilizer and agrichemical industry and to the expected commercial opportunity and impact of transgenic plants. The major commercial vehicle for transgenic plants is expected to be the production and sale of seeds. Chemical and pharmaceutical companies are acquiring seed companies; others are establishing relationships with seed companies and/or food and feed processing and distribution companies. A small group of development-stage biotechnology companies with leading technical capabilities are focused on transgenic plants and their commercial applications; several established agrichemical, pharmaceutical, and seed companies also are emphasizing transgenic plants and their commercial opportunities. The commercial opportunity for transgenic plants is projected to increase the annual value of seeds from $50 to $100–150 billion over the next few decades.

13.1 Customer Needs

Transgenic plant products (seed, vegetation, or other), to be commercialized successfully, must meet the needs of customers. Some of these needs will be in established markets, some in new markets. In the established markets, the transgenic plants will need to have significant advantage over established products. This advantage usually will be economic, for example, decreased cost of production per unit or increased value of the product. The transgenic product will gain market share from established traditional products, depending on the size of the advantage of the transgenic product and the effectiveness of the marketing effort. In the case of new markets, the transgenic plants will satisfy a need that is not met by existing products. New markets are very exciting and may have great potential size, but are much less predictable than existing markets. New markets even may require new or modified infrastructures. A major educational effort often is needed for a new market.

What are the needs of potential customers for transgenic plants? The needs of customers are those that are met inadequately, in most cases, by traditional genetic or chemical products in established markets whereas, in other cases, the needs are not met at all by any traditional genetic or chemical product. In many cases of unmet needs, customers will have to be educated to the value or method of use of the transgenic product. The customers for transgenic plant products are farmers, food and feed processors and distributors, home consumers, and users of plant products for nonfood and nonfeed uses. The general needs of these customers in the 1990s and beyond, when transgenic plant products may play a major role, will be described. Their needs can be classified as improved productivity, improved value-in-use, expanded nonfood and nonfeed markets, and environment-neutral crop production practices.

13.1.1 IMPROVED PRODUCTIVITY

A major need for producers of commodity grains and fiber is increased productivity so they can be the low-cost producer. This need for high productivity is relevant to any commodity, not just grains. For example, in the 1970s, chemical companies invested heavily in process research to develop the most productive (usually the lowest cost) method to manufacture commodity chemicals from oil feedstocks whose cost had increased dramatically. Similarly, grain and fiber farmers need high productivity crop plants to be the low-cost and, therefore, the most profitable producers. Thus, transgenic plants with improved productivity should meet a major current need of corn, wheat, soybean, cotton, and other plant commodity farmers. This key role for transgenic plants in improving productivity in commodity grains and fibers is, of course, only valid in a free market in which subsidization does not distort the normal market factors.

Transgenic plants with potential for improved productivity are being developed in university and commercial laboratories. Examples are plants self-protected against pests to decrease the cost for chemical pesticides, plants with improved efficiency of nutrient utilization and, eventually, with self-nitrogen fertilization to decrease the cost for fertilizer, hybrid and other plants with increased yield to decrease the cost per unit of product, and many others. Hybrid corn, the most profitable segment of the commercial agronomic crop seed business, has been based almost exclusively on continuous year-to-year improvement over 60 years in corn-grain yield (Abelson, 1990). This improvement has been achieved by traditional plant breeding, with major emphasis by the commercial breeders on developing inbred lines that produce hybrids with increased yield. The research director of one of the major United States seed companies (Troyer, 1990) wrote recently that "the primary USA corn breeding objective must continue to be increased grain yield because corn is chosen by users to obtain cheapest starch."

Transgenic plants should enhance the efficiency of improving productivity by facilitating the breeding process and by expanding the genes available to the breeder. The potential for transgenic versus traditional improvement may be likened to the ability of an engineer to design a superior automobile using the vast array of metals, polymers, and composites that are now available as opposed to only traditional materials. Farmers and countries using transgenic plant products with improved productivity for commodity grains and fiber will be more profitable and competitive and will gain world market share at the expense of the less productive farmers and countries (in the absence of protection and subsidization). These commodities, in general, will be produced by large-scale farmers who gain benefits by scale of production, as is the case for commodities in any area.

Possibly the first commercial transgenic crop with improved productivity will be cotton that produces *Bacillus thuringiensis* (*B. thuringiensis*) insect toxin for protection against pest insects. Field trials by Monsanto in 1990 at multiple sites in the United States (Fischhoff, 1991) provided the first evidence that this transgenic *B. thuringiensis* cotton was protected as effectively from insect damage as that given multiple chemical pesticide treatments. The cost of production is expected to be less for purchase of the transgenic *B. thuringiensis* seed than for purchase of the traditional seed plus chemical pesticides and multiple applications of the chemical pesticides. The cost of the transgenic seed will be greater than the traditional seed; this additional seed cost may be 25–50% of the savings in purchased pesticides and their application cost. Other transgenic plants with potential improved productivity are disease-resistant plants and herbicide-tolerant crops. The latter may decrease the cost of weed control by decreasing the total cost for herbicides and their application or by increasing crop yield by increasing the effectiveness of weed control. In the longer term, one of the largest opportunities for improved productivity will be self-nitrogen-fertilizing transgenic crops that eliminate the cost for synthetic nitrogen fertilizer and its application.

13.1.2 IMPROVED VALUE-IN-USE

A very different need of farmers is improved value-in-use in their plant products. Improved value-in-use crops would increase profitability at all steps of production, marketing, and use. Number 2 yellow corn is a commodity (Bridges, 1991); it is the dominant corn of commerce and is the basis for pricing. Number 2 yellow corn is used in many ways, for example, in the United States livestock-feeding industry (55%), grain exporting industry (29%), wet milling industry (13%), and dry milling industry (3%) (Troyer, 1990). Many of these uses can be segmented further. For example, corn is used in livestock feed, including poultry and swine feed. The composition of corn for high value as a poultry feed is different from that for high value as a swine feed. The composition of corn with the most value for wet milling is substantially different from that for livestock feeding. Thus, a plant product of a given composition can have different value, depending on the use to which it is put. At this time, most grain products, such as number 2 yellow corn, are treated in commerce as commodities, with their price unrelated to their value in a specific use. There are a few exceptions to commodity pricing in fruit and vegetable production, in which the processor contracts with a grower to produce a specific plant with high value for that specific market.

Interest is now growing in creating identity-preserved crops to capture the improved value-in-use of the crop. Tom Urban, chairman of Pioneer Hi-Bred International, has emphasized the future importance of identity-preserved crops. Plants with improved value-in-use for a specific end use, for example, corn with a composition that makes it more valuable than number 2 yellow corn for swine feed production, will have its identity preserved throughout its production, marketing, and use. This identity-preserved corn with increased value for swine feed production will be sold at a price based on its value-in-use as a feed for swine and not at a commodity price.

Possibly the most significant major commercial example of an improved value-in-use crop with identity preservation is the rapeseed crop, canola. Canola oil has higher value-in-use as an edible oil than rapeseed oil; its identity is preserved by the name "canola" rather than "rapeseed." Canola is a rapeseed that Canadian scientists, in the 1960s to 1970s, modified by traditional plant breeding to reduce its erucic acid and glucosinolate contents so its oil was suitable for edible uses and the meal was suitable as an animal feed (Downey and Rakow, 1987). Canola oil has become the dominant edible oil in Canada. The United States has recognized canola oil with a designation of GRAS (generally recognized as safe) by the Food and Drug Administration (FDA). Puritan Oil by Procter and Gamble is a canola oil that is being marketed for the more healthful nature of its fatty acid composition and its lighter color as a frying oil than the traditional vegetable oils from corn and soybeans.

There are many other opportunities for improved value-in-use of plant products. Transgenic processes will be more efficacious than traditional breeding processes in producing plants with improved value-in-use. Nutritionally improved feed grain is one example of improved value-in-use, as noted earlier for swine or poultry production. Such transgenic corn plants are being developed by BioTechnica International; the first plants were expected to be in field trials in 1991. Fatty-acid modified crops are another opportunity for corn or soybean oil to be modified to be equivalent or even superior in healthful characteristics to canola. Alternatively, superior oil for industrial uses is being developed by Calgene (Davis, California) (Anonymous, 1991). Starch quantity and quality is another opportunity for higher value-in-use plants. Transgenic crops that eliminate the need for pesticides should decrease consumer concern about the health risk of pesticide residues in food and have a higher value-in-use. Other opportunities for higher value-in-use plants relate to consumer-preference characteristics. Calgene is developing transgenic tomato plants with longer shelflife (Kramer, 1991). This potential product may be one of the earliest examples of a higher value-in-use food crop. Other changes for consumer value include enhanced color, flavor, and texture. The consumer of today is more interested in product value than cost and is prepared to pay a premium for a higher value-in-use food than for a commodity.

The combination of identity preservation and pricing based on value-in-use may provide one of the largest commercial opportunities for transgenic plant products. Proprietariness, as discussed subsequently, is necessary for pricing based on value-in-use. Pricing based on value-in-use will provide farmers the opportunity to produce more profitable crops. Value-in-use is key to the high-profit lower-volume specialty producer whereas productivity is key to the low-profit high-volume commodity producer. The move to identity preservation and value-in-use pricing probably will be the next major advance in crop agriculture. Proprietariness, identity preservation, and transgenic plants are the keys to value-in-use plant products.

Identity-preserved crops, in many cases, will require a reliable, rapid, and low-cost system to enable identification of the crop as it is bought and sold in commerce. A transgenic plant or its seed, with enhanced value for a specific use, probably will be identical in appearance to the commodity plant. For example, corn with enhanced nutritional value as a feed for swine probably will be visually identical to number 2 yellow corn. A marker will be needed to enable the enhanced nutritional value corn to be segregated during processing steps in commerce. Marker systems such as light emission by incorporating the *luxAB* genes for bioluminescence may be useful (Legocki *et al.*, 1986; Koncz *et al.*, 1987). Marker systems will be needed for each of the variants of a given crop whose identity is to be preserved in commerce. This problem has not been solved at this time.

13.1.3 EXPANDED NONFOOD AND FEED MARKETS

Developed-country agriculture suffers from its excess capacity, basically excess arable land that enables it to produce commodity grains in excess of the food and feed needs of the developed world and the ability of the developing world to purchase. New nonfood and nonfeed markets are needed for plant products. Already plants have major established nonfood and nonfeed markets, including materials (e.g., lumber), fiber (e.g., cotton, pulp, and paper), energy (e.g., ethanol for gasohol), industrial chemicals (e.g., acetic acid), and specialty chemicals (e.g., pharmaceuticals or pharmaceutical precursors). Some of these needs are commodities and the opportunity is great, for example, for gasohol. A major need in cases such as gasohol and chemical feedstocks is to modify plants or the processing of plant products to make them economical sources for needs that now use mainly fossil energy. Others of these needs are high value-in-use and low volume, for example, pharmaceuticals or their precursors. At least one plant biotechnology company, Plant Genetic Sciences, has proposed making high value-in-use protein therapeutics in transgenic plants while DNA Plant Technology is making transgenic flowers with the objective of increased value-in-use in the $9 billion floral market (Moffat, 1990a). These nonfood or nonfeed new market opportunities represent a major use for transgenic plants. Commercial activities in this area are minimal at this time because the risk is very high. National and international interests in some of these new nonfood and nonfeed markets, for example, as energy, suggest the need for governments to provide financing that would encourage commercial research and development.

13.1.4 ENVIRONMENT-NEUTRAL CROP PRODUCTION PRACTICES

A growing need of crop production agriculture is the availability of practices that are more environment neutral or compatible with sustainable crop agriculture than existing practices. The extensive use of fertilizers, especially synthetic nitrogen fertilizer, and of agrichemicals has produced contamination, in some cases, of groundwater and soils. Many individuals expect that transgenic plants able to make biodegradable biopesticides and fix nitrogen will facilitate crop production practices that are more environment neutral and compatible with sustainable crop agriculture. However, the use of transgenic plants to meet these needs will be based on economic considerations that relate to productivity or value-in-use of products rather than environment neutrality, unless regulations require the use of the transgenic pesticidal or fertilizing plants rather than chemicals.

Commercial opportunity for transgenic plant products will be greatest in expanded nonfood and nonfeed uses, then in improved value-in-use, and least in improved productivity. It is difficult to assign commercial value to environment-neutral crop production practices other than to incorporate it into each of these opportunities. The expected time to commercialization will be shortest for improved productivity and longest for expanded nonfood and nonfeed uses. Overall, transgenic plants are expected to make major contributions to all these needs. Commercialization in all cases will benefit from strong protection of the intellectual property associated with transgenic plants.

13.2 Proprietariness

Protection of intellectual property rights for novel and useful living organisms, for example, bacteria, plants, and animals, has advanced, especially in the United States, during the 1980s, when the economic potential of biotechnology and transgenic organisms was recognized. Asexually propagated plants were protected by the Plant Patent Act of 1930. Hybrid crops such as corn have built-in proprietariness based on the process of seed production. The major improvement in corn yield over the last several decades arose mainly from the intensive efforts of commercial seed companies that were attracted by the opportunity for economic return protected by the proprietariness of the hybrid production process. However, most crops are not asexually propagated or hybrids; the commercial incentive for crop improvement requires additional methods for proprietariness with these crops.

In 1970, the Plant Variety Protection Act was introduced in the United States to provide improved proprietariness for nonhybrid and nonasexually propagated crops. This act, administered by the USDA, provides certificates for novel plants that could be described and were deposited in a collection. These certificates provide limited proprietariness for improved plants but do not provide the exclusivity of a patent issued by the PTO. Plant variety certificates (PVC) permitted farmers to plant seed and to sell seed to neighbors that was produced from purchased seed that was protected by a PVC. Seeds with major improvements in their value might be sold only once. Further, a competitor could make a modest improvement in a plant covered by a PVC and obtain a new PVC for the modestly improved plant. PVCs do not provide adequate proprietariness for transgenic plants. The cost of development of transgenic plants is much greater than that of plants produced by traditional plant breeding because of the major laboratory- and labor-intensive component of the molecular genetic modification process. This higher cost of development must produce a higher-value transgenic plant product than traditional breeding, and there must be a method to assure the developer a right to obtain an appropriate return on the investment.

During the 1980s, the PTO extended proprietariness first to novel and useful microorganisms, then to novel and useful higher plants, and finally to novel and useful animals. The legal precedent for the patentability of genetically modified plant materials occurred in 1986 (Kleese, 1989) with the issue of a U.S. patent for tryptophan-overproducer mutants of cereal crops (Hibberd *et al.*, 1986). This patent was issued to Molecular Genetics, Inc. The PTO now has issued several patents for genetically modified plants, including transgenic plants, whereas only one plant patent has been issued by the European patent office. A basic difference between the U.S. patent system and that of most of the rest of the world, for example, Europe or Japan, is that in Europe the legal inventor is the first to file an enabling patent application whereas in the United States it is the first individual to invent (Lesser, 1987; Daignault *et al.*, 1988). This difference creates some problems.

The extension of patentability to living organisms has paralleled the ability to describe biological modifications in molecular rather than morphological terms. Such proprietariness is necessary to enable commercial organizations to obtain an equitable return on their research and development risk investment consistent with the added value of the transgenic product. Since many of these transgenic crops will have global markets, it is key that appropriate proprietariness be available, at least in the developed countries of the world. Further, proprietariness of a transgenic crop product will allow each member of the product chain—for example, agribusiness input, farmer, food/feed processor and distributor, and nonfood/nonfeed processor and distributor—to obtain a price consistent with the value of the product at that step. Strong proprietariness is critical to insure aggressive commercial development of transgenic plants and to provide a return to each member of the chain consistent with the value added at that step.

13.3 Technology Needs for Commercialization of Transgenic Crops

Three technology components are necessary for commercialization of transgenic plants and products. These are state-of-the-art germ plasm of host plants, technology for adequate yield plant transformation, and commercially useful genes. The starting material is state-of-the-art germ plasm of the domesticated crop to be improved by molecular genetic modification. This germ plasm must be commercially competitive so the transgenic modification will improve on the germ plasm in commerce. In many cases, this state-of-the-art germ plasm is proprietary to established seed companies or companies called foundation seed companies whose business is to develop and license germ plasm to seed companies that may not develop their own germ plasm. Companies entering the seed business must develop state-of-the-art germ plasm or acquire it.

Technology for transformation of many crop plants is in hand, as described in the chapters of this volume on specific crops. Monsanto pioneered the use of the disarmed *Agrobacterium tumefaciens* plasmid for plants such as soybean, cotton, cucumber, and many others (Fraley *et al.*, 1986). Boyce Thompson Institute (Ithaca, New York) modified electroporation for use as a transformation method for plants. Cornell (Ithaca, New York) invented microprojectile technology for plant transformation (Klein *et al.*, 1988) and licensed the invention exclusively to Du Pont (Anonymous, 1989). Agracetus developed a high-yield variant of the microprojectile process.

These or other processes have produced successful transformation of corn and rice. Of greatest commercial significance for transgenic crops was the press release in January, 1990, by BioTechnica International that it had transformed corn successfully with production of fertile transgenic corn seeds. BioTechnica reported that it was seeking patent protection for its transformed corn, but because of the commercial potential and intensive competitiveness, it elected not to disclose the technology used (Bishop, 1990). Four or five other commercial organizations subsequently reported successful corn transformation between April and December 1990 (Moffat, 1990b). The first and only field introduction of transgenic corn in 1990 was conducted by BioTechnica using corn with marker genes, documenting the validity of their process (Anonymous, 1990a). Transgenic corn is undoubtedly the most competitive commercial target for transgenic crops, and one can expect great intensity in the development work. Although the technologies are available to produce transgenic varieties of most crops, one expects future improvements in the yield of the processes.

The major technology limitation at this time for commercialization of transgenic plants is the limited number of commercially useful genes available. For example, genes available for insect protection are limited to those for *Bacillus thuringiensis* toxins and possibly protease inhibitors. A candidate gene for insect protection is that encoding the enhancing factor discovered at Boyce Thompson Institute (Wood and Granados, 1991). The enhancing factor degrades the peritrophic membrane of an insect, thereby synergizing insect kill by biopesticides that enter through the gastrointestinal system of the insect. The listing of only three different types of insecticidal genes is characteristic of the very small number of useful genes that are available for production of transgenic plants.

This limitation in known genes is common for almost all the proposed commercial targets. The limitation results from the inadequate support of agricultural research for several decades in contrast to the continuous, abundant, long-term support of biomedical research in which many useful genes or gene products are available. Public-sector research in identification of additional genes for use in making transgenic plants is critical to successful commercialization. The U.S. National Research Council 1989 initiative, *Investing in Agriculture,* has led to plans for substantial increases in USDA competitive research funding and should help reduce this limitation (National Research Council, 1989).

Antisense technology is a variant process being used to produce transgenic plants with commercial potential. Calgene has used this approach to provide tomato plants with expanded shelflife, as noted earlier.

13.4 Agribusiness Input Industry and Transgenic Plants

The agribusiness input industries include fertilizer, agrichemical, seed, and equipment companies. All will be impacted by transgenic plants, but the seed industry probably will be affected most and the equipment industry least. During the last 30 years, the fertilizer industry has become a mature commodity business. In the longer term, self-nitrogen-fertilizing transgenic plants will displace much of the need for synthetic nitrogen fertilizer, an over $20 billion industry. This impact will not occur until the early 21st century.

During the last few years, the agrichemical industry has undergone major consolidation; nearly 40 world agrichemical companies have been reduced to less than 20. This consolidation is expected to continue in this maturing but still highly profitable industry. Sales of each agrichemical company are $1–2 billion with total world commercial sales of about $22 billion (Hunter, 1990). Beginning in the mid-1990s, it is projected that transgenic plants, protected against insects and disease, will begin to displace some insecticides and possibly fungicides. In time, most insecticides and fungicides will be replaced by transgenic plants with self-protecting ability or by biopesticides, but it is doubtful that transgenic plants will replace herbicides. Transgenic plants with herbicide tolerance probably will change the use of specific herbicides.

The annual value of world seed use is about $50 billion, including $15 billion of commercial seed sales in 1988 (Fox, 1990). Transgenic plants are expected to increase this value to $100–150 billion by the early 2000s. This increase will come from the value created by improved productivity, improved value-in-use, and expanded nonfood/nonfeed uses. The value of transgenic plants has caused many commercial companies to see the seed business as a major growth opportunity. This recognition is documented by the number of seed companies that have been acquired by established or development-stage companies with strong technology in the area of transgenic plants as well as by some established companies that do not yet have strong technology for transgenic plants (Table 13.1). These strong technology companies recognize that they need established seed companies to provide quality germ plasm for production of transgenic plants as well as to provide a route to market their transgenic plant technology as seed.

TABLE 13.1
Seed Company Acquisitions[a]

Multinational corporation	Country base	Acquired companies
BioTechnica	United States	Horizon, McAllister, J. M. Schultz, Donley, L. Herried
Bristar	United Kingdom	W-L Research, Germains
Calgene	United States	Stoneville Pedigreed, Noble–Bear
CIBA–Geigy	Switzerland	Funks Seed International
Dow/Elanco	United States	United AgriSeeds
Elf Aquitane/Sanofi	France	St. Jeannet Lasserre, Dahlgren, King Agro, Cambier Frere, Arlesa
Hoechst	Germany	Nunhems, Cannets, Hild, KWS (Minority)
ICI	United Kingdom	Garst, Société Europeenne de Semences, Sinclair McGill, Miln Marstets, Contiseeds, Hybrid Wheat Technology from Rohm & Haas, E. J. Funk
ITT	United States	Burpee
Limagrain	France	Ferry–Morse, Shissler, Vilmorin, Nungesser, Tezier, Picard, Nickerson (UK)
Lubrizol	United States	Agrigenetics, Sigco, Lyanville, Jacques
Monsanto	United States	Hartz, Dekalb Hybrid Wheat Program
Orsan	France	Harris–Moran, Clause (50%), Claeys, Wilson, Western Plant Breeders
Rhône–Poulenc	France	Callahan, Clause (50%), Cel-Pril
Sandoz	Switzerland	Northrup King, Stauffer Seed, Zaadunie B.V., Rogers Brothers, Hilleshog, Cokers, Prodes, Musser, Fredonia, Vaughans
Unilever	United Kingdom, Netherlands	PBI, Barenbrug (majority)
Upjohn	United States	Asgrow, O's Gold, Bruinsma

[a]Updated from Kent, 1990.

The top seed companies in the world, ranked in order of 1988 and 1989 revenues in millions (Anonymous, 1990b; Fox, 1990), are listed in Table 13.2. Several have strong programs in transgenic plant technology whereas others are weak and will need to acquire the technology. The seed sales of the largest revenue company is only about $800 million or 5% market share of commercial seed sales, or 1.6% of the value of world seed. Transgenic plants will change the industry. A few major new seed companies will emerge in parallel with the emergence of a few new pharmaceutical companies. Some large established seed companies will merge or be acquired, paralleling the mergers and acquisitions in recent years in the pharmaceutical industry.

A listing made a few years ago includes the top plant biotechnology companies in terms of research and development expenditures. It is assumed that most of these expenditures are for transgenic plants.

TABLE 13.2
Seed Companies Ranked in Order of Revenues (1988–1989)

Rank	Company	Revenue (millions of dollars)
1.	Pioneer Hi-Bred International, U.S.	788
2.	Sandoz, Switzerland	488
3.	Group Limagrain-Blosem, France	360[a]
4.	Upjohn (Asgrow), U.S.	256
5.	ICI England	238
6.	Cargill, U.S.	236
7.	CIBA-Geigy, Switzerland	197
8.	DeKalb, U.S.	190
9.	Sica France Mais, France	170
10.	Takii & Co., Japan	170
11.	KWS AG, Germany	160
12.	Ste. L. Clause S.A., France (Rhône-Poulenc/Orsan joint venture)	159
13.	Sakata, Japan	152
14.	Cebeco, The Netherlands	146
15.	Weibull, Sweden	123
16.	Unilever, United Kingdom	110
17.	Sanofi, France	107

[a]Includes Nickerson Seed Co. acquired in 1990.

TABLE 13.3
Rank of Plant Biotechnology Companies

Company	Research and development expenditures (millions of dollars)
E.I. du Pont de Nemours & Co., Inc., U.S.	15–20
CIBA-Geigy, Switzerland	17
ICI, U.K.	14–17
Sandoz, Switzerland	11–16
DNA Plant Technology Corp., U.S.	15
EniMont, S.P.A.	15
Monsanto Co., U.S.	12–15
Agrigenetics Corp., U.S.	12
BioTechnica International, U.S.	12
Sanofi, France	12
Calgene, U.S.	10
De Danske Sukkerfabrikker	9
Plant Genetic Systems, N.A.	7.5
Pioneer Hi-Bred International, U.S.	7
W.R. Grace/Agracetus, U.S.	7
Agricultural Genetics Co., U.K.	6.1
DeKalb–Pfizer, U.S.	6
KWS AG, Germany	5
Group Limagrain-Blosem, France	5
Upjohn, U.S.	3

The top plant biotechnology companies, based on 1988 and 1989 research and developed expenditures in millions (Teweles, 1989; Fox, 1990), are listed in Table 13.3. Total United States private-sector biotechnology expenditures in 1987 were \$96 million by development-stage and \$104 million by large established companies (Office of Technology Assessment, 1988). Research expenditures are very modest for transgenic plants in comparison with those for biotechnologically generated human therapeutics. The research expenditures are also very modest for transgenic plants in relationship to the large commercial opportunity.

Technology acquisitions have accelerated in recent years. These acquisitions have been driven by the lack of capital for commercialization by the acquired

TABLE 13.4
Transgenic Plant Technology Company Acquisitions

Acquiring company	Acquired company
BioTechnica International	Molecular Genetics, Inc. (plants)
Calgene	Plant Genetics
DeKalb	DeKalb–Pfizer
DNA Plant Technology	Advanced Genetic Sciences
Lubrizol	Agrigenetics
Pioneer Hi-Bred	BioTechnica Canada (canola), Allelix Crop Technologies
W.R. Grace	Agracetus

development-stage companies (Table 13.4). There are now a number of established and a few development-stage companies with strong transgenic plant technology (Table 13.5).

There are alternative commercialization approaches to production and sale of transgenic seed. One is to license the technology for exclusive or nonexclusive use to companies with established seed markets. Monsanto has indicated that it plans to use this route. Du Pont is establishing relationships with seed as well as food or feed companies. However, at this time, most companies view production

TABLE 13.5
Established and Development-Stage Companies with Transgenic Plant Technology

Established	Development stage
American Cyanamid	Agricultural Genetics Co.
CIBA-Geigy	BioTechnica International
DeKalb	Calgene
Du Pont	DNA Plant Technology
Grace	Plant Genetic Systems
Hoechst	
ICI	
Lubrizol	
Monsanto	
Pioneer Hi-Bred International	
Rhône–Poulenc	
Sandoz	
Upjohn	

and sale of seeds as the route for obtaining a maximum return on their transgenic plant technology.

Transgenic plant products are an emerging commercial opportunity for biotechnology. This opportunity is the single largest opportunity in crop agriculture. In time, transgenic plants will displace most of the synthetic insecticide and fungicide world market of over $10 billion and the nitrogen fertilizer market of about $20 billion, but of much greater economic impact will be the increased value-in-use food and feed and expanded nonfood and nonfeed markets.

References

Abelson, P. H. (1990). *Science* **249,** 837.

Anonymous (1989). "Du Pont buys biological 'gun' from Cornell." *Ithaca Journal* **April 29,** p. 1–2.

Anonymous (1990a). *Chem. Engr. News* **68,** 20.

Anonymous (1990b). *AgBiotechnol. News* **7,** 26.

Anonymous (1991). *Chem. Engr. News* **69,** 15.

Bishop, J. E. (1990). *Wall Street Journal,* **January 24,** p. B1.

Bridges, I. G. (1991). *In* "Biological Monitoring of Genetically Engineered Plants and Microbes" (D. R. MacKenzie and S. Henry, eds.), Proceedings of the Kiawah Island Conference, November 1990, pp. 249–255. Agricultural Research Inst., Bethesda, Maryland.

Daignault, R. A., Loertscher, K., Koivuneiemi, P. J., *et al.* (1988). *ASM News* **54,** 612–615.

Downey, R. K., and Rakow, J. F. W. (1987). *In* "Principles in Cultivar Development" (W. R. Fehr, ed.), pp. 437–486. Macmillan, New York.

Fischhoff, D. A. (1991). "AAAS Annual Meeting Abstracts," p. 58. Washington, D.C.

Fox, P. (1990). *AGROW,* **102,** 8–9.

Fraley, R. T., Rogers, S. E., and Horsch, R. B. (1986). *CRC Crit. Rev. Plant Sci.* **4,** 1–46.

Hibberd, K. A., Anderson, P. C., and Barker, M. (1986). U.S. Patent No. 4,518,847. U.S. Patent & Trademark Office, Washington, D.C.

Hunter, D. (1990). *Chem. Week,* **Feb. 14,** 22.

Kent, J. W. (1990). *Farm Chem.* **153,** 34–35.

Kleese, R. A. (1989). *In* "Intellectual Property Rights Associated with Plants." Special Publication 52, pp. 59–64. American Society of Agronomy.

Klein, T. M., Fromm, M., Weissinger, A., *et al.* (1988). *Proc. Natl. Acad. Sci. U.S.A.* **85,** 4305–4309.

Koncz, C., Olsson, O., Langridge, W. H. R., *et al.* (1987). *Proc. Natl. Acad. Sci. U.S.A.* **85,** 131–135.

Kramer, M. G. (1991). *In* "Agricultural Biotechnology: Food Safety and Nutritional Quality for the Consumer" (J. Fessenden MacDonald, ed.), pp. 127–130. National Agricultural Biotechnology Council Report 2. Boyce Thompson Institute/Cornell University, Ithaca, New York.

Legocki, R. P., Legocki, M., Baldwin, T. O., *et al.* (1986). *Proc. Natl. Acad. Sci. U.S.A.* **83,** 9080–9084.

Lesser, W. (1987). *Bio/Technology* **5,** 350–352.

Moffat, A. S. (1990a). *Scientist* **4,** 8–9.

Moffat, A. S. (1990b). *Science* **249,** 630.

National Research Council (1989). "Investing in Research." National Academy Press, Washington, D.C.

Office of Technology Assessment (1988). "New Developments in Biotechnology: U.S. Investment in Biotechnology." U.S. Government Printing Office, Washington, D.C.

Teweles, L. W. (1989). *AgBiotechnol. News* **6,** 19.

Troyer, A. F. (1990). Presentation at XVth Maize and Sorghum EUCARPIA Congress, June 4–8, Baden, Austria.

Wood, H. A., and Granados, R. R. (1991). *Ann. Rev. Microbiol* **45,** 69–87.

Part IV

Concluding Remarks

14

Revolutionary Evolution

Timothy C. Hall

Department of Biology
Texas A&M University
College Station, Texas 77843

The development of human civilization is thought to be coincident with the rise of agricultural practices that permitted settlements where individuals could be spared from hunting and foraging to develop various creative talents. Among these, the development of pottery enabled the storage of agricultural produce, furthering its processing to yield breads and alcohol. Presumably, such items represent early examples of "value-added" commercial stimulation. Medieval agriculture was stimulated by crop rotation, which included renitrification of the soil by purposeful planting of a legume crop or fortuitous colonization by legumes during the fallow year. Dense urban populations characterizing the modern world are possible only because of high agricultural productivity resulting from the directed application of genetic principles in plant breeding programs. The ability to change the characteristics of major crop plants radically through genetic engineering potentially represents the most dramatic revolution yet in agriculture. Ultimately, there seems little reason to doubt that pest-resistant highly productive plants will be derived through recombinant DNA approaches that will provide sufficient nutrition to feed population levels with limits that have nothing to do with the traditional evolutionary constraint of starvation.

14.1 Origins

For those who feel that this prediction of major contributions of plant biotechnology to agriculture is far-fetched, I would recall the scathing skepticism often contained in reviews of proposals submitted by plant scientists in the 1970s

TRANSGENIC PLANTS, VOLUME 2
Present Status and Social and Economic Impacts

to undertake early cloning experiments. Indeed, I recall seeing one review of a 1978 proposal that included experiments to identify and characterize structural genes encoding seed proteins, to obtain insight into their transcriptional regulation and translational control, and to initiate genetic engineering experiments, that stated, "This is undoubtedly the most preposterous proposal I have read, the [principal investigator] wants to improve nutritional value through genetic engineering . . ." Certainly, it was optimistic to hope for the alteration of major crops by techniques not yet developed, but does not progress depend on making the impossible possible?

A prerequisite to the characterization of seed protein genes was cDNA cloning, for which a supply of highly purified mRNA was needed. Even toward the end of the 1970s, few mRNAs had been isolated and shown by *in vitro* translation to yield the protein product desired. Verma *et al.* (1975) had isolated a cellulase mRNA, and mRNA for zein, the major storage protein of corn seed, was isolated and translated by Burr and Burr (1976), Burr *et al.* (1978), and Larkins *et al.* (1976). Separation of zein mRNAs and cDNA cloning were undertaken by Wienand and Feix (1978) and Park *et al.* (1980). Phaseolin mRNA isolation (Sun *et al.*, 1975; Hall *et al.*, 1978) and cloning progressed well, and the isolation of corresponding cDNA and genomic clones permitted confirmation that introns existed in the chromosomal DNA of plants as well as animals (Sun *et al.*, 1981; Slightom *et al.*, 1983). Transfer of genomic DNA to a new host plant and its expression as protein certainly seemed like wishful thinking, although rapid progress toward using the *Agrobacterium tumefaciens* Ti plasmid system as a useful vector was evident by 1980 from papers such as those of Hernalsteens *et al.* (1980) and Chilton *et al.* (1980), which provided rigorous evidence that genes residing within the T-DNA region of the Ti plasmid were transferred into plant nuclear DNA. With John Kemp and collaborators, phaseolin expression was reported in transgenic sunflower cells (Murai *et al.*, 1983). Although doubts had existed about using a highly regulated gene for gene transfer studies, our choice was vindicated by the spatial and tissue-specific expression of phaseolin in seed of fertile transgenic tobacco (Sengupta-Gopalan *et al.*, 1985).

Clearly, progress in plant gene manipulation has been rapid; today, as detailed in the chapters in Parts I and II of Volume 2, transformation and regeneration of many crop species has been reported. For some species, progress has been more rapid than might have been anticipated; others have proven recalcitrant. Unfortunately, transformation and regeneration of many important dicots and essentially all monocot crop plants can be included among the latter. Some of the claims for success with major crops contain license more appropriate to poetical works than to scientific literature. In this regard, Potrykus (1990) has provided a useful critique of 21 published methods for transformation, along with a list of tests needed to confirm claims for success. Among these, demonstration by Southern hybridization that the inserted sequence is present in unrestricted high molecular weight DNA is critical. Recent progress with rice (Shimamoto *et al.*, 1989) and corn (Gordon-Kamm *et al.*, 1990) is encouraging, and the next step is substantiation that the inserted sequence is present, intact, and functional in second generation plants.

14.2 Funding for Plant Molecular Biology

Pragmatically, completion of the rigorous tests needed to prove stable transformation of new plant species poses a problem for plant scientists dependent on competitive grants. Most grants are for 3 years or less, and a competitive renewal showing progress and publications typically must be submitted after 2 years, often all too short a time to carry a transformant through to the second generation. Although plant science has improved its image greatly over the past two decades, we still lack the intensity and length of funding available to researchers who can claim that their work has health-related importance. Although the U.S. Department of Agriculture (USDA) has the responsibility to further plant science, its attempts have been pedestrian; even the new "initiative" program is too small and continues to reflect the repressive political power of traditional agriculture. This is all the more remarkable given the leadership role played by the agricultural industry in furthering plant biotechnology.

University administrations also are frequently guilty of conservative behavior in support of competitive research, despite verbal expressions of a desire to attain "world class" status. All too often defensive responses are given by administrators to newspaper charges of focus on research in preference to teaching when it is obvious that competition for entry to undergraduate as well as graduate programs is highest at institutions renowned for their research.

There can be little doubt that even the limited infusion of funding for plant molecular biology that has taken place would not have happened without the competition from industry. Much credit is due to those entrepreneurs who obtained venture capital funding and made it available to university researchers as well as to various startup companies. In the negative vein of the reviewer of the grant application mentioned earlier, critics of plant biotechnology point to the current lack of molecularly engineered crops and to the demise of many small plant biotechnology companies. Such pessimism is unwarranted: major progress toward field production of herbicide-resistant and pathogen-resistant transgenic plants is evident, despite unreasonable hindrances. The surviving small companies have innovative ideas that will flourish, providing reasonable progress is made in transformation of economically important plants. The commitment by major industrial investors is also likely to be successful, although unduly optimistic fiscal estimates of returns made by business "experts" certainly provide formidable challenges for bench scientists.

14.3 Signal Transduction and Gene Regulation

In parallel with the animal sciences, from which we derive much of our technical insight, detailed understanding of the interactive role of the various motifs present in promoter elements will be forthcoming. The documentation by Benfey and Chua (1990) of several regulatory elements in the 5′ region flanking the 35S promoter of cauliflower mosaic virus (CaMV) is impressive and we (Bustos *et al.*, 1989; 1991a, 1992) have evidence that multiple short sequences contribute to at least five regions (three positive and two negative) that control transcription of the phaseolin gene. Many of these elements are common to other genes that are constitutively or developmentally regulated. It seems certain that spatial permutation of common elements, and the resulting combinatorial conformation of *trans*-acting factors, will be a major means of controlling gene expression. Evidence continues to accumulate for contributions of the poly(A) tail and 3′ untranslated regions to translation and gene expression in animals (Jackson and Standart, 1990) and plants (Ingelbrecht *et al.*, 1989); increased knowledge concerning the role of these regions can be anticipated.

Presently, attention is focused on proteins as the *trans*-factors that directly interface with *cis* elements of DNA sequences. It is not unreasonable that other molecules, some possibly very small, may be shown to interact directly and functionally with nuclear DNA. The dramatic effects of plant growth regulators long have been recognized. The demonstration of transcriptional stimulation of promoter elements of the *Em* gene by abscisic acid (ABA) by Marcotte *et al.* (1989), along with the identification of a leucine zipper protein involved in ABA signal transduction (Guiltinan *et al.*, 1990), represents a seminal contribution that predicts great advances in understanding how plant growth regulators function. Hopefully, cloning the genes encoding the biosynthetic pathways for the major growth regulators also will lead to added understanding of plant development (Hall and Thomas, 1987). It is tempting to speculate that the expression of individual members of a gene family in specific tissues leads to changing hormone ratios (for example, of auxins to cytokinins) necessary for differentiation. Such endogenous changes would explain why regimes for exogenous application of growth regulators to induce regeneration are complex and generally determined empirically.

Many aspects of plant development are controlled by phytochrome (Furuya, 1987), a photoreceptive protein existing in an inactive red light-absorbing form and an active far-red light-absorbing form. In oats and other monocots, phytochrome controls transcription of *phy* genes through a negative feedback system; unfortunately, *phy* genes of dicots appear to lack strong autoregulation (Lissemore *et al.*, 1987) and no phytochrome transcripts were detected in tobacco plants transformed with an oat *phy* gene (Keller *et al.*, 1989). These facts have thwarted attempts to understand molecular mechanisms of signal transduction in an important system, but Bruce *et al.* (1989) have been able to mimic photoregulation of an oat *phy* gene promoter by transient expression in rice. Such advances bode well for the future elucidation of this and other important developmental processes.

Differences in gene expression in monocots and dicots, such as those mentioned for phytochrome, have been encountered several times. Examples include attempts to express maize *adh* (Llewellyn *et al.*, 1985), corn zein (Matzke *et al.*, 1990), and sorghum kafirin (R. C. DeRose and T. C. Hall, unpublished results, 1990) in transgenic tobacco. It has been suggested that the promoter elements of some monocot genes lack regions essential for interaction with dicot *trans*-acting factors (Llewellyn *et al.*, 1985). However, some monocot genes (e.g., barley β-hordein, Marris *et al.*, 1988; wheat glutenin, Robert *et al.*, 1989) have been expressed correctly at the protein level in transgenic tobacco. Understanding the basis for these incongruities is important for plant biotechnology, although pragmatic approaches for expressing monocot coding sequences from dicot promoters (as in the case of zein expressed from the phaseolin promoter, Hoffman *et al.*, 1987) appear suitable for circumventing such difficulties.

Matzke *et al.* (1989) have documented important instances of gene inactivation in sequentially transformed tobacco plants. Inactivation of resident chalcone synthetase (*chs*) genes in petunia following transformational insertion of additional *chs* genes also has been reported (Napoli *et al.*, 1990). The effects appear to parallel classical epigenetic suppression and typically are associated with methylation of genomic DNA encoding the trait studied. Although direct evidence is yet to be adduced for a causal relationship between methylation and suppression of gene expression in these instances, the evidence thus far clearly warrants detailed investigation of the phenomenon. Since several rounds of transformation may be envisaged (if not essential) for the introduction of multiple beneficial traits into a given crop line by genetic engineering, frequent suppression of a recently introduced trait by subsequent transformation would be a major complication. Such concerns would be even greater if suppression of the introduced trait was induced significantly by subsequent rounds of sexual crosses. Both classical and molecular studies show that the amount of a given gene product is rarely directly proportional to the number of gene copies present; it seems possible that a control system (or systems) exists to prevent overly high levels of expression of a given sequence. Although the out-titration of a necessary *trans*-acting factor provides one explanation for these observations, it is not very convincing as a generally applicable mechanism. Another down-regulating influence on gene expression that may be allied to these effects is the effect of chromosome position on gene expression. Individual investigators have ascribed different levels of importance to position effects on gene expression in transgenic plants. At least some of the differences may relate to the genes being studied:

expression from highly regulated genes such as those for the seed proteins phaseolin (Sengupta-Gopalan *et al.*, 1985) and soybean conglycinin (Beachy *et al.*, 1985) may be less subject to such effects than ubiquitously expressed sequences. Possibly, such highly regulated sequences provide substantial constraints on chromatin structure, whereas other sequences are more readily susceptible to methylation or other modifications. Much circumstantial, but little or no rigorous, evidence is available that gene insertion occurs at chromosomal sites that are more receptive than others; such sites are assumed widely to be transcriptionally active. Gene insertion, and the other phenomena just discussed, may alert the cellular organization (policing agencies?) to the existence of strange nucleic acid, motivating silencing or correctional events.

This line of thought leads to a consideration of the ruthlessness with which biological systems deal with nonself events. When mechanisms involved in such control systems fail to work correctly, results are usually very unfortunate, as in the many autoimmune diseases now recognized. DNA correction systems are rigorous; erroneous tracts typically are signaled for subsequent removal. At the cellular level, ubiquitins stand ready to tag foreign or deformed molecules. Cell damage typically is not tolerated, the maimed tissues often being isolated (as in necrotic lesions induced by physical or pathogen-derived damage) and removed by autocatalytic degradation signaled or stimulated by carbohydrates released from injured cell walls. Such signals also may induce the synthesis of lectins or protease inhibitors that kill or deter insect or pathogenic predators (Ryan *et al.*, 1991). Further understanding of cellular responses to physical and pathogenic assaults can be predicted confidently to be among important advances to be reported in the coming decade.

14.4 Macromolecular Modification and Targeting

Protein engineering has been listed among the many important opportunities afforded by biotechnology. Immediate applications are most likely to be found in industrial enzymology, but the value of protein engineering in plant science is likely to grow as the crystal structures of more plant proteins are determined. Detailed structural knowledge may prove to be essential for apparently straightforward purposes, such as improvement in seed protein nutritional value. Although Altenbach *et al.* (1989) were able to increase the sulfur content of transgenic tobacco seed by expression of the methionine-rich Brazil nut protein, analogous attempts to modify seed proteins such as phaseolin by insertion of coding sequences rich in methionine or cysteine (Hoffman *et al.*, 1988) caused little modified protein to accumulate. In the case of phaseolin, elucidation of the crystal structure (Lawrence *et al.*, 1990) revealed that the insertion used by

Hoffman *et al.* (1988) fell within a conserved α-helix, probably destabilizing the structure. Future modifications based on an understanding of the protein structure and physiological properties probably will result in significant enhancement of nutritional value without detrimental effects on stability or function. An alternative approach is that of Kim *et al.* (1990), who have designed an entirely "artificial" protein that has excellent amino acid balance. It will be instructive to learn how well it can be accumulated and stored and whether it has desirable functional qualities and taste.

The seed-storage protein phaseolin also provides an example of the potential role of carbohydrate adducts in protein function. Modification of its amino-acid sequence to eliminate asparagine residues that are *N*-glycosylated resulted in greatly decreased levels of phaseolin in transgenic tobacco seed, compared with those resulting from the wild-type sequence (Bustos *et al.*, 1991b). Although no immediate effects on protein targeting were apparent in these studies for non-glycosylated phaseolin, or in those of Voelker *et al.* (1989) for phytohemagglutinin, comparison with the differing properties of various protein glycoforms (Rademacher *et al.*, 1988) in animal systems suggests that it is premature to assume the absence of such effects in plants. Indeed, subtle yet important consequences may become evident as expertise in plant cell biology increases.

14.5 Blooming Biotechnology

The insightful studies by Goldberg and colleagues (Koltunow *et al.*, 1990) on tissue-specific genes have been rewarded richly by their application to the induction of male sterility by expressing ribonuclease in tapetal cells (Mariani *et al.*, 1990). Changes in flower color and morphology induced by recombinant techniques are contemplated (Napoli *et al.*, 1990; Schwarz-Sommer *et al.*, 1990) that are likely to prove both asthetically pleasing and fiscally rewarding.

The production of biologically active compounds in plants promises to be an exciting new opportunity (Hall *et al.*, 1990). Up to 200 nmol per gram of seed of the neuropeptide Leu-enkephalin have been expressed in transgenic *Brassica napus* plants (Vandekerckhove *et al.*, 1989). Hiatt *et al.* (1989) were able to obtain functional antibody in progeny of sexual crosses between transgenic tobacco plants expressing either γ or κ immunoglobulin chains. These successes represent pioneering examples of the variety of novel applications awaiting the further development of plant biotechnology.

The chapters in this treatise attest to the rich diversity of topics being addressed by plant scientists. Undoubtedly, many other novel concepts will be derived that will expand the role of plant biotechnology. Indeed, the development of biotechnology is likely to parallel that of computer technology. The initial steps are rather simplistic but ultimately will proliferate in complexity, importance, and value until life without such technologies would be hard to imagine.

References

Altenbach, S. B., Pearson, K., Meeker, G., Staraci, L. C., and Sun, S. S. M. (1989). *Plant Mol. Biol.* **13,** 513–522.

Beachy, R. N., Chen, Z.-L., Horsch, R. B., Rogers, S. G., Hoffmann, J., and Fraley, R. T. (1985). *EMBO J.* **4,** 3047–3053.

Benfey, P. N., and Chua, N.-H. (1990). *Science* **250,** 959–966.

Bruce, W. B., Christensen, A. H., Klein, T., Fromm, M., and Quail, P. H. (1989). *Proc. Natl. Acad. Sci. U.S.A.* **86,** 9692–9696.

Burr, B., and Burr, F. A. (1976). *Proc. Natl. Acad. Sci. U.S.A.* **73,** 515–519.

Burr, B., Burr, F. A., Rubenstein, I., and Simon, M. N. (1978). *Proc. Natl. Acad. Sci. U.S.A.* **75,** 696–700.

Bustos, M. M., Guiltinan, M. J., Jordano, J., Begum, D., Kalkan, A. A., and Hall, T. C. (1989). *Plant Cell* **1,** 839–853.

Bustos, M. M., Kalkan, F. A., Begum, D., Battraw, M. J., and Hall, T. C. (1991a). *EMBO J.* **10,** 1469–1479.

Bustos, M. M., Kalkan, F. A., VandenBosch, K. A., and Hall, T. C. (1991b). *Plant Mol. Biol.* **16,** 381–395.

Bustos, M. M., Kalkan, F. A., Begum, D., Battraw, M. J., and Hall, T. C. (1992a). *In* "Biotechnology and Nutrition" (S.-D. Kung and D. Bills, eds.), Butterworths, Stoneham, Massachusetts, pp. 213–235.

Chilton, M.-D., Saiki, R. K., Yadav, N., Gordon, M. P., and Quetier, F. (1980). *Proc. Natl. Acad. Sci. U.S.A.* **77,** 4060–4064.

Furuya, M. (1987). "Phytochrome and Photoregulation in Plants." Academic Press, Tokyo.

Gordon-Kamm, W. J., Spencer, T. M., Mangano, M. L., Adams, T. R., Daines, R. J., Start, W. G., O'Brien, J. V., Chambers, S. A., Adams, W. R. Jr., Willetts, N. G., Rice, T. B., Mackey, C. J., Kreuger, R. W., Kausch, A. P., and Lemaux, P. G. (1990). *Plant Cell* **2,** 603–618.

Guiltinan, M. J., Marcotte, W. R., Jr., and Quatrano, R. S. (1990). *Science* **250,** 267–271.

Hall, T. C., and Thomas, T. L. (1987). *In* "Hormone Action in Plant Development: A Critical Appraisal," Proceedings of the 10th Long Ashton Symposium (G. V. Hoad, J. R. Lenton, M. B. Jackson, and R. K. Atkin, eds.), pp. 287–297. Butterworths, London.

Hall, T. C., Ma, Y., Buchbinder, B. U., Pyne, J. W., Sun, S. M., and Bliss, F. A. (1978). *Proc. Natl. Acad. Sci. U.S.A.* **75,** 3196–3200.

Hall, T. C., Bustos, M. M., Anthony, J. L., Yang, L. J., Domoney, C., and Casey, R. (1990). *Ciba Foundation Symp.* **154,** 177–197.

Hernalsteens, J. P., Holsters, M., van Montagu, M., and Schell, J. (1980). *Nature (London)* **287,** 654–656.

Hiatt, A., Cafferkey, R., and Bowdish, K. (1989). *Nature (London)* **342,** 76–78.

Hoffman, L. M., Donaldson, D. D., Bookland, R., Rashka, K., and Herman, E. M. (1987). *EMBO J.* **6,** 3213–3221.

Hoffman, L. M., Donaldson, D. D., and Herman, E. M. (1988). *Plant Mol. Biol.* **11,** 717–729.

Ingelbrecht, I. L. W., Herman, L. M. F., Dekeyser, R. A., Van Montagu, M. C., and Depicker, A. G. (1989). *Plant Cell* **1,** 671–680.

Jackson, R. J., and Standart, N. (1990). *Cell* **62,** 15–24.

Keller, J. M., Shanklin, J., Vierstra, R. D., and Hershey, H. P. (1989). *EMBO J.* **8,** 1005–1012.

Kim, J. H., Cetiner, M. S., Blackmon, W. J., and Jaynes, J. M. (1990). *J. Cell. Biochem. Suppl.* **14E,** 349.

Koltunow, A. M., Truettner, J., Cox, K. H., Wallroth, M., and Goldberg, R. B. (1990). *Plant Cell* **2,** 1201–1224.

Larkins, B. A., Jones, R. A., and Tsai, C. Y. (1976). *Biochemistry* **15,** 5506–5511.

Lawrence, M. C., Suzuki, E., Varghese, J. N., Davis, P. C., Van Donkelaar, A., Tulloch, P. A., and Colman, P. M. (1990). *EMBO J.* **9,** 9–15.
Lissemore, J. L., Colbert, J. T., and Quail, P. H. (1987). *Plant Mol. Biol.* **8,** 485–496.
Llewellyn, D., Dennis, E., and Peacock, J. (1985). *In* "Molecular Form and Function of the Plant Genome" (L. van Vloten-Doting, G. S. Groot, and T. Hall, eds.), pp. 593–608. Plenum Press, New York.
Marcotte, W. R., Jr., Russell, S. H., and Quatrano, R. S. (1989). *Plant Cell* **1,** 969–976.
Mariani, C., De Beuckeleer, M., Truettner, J., Leemans, J., and Goldberg, R. B., (1990). *Nature (London)* **347,** 733–741.
Marris, C., Gallois, P., Copley, J., and Kreis, M. (1988). *Plant Mol. Biol.* **10,** 359–366.
Matzke, A. J. M., Stöger, E. M., Schernthaner, J. P., and Matzke, M. A. (1990). *Plant Mol. Biol.* **14,** 323–332.
Matzke, M. A., Primig, M., Trnovsky, J., and Matzke, A. J. M. (1989). *EMBO J.* **8,** 643–649.
Murai, N., Sutton, D. W., Murray, M. G., Slightom, J. L., Merlo, D. J., Reichert, N. A., Sengupta-Gopalan, C., Stock, C. A., Barker, R. F., Kemp, J. D., and Hall, T. C. (1983). *Science* **22,** 476–482.
Napoli, C., Lemieux, C., and Jorgensen, R. (1990). *Plant Cell* **2,** 279–289.
Park, W. D., Lewis, E. D., and Rubenstein, I. (1980). *Plant Physiol.* **65,** 98–106.
Potrykus, I. (1990). *Bio/Technology* **8,** 535–542.
Rademacher, T. W., Parekh, R. B., and Dwek, R. A. (1988). *Ann. Rev. Biochem.* **57,** 785–838.
Robert, L. S., Thompson, R. D., and Flavell, R. B. (1989). *Plant Cell* **1,** 569–578.
Ryan, C. A., Farmer, E. E., Pearce, G., and McGurl, B. (1991). *J. Cell. Biochem. Suppl.* **15A,** 10.
Schwarz-Sommer, Z., Huijser, P., Nacken, W., Saedler, H., and Sommer, H. (1990). *Science* **250,** 931–936.
Sengupta-Gopalan, C., Reichert, N. A., Barker, R. F., Hall, T. C., and Kemp, J. D. (1985). *Proc. Natl. Acad. Sci. U.S.A.* **82,** 3320–3324.
Shimamoto, K., Terada, R. T., Izawa, T., and Fujimoto, H. (1989). *Nature (London)* **338,** 274–276.
Slightom, J. L., Sun, S. M., and Hall, T. C. (1983). *Proc. Natl. Acad. Sci. U.S.A.* **80,** 1897–1901.
Sun, S. M., Buchbinder, B. U., and Hall, T. C. (1975). *Plant Physiol.* **56,** 780–785.
Sun, S. M., Slightom, J. L., and Hall, T. C. (1981). *Nature (London)* **289,** 37–41.
Vandekerckhove, J., Van Damme, J., Van Lijsebettens, M., Botterman, J., De Block, M., Vande-Wiele, M., DeClercq, A., Leemans, J., Van Montagu, M., and Krebbers, E. (1989). *Bio/Technology* **7,** 929–932.
Verma, D. P. S., Maclachlan, G. A., Bryne, H., and Ewings, E. (1975). *J. Biol. Chem.* **250,** 1019–1028.
Voelker, T. A., Herman, E. M., and Chrispeels, M. J. (1989). *Plant Cell* **1,** 95–104.
Wienand, U., and Feix, G. (1978). *Eur. J. Biochem.* **92,** 605–611.

Index

ISBN 0-12-428782-4

90051

9 780124 287822